BEEF
CATTLE
PRODUCTION

Prentice-Hall, Inc.,
Englewood Cliffs, New Jersey O7632

BEEF CATTLE PRODUCTION

John F. Lasley
University of Missouri, Columbia

Library of Congress Cataloging in Publication Data

LASLEY, JOHN FOSTER, (date)
   Beef Cattle production.

   Includes index.
   1. Beef cattle.   2. Beef cattle — United States.
I. Title.
SF207.L37   636.2'13   80–13214
ISBN   0–13–072629–X

Editorial/production supervision by Ellen W. Caughey
Interior design by Robert Schorr and Ellen W. Caughey
Cover design by Maureen Olsen
Art direction by Linda Conway
Manufacturing buyer: John B. Hall

Printed in the United States of America

10 9 8 7 6 5 4 3 2 1

Prentice-Hall International, Inc., *London*
Prentice-Hall of Australia Pty. Limited, *Sydney*
Prentice-Hall of Canada, Ltd., *Toronto*
Prentice-Hall of India Private Limited, *New Delhi*
Prentice-Hall of Japan, Inc., *Tokyo*
Prentice-Hall of Southeast Asia Pte. Ltd., *Singapore*
Whitehall Books Limited, *Wellington, New Zealand*

CONTENTS

PART**TWO**

PHYSIOLOGY OF REPRODUCTION / 57

PART**THREE**

GENETICS OF BEEF CATTLE BREEDING / *125*

The author has spent about half a century producing, breeding, and studying beef cattle production in both midwestern and southwestern ranges in the United States. In recent years, the efficiency of beef production has been increased by the use of scientific methods. This book outlines the best of these methods for beef production.

Methods of beef cattle production vary within the United States. One of the objectives of the book was to limit its size; it was sometimes difficult, therefore, to outline methods of production for all areas. Where appropriate, however, reference was made to the varied methods. Some methods of care and management, such as the treatment and control of diseases, are subject to change as the drugs used are banned periodically by federal law. Producers should keep in close contact with their local veterinarian and extension personnel for latest recommendations.

References for further reading are given in some sections throughout the text if the reader wishes to make a more extensive search of a subject. A chapter on cattle behavior was included as behavior is so important to efficient production and management of the herd. Because the literature is somewhat limited, information given was from the author's own experience and observations.

The author is indebted to state extension specialists for up-to-date information on beef cattle production. The added advice and assistance of Homer Sewell, John Massey, Jim Ross, Victor Jacobs, and Glenn Grimes is gratefully acknowledged, as is the preparation of several line drawings by students Jerry Fry and Rosalind Terry. Lastly, many thanks to Dr. A. A. Case who generously supplied photographs of certain poisonous plants from his extensive collection.

JOHN F. LASLEY

PART ONE

GENERAL VIEW

CHAPTER 1

origin
and importance
of cattle
in the world

In the ancient world people were nothing but savages until they began to cultivate crops and domesticate animals. The development of agricultural practices changed their life from that of nomads, or wanderers, where they depended on their hunting skills and their ability to find and gather wild fruits and other plants, to one in which they ceased such wandering and settled down in villages.

Archaeological evidence suggests that the reaping and milling of wild cereals and the possibility of purposeful cultivation of some plants must have been practiced in Palestine 10,000 or more years ago. The reason for this belief is that mortars and pestles as well as flint sickles have been found there at some excavation sites. It is believed that the domestication of food animals followed later, with goats probably being domesticated 8,000 to 9,000 years ago followed by the domestication of cattle, sheep, and pigs.

Plant cultivation and food animal domestication probably began in the hills of southwestern Asia in the Zagros, Lebanese, and Palestinian mountains. The first agriculturalists and stock raisers were probably of the Mediterranean race, similar to those people now living in that area.

1.1 Domestication of Cattle

Cattle belong to the genus *Bos*, being even-toed, hollow-horned, ruminant quadrupeds. They belong to the taurine group, which includes *Bos taurus* (ordinary cattle) and *Bos indicus* (humped zebu cattle). They have many relatives, as shown in Table 1-1.

Cattle were probably first deliberately domesticated in western Asia by nonmigratory farmers (E. Isaac, *Science* 137:195–204, 1962). The time of domestication is thought to be about 5,000 to 6,000 years ago. Zoological studies suggest that all present-day types of domestic cattle came from one ancestral strain because they are fertile in crosses and because they also produce fertile hybrids. The ancestral strain from which they are thought to be descended is the *Bos primigenius Bojanus*, or wild urus, which survived in Europe until the late middle ages, with the last individual dying in 1627. The urus formerly ranged in a widespread area of Asia, Europe, and North Africa. Fossil remains in Europe indicate the urus was a large, long-horned, powerful animal standing 6 feet tall at the shoulder. The urus is also often referred to by various writers as the auroch, uri, or ur.

Many writers have speculated as to why cattle were first domesticated. It is conceivable that the dog and pig, because they were more or less scavengers, followed people in their wanderings and were eventually domesticated. Since the wild urus did not do this, it is possible that people followed them from hunting ground to hunt-

TABLE 1-1   Zoological classification of genus *Bos*

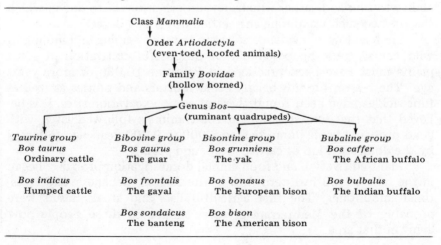

ing ground, and as a result the initiative for the domestication of this species must have come from people. But why did people domesticate the urus? Many zoologists now believe that the first uruses were maintained in the captive state in order to have a supply readily available for sacrificial purposes, mainly because of the size and shape of their horns (C. A. Reed, *Science* 130:1629–1639, 1959). Much of the ancient world worshiped the lunar mother goddess, and it is possible that the urus was selected for sacrificial purposes because of its large, curved horns, which resembled the lunar crescent. This theory sounds reasonable since most ancient people made regular sacrifices to their gods. Even today cattle are considered sacred or holy in some parts of the world. If this theory is true, other uses such as for milk, meat, and draft purposes were a by-product of this religious domestication.

What is the process whereby the wild urus was transformed into a domesticated animal?[1] It has been suggested that at first the captured wild animals were kept in small enclosures so they were handy when needed for sacrificial purposes. Under captivity the wild urus was free to multiply and was probably encouraged to do so. This could have resulted in a certain amount of inbreeding, with the pairing of different recessive genes, or even new mutations, affecting body color and type. Since such animals were protected to a certain extent from predators, they were free to reproduce, and with selection for different colors by their captors or owners, distinct types and colors that bred relatively true could have been developed. At least we know such has been the case with rabbits, mink, and foxes under domestication. Much later, selection and breeding for a specific purpose such as milk production, meat production, and draft type were practiced.

1.2   Importance of Cattle in Ancient Times

Cattle have been held in high esteem by people since ancient times. They were used as a source of food or clothing, as draft animals, or for sacrificial purposes, as mentioned previously.

Greek and Roman mythology make mention of cattle, and in Egypt and other countries the cow was held to be sacred, as it is

[1] *Domestication* may be defined as the condition where the breeding, feeding, and care of animals, in some degree, are subject to continuous control by man.

today in India. In biblical times the wealth of people was largely determined by the size of their herds and flocks. When the Children of Israel made their exodus from Egypt after a period of captivity, they headed for the promised land, a land which "flowed with milk and honey." The high esteem in which cattle were held is further verified in passages in the Old Testament in which it is said that they should not "muzzle the ox that treadeth out the corn."

1.3   Cattle in Great Britain

Early importations of cattle into the United States came mostly from Great Britain, although a few, such as the Holstein-Friesian, came from the European continent.

The first domesticated cattle probably were brought to Great Britain about 2000 B.C. by Neolithic tribes which were migrating westward from the European continent. These cattle were thought to be black or red in color and were small, with short horns. The Romans came to Great Britain in the first century A.D., and they brought their own cattle with them. These cattle were large and had a white coat color and a great spread of horns. The Saxons and Norsemen came to Great Britain between A.D. 450 and 850 and brought red or dun-colored, polled cattle with them. These importations are probably the foundation stock for many of the British breeds.

Cattle breeding in Great Britain, in which various pure breeds were developed, centers along the shores of the English Channel. Most of today's modern breeds of improved dairy and beef cattle originated either in Great Britain, upon the mainland opposite Great Britain, or on the islands off the coast of Normandy and Brittany.

Intensified agriculture and excellent grazing land no doubt were responsible for the development of breeds and the improvement of beef cattle in Great Britain. People such as Robert Bakewell, often referred to as the father of animal breeding, the Collings brothers, and Thomas Bates were early breeders of the Shorthorns. Benjamin Tompkins, Jr., William Galliers, John Price, William and John Hewer, and T. W. Carwardine were early breeders of Herefords, whereas Hugh Watson, William McCombie, and several others were the early developers of the Angus breed. Most of these breeds were developed in the early to middle 1800s, with the first herd books for the Shorthorns established in 1833, the Herefords in 1846, and the Angus in 1863.

1.4 American Buffalo on the Western Plains

Cattle are not natives of either North or South America. When the first white people came to the new world, an estimated 70 to 100 million buffalo, or bison, were roaming the North American continent. The buffalo ranged from Florida to the Rockies and from Mexico to Canada, but the majority were found on the prairies of the Great Plains. The American buffalo (*Bos bison*) were relatives of domesticated cattle as shown by the fact that matings between them have produced viable offspring which are called Cattalo or Beefalo. Cattle and the American bison have the same number of chromosomes, and the shape and structure of their chromosomes are similar. Buffalo were one of the main sources of food for the thousands of Indians on the western plains, and when the buffalo migrated to new areas at different seasons of the year, the Indians migrated with them (Figure 1.1).

As far as is known, the Indians made no attempt to domesticate the buffalo. This may have been because the Indians differed from ancient Europeans in their religious beliefs or because buffalo were

Figure 1.1.   The American bison on the Great Plains.

so plentiful and easy to find that there was no need to domesticate them.

The Indians killed a few buffalo from time to time to meet their needs for meat, but they did not wantonly destroy the great beasts as the white people did. When the Indian braves killed buffalo, their families feasted on the fresh meat. Any surplus meat was not wasted but was cut into strips and dried in the sun to produce a product known as *pemmican* (sometimes called *jerky*) which could be kept indefinitely. The buffalo hides were also valuable as they could be dried and cured and made into tepees, saddles, clothing, sleeping robes, and many other necessities. The hooves and horns were used for making household utensils such as spoons and cups. This was the situation which existed in 1540 when Coronado entered the southwestern portion of what is now called the United States in his search for the seven cities of Cibola (A. H. Sanders, *The Cattle of the World*, National Geographic Society, Washington, D.C., 1926).

The discovery of a new process for making serviceable leather from buffalo hides in about 1870 almost led to the extinction of the American buffalo. Hunters from many walks of life began slaughtering them by the thousands for their hides. The huge herds in Texas were gone by 1878, and by 1885 only a few buffalo were left in isolated valleys in the west. The plains where the buffalo once roamed were white with buffalo bones, but even these proved to be valuable since they could be made into fertilizer or used to make carbon for refining sugar. By 1905, it was estimated that only 324 buffalo were left in the United States and Canada. In 1907, 15 of the finest buffalo in the New York Zoological Park were shipped by rail to the Wichita Mountain Wildlife Refuge near Lawton, Oklahoma to establish a herd and to prevent the complete extinction of this species. This small herd grew and prospered in subsequent years until it numbered about 1,000 head, with surplus animals being sold almost yearly to prevent the overstocking of the range in the refuge. Many buffalo are now also produced in private herds in many parts of the United States.

The elimination of the buffalo from the western range had far-reaching consequences. It brought about a change in the hunting habits of the Indians and in many cases left them starving and destitute. It also opened the western range country to white ranchers who stocked the vacated range land with cattle. The ranchers were soon followed by settlers who plowed the prairies and seeded them to crops. The encroachment of white people on land which the Indians rightly considered their own led to the Indian wars, the loss of many lives on both sides, and eventually the defeat of the Indians and their confinement to reservations where most of them live today. Thus, the range where both the buffalo and Indians once freely roamed became the great farms and ranches of today.

1.5   Growth of the Cattle Industry in the United States

Columbus brought the first cattle to the West Indies on his second voyage to the New World in 1493. These cattle were of Spanish breeding and were not of desired beef type by modern standards. It is believed that the Spanish cattle were first brought to the mainland of Mexico in 1521 from Santo Domingo by Governor General Villalobos. They probably were descendants of the cattle first brought to the western hemisphere by Columbus. Other importations of Spanish cattle followed, and the number of cattle in Mexico increased greatly in the next few years, with many large ranches being established.

Coronado and other Spanish explorers took cattle, horses, and sheep with them when they explored the country north of the Rio Grande. Some of these cattle and horses escaped along the way and may have been the ancestors of the wild Spanish cattle and horses that eventually roamed the western ranges. It is estimated that the first herds of cattle in what is now known as Texas were there in 1583. The Spanish cattle, although not of good beef type, had great endurance and energy and the ability to resist diseases. They could also hold their own in battles with predators and seemed to know how to find grass and water. In any event, the Spanish cattle rapidly increased in numbers in Mexico and the region north of the Rio Grande. The Spanish began their colonization of the southwestern United States in about 1810. By 1820, the region was populated with Mexican cattle, and with ranches of extremely large size.

Colonists coming from France, England, and other European countries brought cattle with them when they landed on the east coast of what is now the United States. The Jamestown colony first imported cattle in May 1611. Other importations of cattle were made from time to time, with one of the most important ones being made from England to New England in 1624. Cattle from Denmark were introduced into New Hampshire in 1638. By 1655, cattle were being driven to the Boston market, and by 1670 winter-fed cattle were being marketed there. Other colonists soon followed suit in the cattle feeding and raising business, and the business was firmly established in the area by the time of the Revolutionary War.

Goff and Miller imported Shorthorn cattle from Great Britain to Virginia in 1783 and started the business of fattening them on corn and bluegrass. Beef cattle soon found their way over the Blue Ridge Mountains to the Ohio Valley and finally into the midwestern region of what is now the states of Illinois, Indiana, and Missouri. Other importations from England in the early 1800s started a great business on these farms. Some of the English cattle moving westward with their owners eventually mixed with those of Spanish origin. The Texas Longhorn, which was mostly of Spanish breeding, may have a trace of the blood from English cattle in its makeup.

During the Civil War cattle greatly increased in numbers in the southwestern United States. Few of them were marketed, however, because most of the population's efforts were directed to fighting the war. After the Civil War there developed a great demand for beef in the thickly populated regions of the north and east, resulting in high prices for beef. The abundance of beef cattle of Longhorn breeding, especially in Texas, naturally led to an attempt, sooner or later, to drive these animals on foot to points where they could be within reach of the markets (Figure 1.2). This led to the era of the trail drives, which have been the subject of many books, movies, and television series in recent years. The trail drives covered from 500 to 1,700 miles and required 2.5 to 5.5 months for their completion. The average distance traveled per day was 15 to 20 miles. Overstocking of the southwestern ranges (see Table 1-2) with a resultant scarcity of grass and a deterioration of the native grass supply together with the discovery of rich grazing lands in western Nebraska, Wyoming, and other states resulted in trail drives where ranchers moved their entire breeding herds to these areas.

The long-legged Longhorns were well adapted to these trail drives, and most ranchers during this era had little or no desire to improve their beef qualities through the introduction and use of the British breeds of bulls. With the extension of the railroads farther west and southwest, the need for trail drives ended, and ranchers eventually focused their attention on the improvement of beef qualities in their herds. Shorthorn bulls first found widespread use on the western range for crossing with Longhorn cows. They were later largely replaced by Hereford bulls, which found great favor in these regions.

The boom of the western cattle business after the Civil War was almost an overnight affair. It attracted people and investors from many different walks of life. It finally ended, however, because of a combination of circumstances. Among these were drought conditions combined with overstocking of the ranges which resulted in less

Figure 1.2.   Routes of the trail drives in the late 1800s.

forage being produced for grazing purposes and extremely heavy winter losses of cattle. The demand for feeder cattle also decreased, and the "bust" which came in 1887 caused many large ranchers to go broke. Many of them never recovered from their losses. Another reason for a decline in the western cattle business was the occurrence of the disease often called *Texas fever* but more properly called *cattle tick fever.* In natural infections (W. M. Mackellar, "Cattle Tick Fever: Keeping Livestock Healthy," *U.S.D.A. Yearbook of Agriculture,* Washington, D.C., 1942, pp. 572–578) the disease was spread only by the bite of the cattle tick, *Boophilus annulatus.* Many affected animals died or became extremely emaciated. Animals responsible for its spread were apparently healthy, whereas those with the disease usually did not transmit the infection to others. Thus, cattle coming up the trail from the south and southwest were apparently healthy but introduced the disease into susceptible native cattle farther north. The disease caused enormous losses so that many

TABLE 1-2   Grazing capacity of the southwestern range.[a] Overstocking in
early years led to severe restriction of carrying capacity of the
range.

Name of roundup ground	1916		1946	
	Date branded	No. calves branded	Date branded	No. calves branded
Bear Canyon	4-20-16	1,262	4-21-46	121
Sontag	4-23-16	945	4-24-46	129
Blue River	4-24-16	309	4-29-46	148
Rocky	4-25-16	1,116	5-1-46	349
Brush Corral	4-30-16	738	5-9-46	106
Kidde Canyon	5-1-16	448	5-15-46	62
Freezeout	5-2-16	815	5-16-46	78
Sweetmeat	5-3-16	1,055	5-17-46	155
Ash Creek	5-4-16	1,078	5-19-46	221
Dead Man	5-6-16	462	5-20-46	146
Totals		8,228		1,515

[a]Information from the San Carlos Indian Reservation, San Carlos,
Arizona.

states passed laws regulating the movement of southern cattle into the
northern areas in order to prevent the occurrence of the disease. Still
another factor involved in the decline of the cattle business was that
much of the range country was now being settled under the home-
stead law. The homesteaders naturally settled where the water
supply was ample, and this was around the watering places used by
range cattle. These watering places were few and far between in
many regions. This caused a great deal of friction between the homesteaders and the cattle producers and eventually caused a decline in
the amount of free range.

The production of beef cattle has become of great importance
in the United States in recent years, and many persons depend on it
for an income. This will be discussed in more detail in later chapters.

1.6   Importance of Cattle Today

Cattle of today are useful to people in many ways in different regions
of the world. They supply beef and veal when slaughtered as well as
leather for shoes and other products. Various glands of the body
supply hormones and vitamin extracts, and even the bones are
processed into fertilizers and mineral supplements to feed live-

stock. Offal and certain nonedible glands and organs are made into the meat scraps and tankage which are widely used as protein supplements to the feed grains for poultry, swine, and other single-stomached animals. Dairy cattle supply milk, butter, cheese, and other products for human consumption. About one third of the cattle in the world are still used as a source of power as working animals for drawing plows, carts, and other vehicles.

Cattle are not as efficient as swine in converting feed to human food. Swine convert about 20 percent of the feed they eat to human food, whereas dairy cattle convert about 17 percent and beef cattle about 5 percent. Swine are more efficient, however, because they consume concentrated high-energy feeds such as the grains, whereas cattle utilize larger amounts of roughages which are much less concentrated in energy and high in fiber. The fiber is digested by microorganisms in the rumen of cattle; these microorganisms are not found in large numbers in the digestive tract of swine. Beef cattle are especially useful in the utilization of forages and grasses in semiarid regions which otherwise would be of limited use to people. The efficiency of production of all classes of livestock can be improved through the use of improved breeding and management methods, a fact that has received a lot of attention in recent years.

Dairy cattle, as a general rule, are raised in large numbers for the production of milk and milk products in regions of high human populations (see Table 1-5). They supply beef as a by-product since surplus calves and cows are slaughtered for meat. In fact, all cattle, regardless of whether they are dairy or beef type, either die a natural death or end up on the butcher's block as meat for the table. Beef cows for the production of feeder calves are usually found in large numbers in sparsely populated areas such as the range regions of the United States and the pampa plains of South America.

Most of the cattle in the world today belong to two different species, the *Bos taurus* and the *Bos indicus.* Water buffalo, which are distant relatives of cattle, are found in large numbers in certain tropical regions such as India, Thailand, and China. Buffalo are used in the tropics not because they are more heat resistant than cattle, but because they are better able to utilize the highly fibrous tropical vegetation. In fact, buffalo are even less heat resistant than most European breeds of cattle and have no sweat glands except around the muzzle. They must be cooled with water in hot weather or allowed to wallow in mud or water if they are to produce efficiently. In China, buffalo are used mainly for work, but elsewhere they are used for milk production, with surplus animals slaughtered for meat.

European cattle originated in western Europe and are better

adapted to the temperate zone. Zebu cattle probably originated in India and were introduced into south Asia and central Africa (R. W. Phillips, *Sci. Am.* 198:51, 1958). More recently zebus have been introduced into Brazil and the Gulf Coast states of the United States where conditions are semitropical. Zebus are better adapted to the tropics because they have been selected under these conditions for thousands of years by both artificial selection (by people) and natural selection (by nature), especially the latter.

1.7 Cattle Production in the Different Continents of the World

The total population of cattle and buffalo in the world is more than 1 billion head (Table 1-3). The most cattle are found in Asia, followed by South America and North America in that order. The greatest number of cattle per person are found in Oceania, with the smallest number per person in the Far East.

1.7.1 Asia

According to U.S. Department of Agriculture figures, Asia had an average of 475 million head of cattle and buffalo during the period from 1974 to 1976. This is more than twice the number found in either North or South America. Among the nations of the world, India ranks first with about 240 million head of cattle and buffalo (Table 1-4). The density of cattle in India is extremely high as compared to other Asian countries, because cattle are considered sacred

TABLE 1-3   Number of cattle on different continents of the world[a]

Continent	Estimated head (thousands)
Asia	475,844
South America	211,329
North America	171,020
Europe	131,748
USSR	112,500
Oceania (Australia & New Zealand)	38,750

[a]*Agricultural Statistics*, U.S. Department of Agriculture, Washington, D.C., 1978, p. 304.

TABLE 1-4   Leading cattle and buffalo producing countries of the world[a]

Country	Estimated head (thousands)
India	242,690
United States	116,265
USSR	112,500
Brazil	99,500
China (mainland)	92,805
Argentina	56,750
Australia	29,500
Mexico	28,800
Colombia	24,385
France	23,800
Turkey	15,300
West Germany	14,756
United Kingdom	13,550
Union of South Africa	13,500
Poland	12,360
Uruguay	10,000

[a]*Agricultural Statistics,* U.S. Department of Agriculture, Washingington, D.C., 1978, p. 304.

by the Hindus and usually are not slaughtered but are allowed to roam the countryside until they die a natural death. Thus, even though cattle are present in large numbers in India, they are of little economic importance since they are not used for production. In Japan and China as well as in other Asian countries, cattle are kept within the limits that the land can support because there is a great demand to use the land for the direct production of foods without processing them through animals, which would result in a lower efficiency in human food production. In some parts of Asia cattle are not produced in large numbers because of extreme cold or other conditions which do not favor their production.

1.7.2   South America

Cattle numbers total about 210 million head in South America, which ranks second to Asia in this respect (Table 1-3). On this continent, beef cattle are used to graze on the interior lands of Brazil and the belts of pastureland east of the Andes. They are also produced in large numbers in the pampa region. Thus, cattle ranching, or beef production, is very important in Argentina, Uruguay, and southern

Brazil. Brazil and Argentina rank fourth and sixth, respectively, among the nations of the world in total numbers of cattle.

Argentina produces large numbers of beef cattle in the pampa region. This area was formerly largely grassland, but it now produces a large amount of alfalfa and grain which is used to fatten cattle. The eastern part of the pampa plain is humid and supplies good grazing throughout the year. Grazing conditions are especially favorable in the spring and again in the fall. The British breeds of cattle are largely produced in this area. Argentina exports about 30 to 40 percent of the beef it produces as chilled or canned beef. The United Kingdom, Germany, and other countries of northwestern Europe are the chief markets.

Uruguay produces many beef cattle, but they are raised mostly on pasture, without being fattened on grain. About 50 to 60 percent of the beef produced in this country is exported.

1.7.3   North America

The United States is the chief cattle producer in North America and is second among the nations of the world in this respect. Beef production is a major source of income, especially in the range areas as well as in the rougher lands all over the United States. An abundance of feed grains, particularly in the Corn Belt, favors the feeding and fattening of cattle in feed lots. In recent years more and more cattle are being fed in the West and Southwest, most notably in regions of alfalfa and sorghum production.

Beef cattle are produced in large numbers in southern Canada, especially in the Frazer River Valley of British Columbia. Most of northern Canada raises small numbers of cattle because of the extreme cold and poor grazing conditions.

Mexico produces a considerable number of cattle and ranks eighth among nations of the world with about 28 million head. Beef cattle production is important on the semiarid grazing lands and especially in northern Mexico, which borders some of the southwestern states of the United States.

1.7.4   Africa

Cattle are of great economic importance to European farmers who have settled eastern and southern Africa. They are also of great economic and social importance to the natives and are used by them as a kind of currency, with a person's wealth being measured to a great extent by the number of cattle owned. Thus, there is more interest in total numbers in a particular herd than in quality, which does not favor the improvement of the cattle through the application of breeding methods. Some tribes draw blood from their cattle at various intervals and drink it fresh or mix it with other foods as a source of nutrients.

The cattle industry in many parts of Africa is handicapped by disease, lack of feed during the dry winter months, and the scarcity of water. The quality of cattle is generally poor, especially among the native breeds. The total number of cattle in Africa is estimated at a little over 150 million head; this continent ranks fourth in cattle numbers among the continents of the world.

1.7.5   Europe

The cattle population of Europe is particularly heavy in Great Britain, Eire, Denmark, the Netherlands, Belgium, northwestern France, southern Germany, the Swiss Mittelland, the Po Valley of Italy, Austria, western Czechoslovakia, and the northwestern Iberian Peninsula (W. Van Royen, *The Agricultural Resources of the World*, Prentice-Hall, Englewood Cliffs, N.J., 1954). The density of cattle is low in countries around the Mediterranean and countries of the East. Cattle in these areas of low density are used mainly for draft purposes, with meat and milk by-products of little economic importance. In the areas of heavy cattle populations, most dairy cattle are produced mainly because of a lack of surplus land for grazing purposes. Surplus dairy stock furnish beef and veal in these areas. The raising of beef cattle strictly for beef production is important in only a few areas of Europe, such as in parts of Great Britain. It was here that most of the British breeds of beef cattle were developed.

About 130 million head of cattle are found in Europe, which is somewhat less than the number found in Africa.

1.7.6   Oceania

Cattle numbers on this continent (Australia and New Zealand) total about 38 million head. Sheep are of greater importance than cattle in Australia, although beef cattle production is also of great economic value. Beef cattle are produced on the grazing lands of the interior, which are covered largely by native grasses. These lands support mostly breeding cattle, and few are fattened there. The coastal regions of eastern Queensland receive a more even distribution of rainfall so that improved pastures and feed crops are produced. In this region, breeding cattle are maintained, and surplus cattle are fattened. Cattle produced on the inland grazing areas are often brought long distances to the coastal regions for fattening. Many of these cattle are trailed long distances on foot, although some are transported by rail. Approximately one fourth of Australia's yearly beef supply is exported to other countries.

Most of the cattle in New Zealand are produced for dairy purposes, with beef production involving mostly surplus dairy calves and cows.

1.8   Beef Cattle Production in the United States

The United States may be divided into four major regions: the Northeast, the South, the north central region, and the western region.

Cattle numbers per square mile are the greatest in the north central region and the least in the western region (Figure 1.3). The large concentration of beef cattle on farms in the north central region does not necessarily mean that they are raised there since this region includes the Corn Belt states where many feeder cattle are shipped in from the range to be fattened for slaughter.

Figure 1.3.   Regions of beef cattle production in the United States.

The percentage of beef cattle on farms in the north central region as compared to the United States as a whole has remained fairly constant during the past 20 years. Beef cattle numbers on this basis have declined slightly in the Northeast but have shown increases in the other regions, although these increases have been small. The number of beef and dairy cattle by states in the United States, ranked according to numbers, is shown in Table 1-5.

The concentrations of dairy and beef cattle in the different regions of the United States are quite different (Table 1-5). Dairy cattle numbers per square mile are the highest in the eastern one half of the United States, especially in states of the North and East where the human population is concentrated. California also has a high concentration of dairy cattle per square mile, and here, too, the human population is dense and the intensity of agricultural operations great.

The number of beef cattle per square mile is the greatest in the Corn Belt states and other states of the north central and southern regions. These are regions of greatest forage production per acre as well as regions of greater grain production. Beef cattle numbers per square mile (Figure 1.4) are the greatest in Iowa, Missouri, Kentucky, and Oklahoma. The numbers per square mile are less in the eastern coastal states, probably because of the greater density of the human population and dairy cattle and less available land for grazing purposes. The number of beef cattle per square mile in the western region of the United States is low because of semiarid conditions and a resultant low forage production per acre except when irrigation is practiced.

The number of beef cattle per 100 persons in each state in the United States is shown in Figure 1.5. The numbers per 100 persons is still very high in the Corn Belt states, but they are also very high on this basis in the range states, showing that beef cattle production is of greatest economic importance in these regions where intensive agriculture cannot be practiced.

Increases in cattle numbers have not kept pace with increases in the human population (Figure 1.6). Dairy cattle numbers have remained fairly constant in the last 50 years and actually have shown a decline in the past 10 years. In spite of this decrease in dairy cow numbers, there is no scarcity of milk and other dairy products, even though there have been large increases in the human population. This results from the fact that the production per cow has greatly increased, probably through the use of improved methods of breeding, feeding, and management. It would be well for beef cattle producers to study the more efficient methods used by dairy cattle producers in the past decade.

TABLE 1-5 Estimated number of beef, dairy, and all cattle by states in the United States[a]

State, ranked for all cattle	Head (thousands)[b]		
	Beef	Milk	All cattle
Texas	6,480	320	15,600
Iowa	1,931	392	7,500
Missouri	2,700	300	6,600
Nebraska	2,142	148	6,550
Kansas	1,978	143	6,450
Oklahoma	2,673	117	6,400
California	1,010	810	5,000
Wisconsin	350	1,812	4,550
South Dakota	1,905	156	4,500
Minnesota	751	890	4,430
Kentucky	1,365	285	3,450
Illinois	842	244	3,400
Colorado	1,040	75	3,250
Montana	1,614	26	3,150
Tennessee	1,268	212	3,100
Florida	1,419	196	2,920
Alabama	1,238	90	2,850
Mississippi	1,317	117	2,723
Arkansas	1,185	89	2,385
North Dakota	1,150	119	2,380
Georgia	1,037	129	2,370
Ohio	500	390	2,305
Indiana	617	213	2,225
New York	130	916	1,915
Pennsylvania	208	706	1,960
Louisiana	952	138	1,880
Idaho	636	144	1,875
New Mexico	644	31	1,650
Michigan	195	421	1,650
Virginia	621	159	1,650
Wyoming	748	12	1,580
Oregon	620	90	1,440
Washington	393	179	1,375
Arizona	312	68	1,280
North Carolina	425	153	1,130
Utah	362	79	927
South Carolina	314	54	725
Nevada	336	14	651
West Virginia	244	39	555
Maryland	79	141	460
Vermont	14	194	346
Hawaii	89	14	245
Maine	12	60	141
New Jersey	14	45	110
Connecticut	7	54	110

TABLE 1-5   *Continued*

State, ranked for all cattle	Head (thousands)[b]		
	Beef	Milk	All cattle
Massachusetts	9	55	107
New Hampshire	6	33	72
Delaware	6	12	33
Rhode Island	1	6	12
Alaska	3	2	9

[a]*Agricultural Statistics,* U.S. Department of Agriculture, Washington, D.C., 1976, pp. 298–299.
[b]Beef and milk cattle include only cows and heifers that calved.

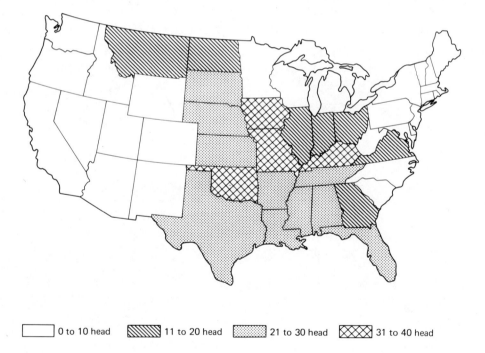

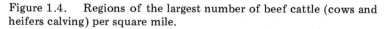 0 to 10 head   11 to 20 head   21 to 30 head   31 to 40 head

Figure 1.4.   Regions of the largest number of beef cattle (cows and heifers calving) per square mile.

21

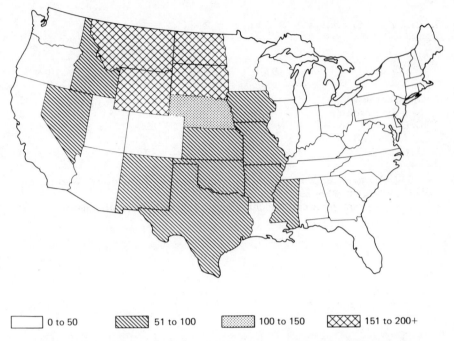

| | 0 to 50 | | 51 to 100 | | 100 to 150 | | 151 to 200+ |

Figure 1.5.   Number of beef cattle (cows and heifers that calved) per 100 human population in each state of the United States.

Figure 1.6.   The increase in human and cattle populations in the United States since 1900.

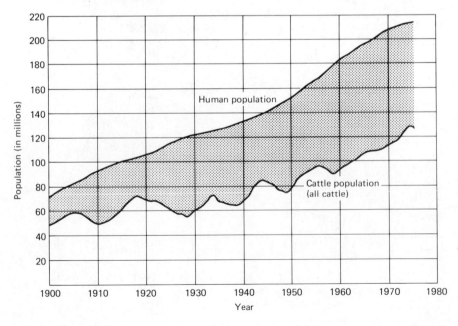

Beef cattle numbers have shown a tendency to increase gradually in the United States, especially in the last 20 to 30 years (Figure 1.6). Beef cattle numbers have shown definite cyclical trends over the years, with periods of large increases in numbers usually followed by periods of large declines. This cyclical trend is due to supply and demand and will be discussed more fully in later chapters. Unless there is a definite decline in the demand by the human population in the United States for beef and beef products, the future of the beef cattle business looks very bright indeed. However, it also appears that greater efficiency in production of beef through attention to all aspects of breeding, feeding, and management will be required to feed our rapidly increasing human population.

1.9 Beef Cattle Research

The population explosion and the diminishing amount of land available for agricultural purposes have served to emphasize the need for greater efficiency in livestock production. The demand for more lean and less fat on the part of the consumer has also influenced the type of research done in recent years. Research objectives have included physiology of reproduction, genetics, nutrition, meats, production and management, and the control of disease.

Research in physiology of reproduction has included studies of superovulation, synchronization of estrus and ovulation, induced multiple births, and induced parturition.

Genetic studies have involved selection for desirable traits such as rapid and more efficient gains and superior carcass quality and quantity. Crossbreeding and crossbreeding systems have also been studied. Research has been directed toward identifying and eliminating genetic defects. Genetic principles developed by this research have found good acceptance and use by beef cattle producers in recent years. Exotic breeds of cattle have been introduced into the United States to determine their possible use in crossbreeding for more efficient beef production. Cow size and its relationship to efficiency of production have also received attention.

Nutrition studies have been directed toward designing more efficient and complete rations for growing and fattening steers and for breeding bulls and brood cows and their calves. Much research has been done toward locating and testing certain feed additives such as diethylstilbestrol, rumensin, and others which stimulate more rapid and/or more efficient gains. The use of simple nitrogen compounds as a part of the ration to replace proteins has also been studied. Such rations have even included chicken litter (manure). Manure from beef cattle feedlots has also been recycled and tested as a possible source of nutrients for rations of beef animals.

Meats research has been directed toward producing a better-quality product for the consumer. Research on producing more tender meat with a better flavor has been conducted, some of this in cooperation with animal breeders and production and management specialists. Methods of retaining better-quality meat with a better flavor when stored for considerable periods of time have been studied. Estimating carcass quality and quantity in the live animal through the use of ultrasonics, the 40^k counter, and other methods have been developed.

Production and management research has also emphasized more efficient methods of production. More efficient methods of wintering brood cows, creep-feeding calves, and fattening slaughter cattle have been investigated as well as the most desirable size and conformation for the greatest efficiency. The cost of production when cows are confined to small lots the year round and all of their feed brought to them has been compared with more conventional methods where cows are grazed or allowed to forage in fields and pastures throughout the year. More recent research has been directed toward solving pollution problems, especially near large feedlots where thousands of animals are fattened each year. It has been estimated that a fattening steer can produce 10 to 12 tons of manure per year. Some research has been done to produce methane (natural gas) from cow manure, and it has been found that feedlots feeding 100,000 head of cattle per year can produce enough methane to supply the energy needs of a city of 30,000 population.

Methods of keeping livestock healthy through development of vaccines and treatments of various cattle diseases have been developed. Research of this kind is almost continuous because new or previously unrecognized diseases are discovered and become important periodically, and methods of treatment and control must be developed.

1.10   Future of the Beef Cattle Industry

The future looks bright for beef cattle production. Many acres of land all over the world cannot be farmed because they are rough and rocky or there is too little rainfall for crop production. Much of the land in the world can never be used for anything but grazing. Little energy is required for beef production by grazing as compared to crop production, and this may be of importance.

In the future beef cattle will be produced mainly on grass because of the increasing need to use grains and concentrates for feeding the human population. Methods of production and animals best able to produce beef on forage alone will be emphasized. Thus, more efficient production, including breeding, feeding, management, and disease control, will be of greatest importance in the future.

STUDY QUESTIONS

1. In what part of the world did plant cultivation and animal domestication begin?

2. To what genus and family do cattle belong?

3. When were cattle first domesticated, and what is their probable ancestor?

4. Why were cattle first domesticated? Define *domestication*.

5. Explain the possible reason for the development of many coat colors in cattle after domestication.

6. Trace the possible sources of cattle introduction into Great Britain in early times.

7. Why is Great Britain noted for breeding and producing excellent breeds of livestock?

8. Are cattle native to North and South America? Are American bison related to cattle?

9. How did the American Indians make use of the bison?

10. Did the American Indians try to domesticate bison? Explain your answer.

11. What was responsible for the near extinction of the American bison? What use was made of buffalo bones on the prairies?

12. What were the consequences of the removal of the bison from the ranges?

13. When were domesticated cattle introduced into the New World and by whom?

14. How did Spanish-type cattle become established in the western and southwestern United States?

15. Why did range cattle greatly increase in numbers during the Civil War?

16. What was responsible for the development of cattle trail drives in the Southwest after the Civil War?

17. What factors were responsible for the decline and end of the cattle trail drives?

18. In what ways are cattle of today of importance to the people of the world?

19. Which are most efficient in the conversion of feed to meat, cattle or swine? Why the difference?

20. What country of the world has the largest number of cattle? Why are cattle numbers so high in this country?

21. Most of the cattle in the world today belong to what two species? Describe these species.

22. What are the advantages and disadvantages of buffalo in tropical regions of the world?

23. What countries in South America are the leading cattle producers?

24. What country in North America is the leading cattle producer?

25. What handicaps do cattle have to cope with in Africa?

26. How much of Australia's beef supply is exported? Why is it exported?

27. In what parts of the United States are dairy cattle kept in largest numbers? Why?

28. In what regions of the United States are beef cattle numbers per 100 humans the highest? Why?

29. In what regions of the United States are beef cattle numbers per square mile the highest?

30. Have increases in cattle numbers in the United States kept pace with increases in the human population?

31. Do human population growth curves in the United States show cyclical trends as do increases in cattle numbers? Explain your answer.

32. In recent years what have been the main research objectives in (a) physiology of reproduction, (b) genetics, (c) nutrition, (d) meats, (e) production and management, and (f) veterinary medicine in beef cattle?

33. Is the future of beef cattle production in the United States doubtful? Explain.

34. Can the energy shortage have an effect on the future of beef cattle production in the United States? Explain.

35. Why is increased efficiency of beef production in the United States so important?

breeds of beef cattle

A breed of livestock may be defined as a group of individuals possessing certain common characteristics as a result of selection and breeding which distinguish them from other groups within the same species. Many centuries of domestication have produced numerous breeds of cattle all over the world.

Settlers from Europe who migrated to the temperate zone regions such as America, New Zealand, and Australia took their native cattle with them. These belonged to the species *Bos taurus*. Later when there was a greater emphasis placed on the use of purebred cattle, other importations of pure breeds were made, and an effort was made to further improve the pure breeds in certain areas.

The zebu-type cattle (*Bos indicus*) probably first originated in tropical or subtropical regions largely through natural selection for traits which adapted them to those regions. In recent years, zebu cattle have been introduced into many tropical or subtropical regions such as the Gulf Region of the United States and parts of central South America (R. W. Phillips, *Sci. Am.* 198:51, 1958). At least thirty-two different breeds of Brahman cattle are known in India, some of which are of dairy type and some of dual-purpose type (R. L. Kaura, *Indian Breeds of Livestock*, Prem Publishers, Golaganj, Lucknow, 1961).

Each breed of beef cattle today has its own registry association. In general, the breed association establishes an office, employs a secretary, receives and approves applications for registration according to certain requirements, and issues registration certificates. In some instances, the breed association develops a standard of perfection or official scorecard for the breed and adopts rules for advanced registry requirements.

Each breed association has as its goals the improvement and promotion of the breed. This is done by sponsoring breed shows and sales where prizes and premiums are offered for the winners. They also print and circulate booklets and leaflets for the purpose of advertising the breed and often print a breed magazine. The breed magazine, however, is sometimes printed by private individuals. The major source of income of the breed associations is from registration fees.

Breeds of beef cattle differ genetically, especially for certain of the simply inherited traits (qualitative traits) such as coat color. True and significant differences among breeds also exist for traits such as rate of gain, milking ability, and carcass quality and quantity, which are determined by many different pairs of genes. These traits are referred to as *quantitative traits*. Genetic differences among breeds exist because the various breeds have been isolated from each other for many years, they may have started from a different genetic background, and selection goals for the breed may have been different. Research indicates, however, that there is still considerable genetic variation within each of the breeds for the important economic traits so that improvement could be made in most of these by selecting and mating superior individuals within the breed which possess these desired traits.

2.1   Kinds of Cattle

Cattle have been selected and developed for many years with a particular purpose in mind: for beef, for milk production, and for draft purposes.

2.1.1   Beef Cattle

Beef cattle are those developed mainly for producing beef. Breeding animals are seldom fed grain and concentrates but subsist largely on roughage supplemented with minerals and proteins in the winter months or when pastures are poor and of low quality. The offspring, especially the steers, are placed on a full feed in the feedlot after weaning to fatten them for slaughter purposes. Generally, beef cows are not extremely heavy milkers, although some produce more milk (or less milk) than others. Beef cattle are well adapted to pasture and range conditions. Beef cattle tend to be rectangular in shape as viewed from the side, and the ideal animal is heavily muscled.

2.1.2   Dairy Cattle

Dairy cattle have been developed largely for milk production. They are usually fed grain and other concentrates in addition to roughage when lactating. They are more angular than beef cattle when viewed from the side, being deeper in the back portion of the body than in the front. In recent years, attention has been paid to the crossing of dairy and beef breeds for beef production, but the main purpose of the dairy breeds has been the production of milk for human consumption.

2.1.3   Dual-Purpose Cattle

Dual-purpose cattle are developed for both beef and milk production. In conformation and type they tend to be intermediate between the beef and dairy breeds. In recent years they have become popular in crosses with the beef breeds for beef production. Dual-purpose breeds have been more popular in European countries than in the United States but are now being used for beef production in some areas of the United States.

2.1.4   Triple-Purpose Cattle

Triple-purpose cattle have been developed for meat, milk, and draft purposes — hence their name. They have not been popular in the United States but have been more popular in regions of the world where machinery is scarce and oxen are still used for draft purposes. They are large at maturity and heavily muscled.

2.2   How Breeds Were Developed

Older breeds of beef cattle were developed in localized areas of a country by a relatively few breeders. When a group of cattle was developed in an area on farms that were productive and attracted the interest of cattle producers in that community, a new breed was soon developed. Since the original base for a breed was rather small, inbreeding and linebreeding were often used, at least until total numbers of individuals within the breed increased to the point where good-quality nonrelated individuals (not related within the last few generations) were available for mating. After a considerable number of animals were available, breeders began to select for certain points of type, coat color, and performance so that individuals within the group appeared more uniform. Still later a breed registry was formed to record the pedigrees of individuals to maintain the purity of the breed.

In recent years new breeds of cattle have been developed by first crossing two or more of the older breeds and then selecting within the cross for certain traits. For example, the Santa Gertrudis was developed by the King Ranch in Texas by crossing the Brahman and the Shorthorn breeds. The Brangus breed was developed by crossing the Brahman and Angus breeds. Breeds possessing Brahman blood were developed for the semitropical climates or arid regions of the United States.

Several new breeds from crosses of two or more older breeds have been developed by some commercial companies, especially for the purpose of using them in a rotational crossbreeding program. Single breeds, for example, the Hays Converter, have been developed by individuals in the same manner.

Old, established breeds from other countries, especially from western Europe, have been imported into Canada and the United States in recent years. These are referred to as exotic breeds. The term *exotic* means they are from a foreign source and not a native breed. Exotic breeds are pure breeds in the country of their origin and include a long list such as the Limousin, Simmental, and Chianina in addition to others. The exotic breeds are a new source of germ plasm for the country into which they are imported, and they may be of value in crossing with the older, standard breeds of cattle such as the Angus and Hereford. The Meat Animal Research Laboratory at Clay Center, Nebraska has an extensive project in which bulls of the exotic breeds are crossed with Angus and Hereford cows to determine their value in such crosses for beef production. The results of these experiments will be discussed in a later chapter.

Frozen semen from bulls of the exotic breeds has been extensively used for artificial insemination. The introduction of individual purebred animals of these various breeds is a slow process because they must be under quarantine for some time before being introduced into Canada and the United States. They are quarantined to prevent introducing new diseases such as foot and mouth disease into these countries. As a general rule, semen from bulls of the exotic breeds is used on cows of another breed, and then the crossbred females produced are bred back to a purebred bull of that breed, gradually increasing the percentage of blood of the exotic breed toward the purebred. Some breed associations consider such an animal a purebred when it reaches 31/32, or more, of the exotic breed. This is a grading-up process and takes several generations. The method of superovulation of the purebred exotic breed of cow and transfer of the fertilized ova to grade cows which nourish them internally and give birth to them has also been used to increase the number of purebreds of the exotic breed.

Following the first introduction of the exotic breeds, prices for individuals are very high because of a small supply. As the number of animals from these breeds increases, prices tend to become much lower.

2.3   Breed Differences

Breeds of beef cattle differ significantly in color, type, and certain traits of performance. Breed differences are genetic differences.

No single breed of beef cattle is superior genetically to all other breeds. One breed, however, may be superior for one trait and average or lower in another. When a new breed is developed by first crossing two or more older breeds, it is hoped that the best characteristics of each of the original breeds can be incorporated and retained in the new breed. The first cross, however, of two or more breeds produces hybrid vigor in traits which are affected by it. These are the traits related to physical fitness. When selections are made within the cross in successive generations, hybrid vigor tends to decline. For this reason it is difficult to maintain superiority for such traits in later generations of the new breed.

2.4   Characteristics of the Different Breeds

It is our purpose here to give a brief description of each breed of cattle which is present in considerable numbers in the United States. The breeds described will include the old standard beef breeds, some new breeds, and some exotic breeds which have enjoyed popularity after being imported into the United States. (See Table 2-1.)

2.4.1 Black Angus

The Black Angus was originally known as the Aberdeen Angus (Figure 2.1). It originated in northeastern Scotland in the counties of Aberdeen, Kincardine, and Farfar. Much of the land there is hilly or mountainous and is used mostly for grazing, but good crops of oats, turnips, and hay are also produced. The area in which the Angus breed was developed lies in a latitude north of Quebec, Ontario, and Manitoba in Canada. The weather is cold and damp most of the time, but the summers are not as hot as in the United States and not as cold in the winter as in our northern states.

Early cattle in the region where the Angus breed was developed varied in color from dun, red, brown, and brindle with white patches.

TABLE 2-1   Numbers of registrations for some major breeds of beef cattle in recent years

Year	Angus	Red Angus	Horned Hereford	Polled Hereford
1970	352,471	N.A.[a]	236,617	160,374
1971	346,196	N.A.	253,832	168,021
1972	350,910	N.A.	260,676	178,391
1973	348,517	6,661	269,890	168,746
1974	350,558	6,288	272,479	207,882
1975	306,495	5,796	251,772	165,772
1976	228,545	6,968	230,144	181,579
1977	264,621	9,993	218,681	173,010

Year	Shorthorn	Charolais	Limousin	Simmental
1970	36,651	45,328	5,660	6,630
1971	32,647	52,963	14,118	18,418
1972	34,917	58,824	25,135	48,625
1973	30,503	60,376	27,898	65,885
1974	N.A.	56,378	33,872	96,869
1975	29,185	45,232	29,913	83,540
1976	25,587	45,351	24,742	72,095
1977	26,648	41,348	23,137	62,001

[a]N.A., not available.

Figure 2.1.   The purebred Black Angus bull. (Courtesy American Angus Association)

Some were polled and some were horned. The polled characteristic is thought to have been introduced by a polled Scandinavian stock at a very early date. Two strains of cattle are considered to be directly involved in the formation of the Black Angus breed. Both were largely black and polled, although neither strain was uniform in type. One strain was known as the *Buchan humlies* (humbled refers to polledness) and the other as *Angus doddies*.

Angus were first introduced into the United States in 1873. Later importations followed, and the American Aberdeen Angus Breeders Association was established in 1883.

The modern Angus is black and polled, although occasionally red calves are produced. The red coat color is recessive to black. Angus cattle in the United States tend to be smaller than Herefords and Shorthorns and mature quickly. They are especially noted for the excellent marbling of their meat, which is a desirable characteristic. In recent years successful attempts have been made to increase the size and growth of this breed. Angus cows are good milkers and good mothers and are quite popular in the Midwest. They are popular for crossing with other breeds.

2.4.2   Red Angus

The recessive gene for red coat color in the Black Angus has not been eliminated in spite of selection against it for many years (Figure 2.2). When two Black Angus which are carrying the red gene (genotype *Bb*) are mated, they may be expected to produce a red calf about 25 percent of the time. Red Angus are not eligible for registration in the American Angus Breeders Association. Since red is recessive to black in the Angus, red individuals breed true when mated together, and it is relatively easy to develop a Red Angus herd.

The Red Angus Association of America was founded in 1954, and eligible red individuals are now registered in this association. This association is very active in production testing and has adopted a progressive improvement program.

Since the Red Angus has the same ancestry as the Black Angus, its origin and beef qualities are the same.

2.4.3   Hereford

The Hereford breed of beef cattle originated in the English county of Hereford on the southern border of Wales (Figure 2.3). The

Figure 2.2.   The Red Angus bull. (Courtesy Red Angus Association of America)

Figure 2.3.   The Hereford (horned). (Courtesy American Hereford Association)

country is well watered with many rivers and is primarily a grazing district. The surface of the area is undulating in long ridges. The lowlands between these ridges have many pear and apple orchards. The cattle depended more on grass than grain for fattening since the land was less fertile than the native home of the Shorthorns.

The early ancestors of the Hereford were solid red in color with widespread horns. They resembled cattle now existing in the counties of Devon and Sussex. In the early days, Herefords were bred for size in order to produce oxen for draft purposes. Most of the beef produced at that time was from the sale of old oxen and cows which had outlived their period of usefulness. Cattle with a red body and a white face had been in the county of Hereford for many years when breeders began to develop and improve the Hereford breed.

The first herd book was established as a private project by T. C. Eyton of Shropshire in 1846. After two volumes were published it was taken over by Thomas Duckham. After nine volumes were published, the herd book was purchased and published by the Hereford Herd Book Society of Great Britain. About the middle of the nineteenth century the extreme weight of Herefords was changed toward greater refinement, quality, and smoothness. It has been reported that the bull Cotmere (registration number 349), which was the winner of the first Royal Show in 1839, weighed over 3,500 pounds.

About 50 years later the heaviest of the winning bulls weighed about 2,600 pounds. The reason for the change was to produce better beef at an earlier age.

Herefords were first imported into America in 1817 by Henry Clay of Kentucky. Other importations soon followed, but the breed did not gain a good foothold in the United States until about 1870. Herefords found great favor on the ranges of the United States when crossed with the native Texas Longhorns. The first Herefords were taken to Colorado in 1870 and into Texas in 1874. The winters of 1880-1881 and 1886-1887 were extremely severe on the range, and cattle losses were heavy. Losses of all breeds of cattle were severe, but ranchers concluded that Herefords withstood these severe winters better than any other breed. This was probably why Hereford bulls were in great demand for crossing on range cattle, and they are still the favorites with many western cattle producers.

The American Hereford Cattle Breeders Association was organized in 1881. T. L. Miller printed the first volume of the *American Hereford Record* in that year, but the association published later volumes.

The modern Hereford in America varies in shades of red to yellow in color with white on the face and varying amounts of white on the crops, shoulders, and legs. Herefords have good constitution and vigor, are quiet and docile, and fatten easily. The cows are average to good milkers, depending on selection applied to improve this trait in a particular herd. The Hereford produces a good carcass when fattened in the feedlot.

2.4.4   Polled Hereford

The polled Hereford originated in the United States and Canada (Figure 2.4). Single-standard polled Herefords were developed in Canada by Mossom Boyd in 1893 by crossing two Angus bulls with ten purebred Hereford cows. A crossbred polled bull from these matings was then mated to purebred Hereford cows, and many of the calves produced were polled. Selection within this strain produced many polled calves. W. W. Guthrie of Atchinson, Kansas used a naturally polled Hereford X Shorthorn bull on purebred Hereford cows, leading to the establishment of another polled strain.

Warren Gammon and son of Des Moines, Iowa were the first to breed purebred polled Herefords. They sent out letters to many pure-bred Hereford breeders in order to locate purebred polled cattle. From this search they located and purchased seven cows and four

Figure 2.4.   The Polled Hereford. (Courtesy American Polled Hereford Association)

polled bulls. This served as the foundation of the double standard polled Herefords since they could be registered with both the *American Hereford Record* and the *Double Standard American Polled Hereford Record*. The American Polled Hereford Breeders' Association was organized in 1900. The polled Hereford has been greatly improved in the United States and is gaining in popularity.

Polled steers are greatly preferred in the feedlot, and even horned individuals are dehorned when they are calves. Naturally polled cows sometimes produce calves which are horned since the horned condition is recessive to polledness. Polled cattle also sometimes possess scurs (rudimentary horns not connected to the head), which is also hereditary. The mode of inheritance of scurs in polled Herefords appears to be a sex-influenced trait, being dominant in males and recessive in females. Scurs and horns are determined by separate pairs of genes in Herefords.

2.4.5   Shorthorn

The Shorthorn breed of cattle originated on the northeastern coast of England in the vicinity of the Tees River, which is the dividing line between the counties of Durham and York (H. W. Vaughan, *Breeds of Livestock in America*, College Book Company, Columbus, Ohio, 1941). The ancestors of the Shorthorn breed were probably introduced into England from northwestern Europe many centuries ago. At least in the early days there were cattle along the Tees River which resembled in size, shape, and color many of the cattle of Denmark and other regions of northwestern Europe. These cattle were extremely large in size, had a coarse head and short stubby horns, and were very rough and coarse in the body. They were late maturing but fattened rapidly when mature, making very heavy carcasses. Their flesh was coarse grained and was less tasteful than that from smaller breeds. Their coat color was quite varied with a decided lack of uniformity, apparently governed to a great extent by the law of chance. The cows were heavy milkers and gave more milk than most cows of those early days (Figure 2.5).

A certain amount of improvement was made in the ancestors of the Shorthorn breed before the herd book was established in 1822. These early cattle on both sides of the Tees River were then known as Teeswater cattle. Later they were sometimes referred to as Durham cattle. Many bulls in the first volume of the herd book lived before the year 1780, but their pedigree usually included only the

Figure 2.5. The Shorthorn. (Courtesy American Shorthorn Association)

name of their sire, and sometimes their grandsire, and very seldom included the name of the dam. It is assumed that the quality of the dams of these bulls was similar to that of their sires, although this cannot be proved.

Shorthorns were the first breed of beef cattle to be imported into the United States, and they became widely distributed at an early date. The first importations of this breed were made into Virginia in 1783. A second importation was made into New York in 1791, and the first direct importation into the Midwest (Kentucky) was made between 1820 and 1850. The Shorthorns became very popular in the United States, especially those of the Duchess line bred by Thomas Bates. Some Shorthorn bulls were used in the range country but later found very stiff competition from the Hereford breed. Shorthorns appear in largest numbers in the Corn Belt states.

Shorthorns of today are either white, roan, or red in color, although some of them are red and white spotted. The preferred color in modern times is a deep red color, with a dark roan being acceptable. White Shorthorns are not preferred unless they are outstanding in quality. Most Shorthorn breeders have not worried too much about color and have concentrated their efforts toward improving important economic traits.

Shorthorns, as the name indicates, have horns of moderate size and length. This breed is one of the heavier of the old, established beef breeds in the United States. Although they are large, they are considerably smaller than their early ancestors and are much more refined. One of the outstanding traits of the Shorthorn breed is the milking ability of the cows.

2.4.6   Polled Shorthorn

Polled Shorthorns originated in the United States in Ohio, Indiana, Illinois, and Iowa in about 1885, although efforts were directed toward this objective beginning about 15 years earlier. The breed consisted of two strains, the single standard and the double standard.

The first polled Shorthorns were produced by mating native polled cows to purebred Shorthorn bulls. The heifers from this first cross which had no horns were again mated to a purebred Shorthorn bull. This top crossing system of breeding each new generation of polled heifers to a purebred Shorthorn bull was continued until they were eligible for registration in the *American Polled Shorthorn Herd Book*. They were not eligible, however, for registration in the *Ameri-*

can Shorthorn Herd Book. This system of breeding eventually resulted in a strain that produced calves, most of which were polled.
The double standard polled Shorthorns were developed from purebred Shorthorns who produced naturally polled offspring and were
also eligible for registration in both the *American Shorthorn Herd
Book* and the *American Polled Shorthorn Herd Book* — thus the
name *double-standard.*

Polled Shorthorns are similar to horned Shorthorns for type,
color, and other economic traits except they are hornless.

2.4.7   Galloway

The Galloway originated in southwestern Scotland in the region
formerly known as the province of Galloway (Figure 2.6). Early
cattle in this area were horned and predominantly black in color,
although some were red, brown, and brindled. Some even had white
faces. It is not known how the polled trait was introduced, but it
apparently was recognized as desirable by early breeders, and they
selected for it because by 1810 the majority of cattle in this region

Figure 2.6.   The Galloway. (Courtesy American Galloway Breeders
Association)

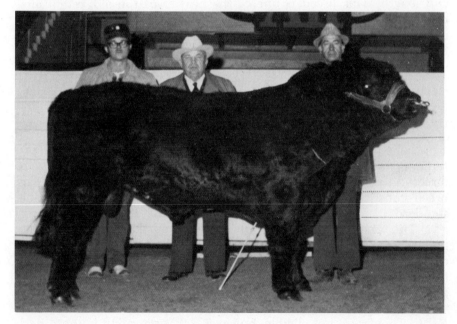

were polled. The Aberdeen Angus and Galloway are very similar in type and other characteristics.

The *Polled Herd Book* in which both the Angus and Galloway were recorded was established in 1862. In 1877, Galloway breeders established their own herd book.

The breed first gained a foothold in America when the Graham brothers imported cattle into Ontario, Canada, in 1853. Other importations followed, and in about 1882 and 1883, several hundred head were imported into the United States. They were popular in the United States for a time but later gave way to the Angus and have never been present in large numbers since that time.

Galloway cattle are smaller in size than Angus cattle and are very early maturing. They are excellent grazers and feeders, are average in milk production, and produce excellent beef. In the winter, Galloways develop a coat of long, wavy hair which at one time was in demand for the making of robes and coats.

2.4.8   Charolais

The Charolais breed is one of the oldest and most important breeds of beef cattle in France and was apparently developed in the region near Charolles (Figure 2.7). Pictures and written statements date white cattle in this area to as early as A.D. 878. The Charolais breed remained in its native district until about 1773 when Claude Mathieu moved from Oye in Saone-et-Loire and settled in Anlezy in Nieure with his white cattle. His success in agricultural operations attracted the attention of others who followed his lead, which resulted in the spreading of the Charolais breed into central and northern France. Because they were well muscled, Charolais oxen proved to be excellent draft animals until World War I. They were widely used for this purpose until supplanted by tractors and other machines. This resulted in more attention toward breeding Charolais only for meat production.

The *Charolais Herd Book* was established in 1919 by combining the herd book of the Nivernais-Charolais breed founded at Nevers in 1864 and the herd book of the pure Charolais breed established at Charolles in 1882. It has been closed since 1920 and accepts registrations only on the grounds of progeny. Total registrations in France are around 60,000 to 70,000 head, with yearly registrations of 25,000 to 30,000 head.

The Charolais is a large breed. The breed standard calls for a creamy white color; a fairly small, short head with a broad fore-

Figure 2.7   The Charolais. (Courtesy American International Charolais Association)

head; well-developed jaws; a broad muzzle; long, white, round horns;
a deep chest; round ribs merging into the shoulders; a horizontal
muscular back; very wide and thick loins; hips which are slightly
recessed but very wide at the rump; a well-rounded thigh that is fairly
short; and well-balanced and not overthin legs.

The Charolais breed has spread to South America, Mexico, the
Philippines, and southern Africa as well as other parts of Europe. In
recent years it has gained in popularity in the United States. This
breed is very hardy and does well in semitropical areas.

2.4.9   Brahman

Zebu or "humped cattle" belong to the species *Bos indicus* even
though they cross readily with European cattle of the species *Bos
taurus* (Figure 2.8). Offspring from such matings are also fertile.
Archaeological evidence suggests that zebu cattle were domesticated
in Egypt and Asia Minor about 2000 to 3000 B.C. They later spread
to India and Africa.

Most of the Brahman cattle as we know them today are of

Figure 2.8.   The American Brahman. (Courtesy American Brahman Breeders Association)

Indian origin. Thirty-two breeds of cattle have been described in India (R. L. Kaura, *Indian Breeds of Livestock*, Prem Publishers, Golaganj, Lucknow, 1961). Cattle in this country are the main source of power for agricultural operations and play a major role in the maintenance of soil fertility. They also provide milk and milk products, which are the main sources of animal proteins in the predominantly vegetarian diet of the Indian people. Cattle are of no great economic importance as meat producers in India because they are generally held sacred by the masses.

The first Brahman cattle were imported into South Carolina in 1849; importations into Louisiana were made in 1854. Other importations were made into Texas in 1885, 1895, 1906, and 1910. The American Brahman Breeders Association was organized in 1924, and a herd book was established.

Gray is the most popular color for Brahmans in the United States, although red is preferred by some breeders. Other colors sometimes occur but much less frequently than gray or red. The red color in American Brahmans comes mainly from the Gir, Sahiwal, and possibly the Sindhi breeds of India.

The Brahmans are especially noted for their ability to withstand high temperatures. This ability is attributed to an abundance of loose skin which increases the body surface exposed to cooling, the presence of sweat glands, and the production of less internal body heat in warm weather than the European breeds. Brahmans also have a certain amount of resistance to insects. The short, thick hair of the Brahmans prevents penetration by insects. Their skin is difficult to penetrate because of its thickness and a well-developed subcutaneous muscle which enables them to dislodge insects by shaking the skin. Brahmans also secrete an oily substance through the skin which is an effective insect repellent. They are also more resistant to certain diseases such as cancer eye, pinkeye, and anaplasmosis than the British breeds.

Indian breeds of Brahman cattle are resistant to heat and high humidity and are better able to withstand attacks of the major cattle plagues and tick-borne diseases than cattle developed in other parts of the world. They also do well where grazing is sparse. Although Brahman cattle are relatively poor producers of milk and meat, performance can be greatly improved through better management and feeding and by selection and breeding.

The greatest usefulness of the Brahmans in the United States has been in crossbreeding with native cattle or cattle of the European breeds and the formation of new breeds, particularly adaptable to semitropical or semiarid conditions.

2.5   New American Breeds

New breeds of beef cattle first developed in the United States were based on crosses among the European and Brahman breeds. Such crosses have introduced the hardiness, heat, and disease resistance of the Brahmans into the new breed as well as the beef and performance qualities of the European breeds. After the first cross, or later crosses, selection within the new breed has been directed toward maintaining the desired qualities of both parental breeds, and in some instances there has been an attempt to control a particular proportion of blood from each parent breed. The new breeds have been bred and developed mainly for the southern and southwestern states.

2.5.1   Santa Gertrudis

The Santa Gertrudis is the first true breed developed in the United States (Figure 2.9). It was developed on the King Ranch of south Texas because the European breeds did not thrive in the semi-

Figure 2.9.   The   Santa   Gertrudis.   (Courtesy   Santa   Gertrudis
Breeders International)

tropical environment of that area. The King Ranch was founded by
Captain Richard King in 1853. From that time and until about 1880
the King Ranch was stocked with Longhorn cattle which were very
hardy but did not possess desirable carcass quality and quantity. The
King Ranch began crossing Shorthorns and Brahmans in about 1910
with the hope of combining the desired qualities of both breeds in
their offspring. A crossbred bull calf named Monkey (because of his
antics) was born in 1920. He had a gentle disposition, was red in
color, and was beefy. He weighed 1,100 pounds by the time he was
12 months old (Anonymous, *Santa Gertrudis Breeders International
Recorded Herds,* Vol. 2, Santa Gertrudis Breeders International,
Kingsville, Texas, 1959). Later, when used for breeding, he proved to
be a superior sire, producing more than 150 superior sons before his
death in 1932.

Through the use of linebreeding and inbreeding and mass selection, the characteristics of Monkey and his descendants were developed into a new strain first recognized as a breed by the U.S.
Department of Agriculture in 1940. The name of the breed was
derived from the Spanish land grant named Santa Gertrudis on which
the headquarters of the King Ranch are now located.

The Santa Gertrudis Breeders International was formed officially
in 1951 with a charter membership of 160 breeders. The association

proceeded to classify all Santa Gertrudis cattle into two categories, certified purebreds and accredited. Certified purebreds were those from multiple- or single-sire herds which had four or more top crosses and a high standard of excellence. Accredited herds (females only) were multiple- or single-sire herds with at least three top crosses and which passed a minimum standard of excellence.

The standard of excellence for the breed lists a solid red coat color with red-pigmented skin, mucous membrane, horns, and hide as the most desirable. A mild temperament is desired, and the hair should be short and straight with a slick coat. The hide most desired is loose and thin with the surface area increased by neck folds and sheath. A large size is desirable, and especially desirable is the ability to make rapid gains on grass alone. The association's program calls for strict classification and performance-testing programs aimed at breed improvement.

2.5.2   Beefmaster

The foundation herd which eventually led to the establishment of a new breed of beef cattle known as Beefmasters was established in 1908 by Edward C. Lasater of Falfurrias, Texas (Figure 2.10).

Figure 2.10.   The   Beefmaster.   (Courtesy   Beefmaster   Breeders Universal)

The ranch where this breed was developed consisted of about 400,000 acres stocked with about 20,000 head of cattle. The large range herd consisted mostly of Herefords and Shorthorns, but in 1908 Brahman bulls were purchased and mated to some of these cows. Top crosses of the Brahmans were continued until finally a nearly pure strain of Brahmans was developed.

After Ed Lasater's death in 1930, his son, Tom, took over the management of the Brahman and registered Hereford herds his father had developed. Tom Lasater began a crossbreeding program by mating Herefords and Brahmans reciprocally. He also developed some Shorthorn X Brahman crosses by mating Shorthorn bulls to Brahman cows. Many of the bulls from the Shorthorn X Brahman cross were mated to Brahman X Hereford cows, giving a three-way cross. The calves from such a cross were superior to any of the calves from the two-breed crosses. Upon observing this, Mr. Lasater started an effort to convert his entire herd to the Hereford X Brahman X Shorthorn cross which finally became the new Beefmaster breed.

In developing the breed, selection was for the best animal for commercial production under range conditions. Little attention was paid to selection for polledness and coat color, and no attempt was made to maintain a certain percentage of blood from any of the three breeds. The Beefmaster is probably a little less than one-half Brahman with the remainder equally divided between Herefords and Shorthorns. Selection was, and is, based on disposition, fertility, weight, conformation, hardiness, and milk production. Beefmasters were recognized as a breed by the U.S. Department of Agriculture in 1954. The Beefmaster Breeders Universal was organized in 1961. Headquarters are now located in San Antonio, Texas.

2.5.3   Brangus

The Brangus is strictly an American breed which has been developed by blending the blood of the Angus and Brahman breeds. Brangus cattle are three-eighths Brahman and five-eighths Angus (Figure 2.11). The breed traces its origin to the efforts of Frank Buttram, who purchased a ranch in northeast Oklahoma in a region famous for its limestone soil and bluestem grass. This ranch, later known as the "home of Brangus," was located near Welch, Oklahoma. Mr. Buttram began full-scale operations to develop the breed in 1942. About the time that he began his operation, Tom Slick of San Antonio, Texas, began crossbreeding Brahman and Angus cattle. By 1947 Mr. Slick

Figure 2.11.   The   Brangus.   (Courtesy   International   Brangus
Breeders Association)

set up a ranch to breed for Brangus. Others have followed in the development of the breed in later years.

The American Brangus Breeders Association was organized in
1949 with the association office at Vinita, Oklahoma. The main
office was later moved to Kansas City, Missouri, and the name
changed to International Brangus Breeders Association.

The Association registers only Brangus cattle of the three-eighths
Brahman and five-eighths Angus breeding. Some Brangus are produced by the mating of Brangus to Brangus, but it is possible to develop Brangus quickly by using certain foundation stock. Foundation
stock used for this purpose must meet certain qualifications set by
the association. Although they register only Brangus, the association
requires that all foundation stock necessary to produce the three-
eighths–five-eighths Brangus be enrolled with them. Intermediate
foundation animals, based on their percentage of Brahman blood, are
known as quarter bloods, half bloods, and three-quarter bloods.
Quarter bloods, for instance, result from the mating of half bloods
with purebred Angus, whereas three-quarter bloods are the result of
mating half bloods with pure Brahmans. Quarter bloods mated to
half bloods or three-quarter bloods mated to purebred Angus give
the necessary three-eighths–five-eighths blood and are eligible for
registration.

Brangus cattle are black and polled and are large cattle with the ability to make rapid gains, with many bull calves weighing more than 1,000 pounds at one year of age. They also produce very desirable carcasses. Brangus possess many of the desirable traits of both of the foundation breeds and appear to be gaining in popularity, especially in the southwestern United States.

The standard of perfection for Brangus cattle calls for a solid black color as the most desirable with black-pigmented mucous membranes, skin, and hooves. Red Brangus are also produced; they are also polled.

2.5.4   Charbray

The Charbray was developed in the United States by a crossbred foundation of Charolais and Brahmans (Figure 2.12). The American Charbray Breeders Association was formed in 1949 and is now registered with the American International Charolais Association. Charbrays must trace in ancestry to a purebred Charolais and purebred Brahman cross to be eligible for registration. To be accepted for registration blood proportions must fall within limits of three-

Figure 2.12.   The Charbray.   (Courtesy   American   International Charolais Association)

fourths–one-fourth to seven-eighths–one-eighth Charolais-Brahman blood percentages. Some breeders feel that the three-sixteenths Brahman and the thirteen-sixteenths Charolais blood proportion is the most desirable.

Charbray cattle possess the ability of the Brahmans for heat tolerance, resistance to diseases such as pinkeye and cancer eye and to insects, as well as rustling ability on the range. They also possess many of the qualities of the Charolais, such as high milk production and rapid and efficient gains.

Most Charbrays are white in color, have horns, and are particularly adapted to semitropical regions.

2.5.5   Other Breeds

Other new breeds developed in the United States include the Beefalo,[1] the Barzona, and the Longhorn.

The Beefalo was developed by D. C. Basolo in Tracy, California. A typical Beefalo is three-eighths American bison, three-eighths Charolais, and one-fourth Hereford. Few experimental data are available to compare type, performance, and carcass traits of the Beefalo with other breeds in the United States.

The Barzona was developed by the Bard Ranch at Kirkland, Arizona. The genetic pool used to develop this breed included the Africander, Hereford, Santa Gertrudis, and Angus. The breed is adapted to semiarid regions.

The Longhorns are descendants of the original Texas Longhorns that were so widespread in the United States shortly after the Civil War. Some cattle of this breed are being crossed with others to determine the amount of hybrid vigor produced in the crossbred calves.

2.6   Exotic Breeds of Cattle

In recent years many breeds of cattle have been imported into Canada and the United States. Most of these have been from European countries. It is our purpose here to describe some of the breeds which were introduced first and which have increased in numbers since their arrival.

---

[1] There is some question as to whether or not the Beefalo is correctly referred to as a new breed.

2.6.1 Simmental

The Simmental is a very old breed of cattle which originated in the Simme Valley of Switzerland (Figure 2.13). It was not recognized as a breed, however, until 1862. Simmentals make up about 50 percent of all cattle in Switzerland and have spread to most European countries, South America, and Africa. In some European countries they are called the Pie Rouge. They were introduced into North America in the last few years.

Simmentals were first developed as a dual-purpose type of cattle, but more recently emphasis has been placed on the production of milk and meat. They are red and white in color, although some have an almost solid color. They have a white face similar to that of Herefords. The red coat color varies from dark red to almost yellow. They are well muscled and heavy boned with a good temperament. Simmentals are large, and their crossbred progeny make rapid gains in the feedlot. They also produce a desirable carcass. Simmental cows are good milkers and produce heavy calves at weaning. They cross well with the older, established breeds in the United States such as Angus and Herefords.

The American Simmental Association registers calves from this breed.

Figure 2.13. The Simmental. (Courtesy American Simmental Association)

2.6.2 Limousin

The Limousin originated in central France, and they were used as draft oxen for many centuries (Figure 2.14). They are a large breed but somewhat smaller and finer boned than the Charolais. They are heavily muscled with a small amount of external fat. They are red in color and possess horns.

The North American Limousin Foundation has been formed for this breed. Limousins appear to be growing rapidly in numbers in the United States. They cross well with the older American breeds to produce a steer with excellent carcass quality and quantity.

2.6.3 Maine-Anjou

The Maine-Anjou originated in western France in the provinces of Maine and Anjou. Maine-Anjou cattle have been known as a breed since 1900 (Figure 2.15).

The Maine-Anjou is probably larger than any other French breed. They are considered to be dual-purpose cattle, but presently most emphasis is on beef production. They are above average in milk production.

Figure 2.14.   The Limousin. (Courtesy North American Limousin Foundation)

Figure 2.15.   The Maine-Anjou. (Courtesy American Maine-Anjou
Association)

The Maine-Anjou is a long, rather upstanding breed with a very
long rump and average-sized bones. They are mostly red and white
spotted in color.

2.6.4   Chianina

The Chianina is probably the oldest breed in Italy and represents about 6 percent of the meat-milk-draft (triple-purpose) cattle in
that country. It is probably the largest breed of cattle in the world
(Figure 2.16).

Chianina cattle are named for their place of origin, which is the
Chiana Valley in the province of Tuscany. Their origin may trace
back to the days of the Roman Empire, and they may have been
introduced into France by Roman colonists and may have later served
as a foundation for the Charolais breed of today.

Mature Chianina bulls stand as much as 6 feet high at the withers
and may weigh as much as 4,000 pounds. The hair is white, but the
tongue, palate, nose, tail switch, and patches around the eyes are
black. The skin has black pigmentation except for the underline. The
breed shows about 13 to 24 percent inbreeding.

Figure 2.16   The Chianina. (Courtesy American Chianina Association)

The first importation of Chianina semen into the United States was made by the Italian White Cattle Breeders, Inc. of Monterey, California, in 1971.

2.6.5   Other Breeds

Other breeds imported into North America include the Gelbvieh, Norwegian Red, Scotch Highlander, Marchigiana, Pinzgauer, Blonde d'Aquitane, Murray Grey, Normande, South Devon, Red Poll, Tarentaise, Devon, and Welsh Black. Most of these are present in America in limited numbers. Other breeds are also imported from time to time.

STUDY QUESTIONS

1. Define a breed. How is it different from a species?
2. What are the two species of domestic cattle in the world. Will they cross and produce fertile offspring? Explain your answer.

3. What are the objectives of breed registry associations?

4. What are dual-purpose cattle? Triple-purpose cattle?

5. How were breeds of cattle of today developed?

6. Why is a new breed often developed from the cross of two or more older, established breeds?

7. Where in the United States are the exotic breeds being tested in crosses with the Angus and Hereford?

8. Why do breeds differ genetically for many traits?

9. What are the main differences between Red and Black Angus? Why is there a breed association for each of them?

10. Where did the Aberdeen Angus originate? When was it introduced into the United States?

11. Which is the easiest to breed, pure, Red, or Black Angus? Why?

12. Why were Herefords so popular in the western range country?

13. How do polled and horned Herefords differ? How were polled Herefords developed?

14. What are the distinguishing characteristics of Shorthorns? How were Shorthorns developed?

15. What are the distinguishing features of the Galloway? To what breed of cattle are they related?

16. Discuss the origin of Charolais cattle.

17. Why did the Brahmans become popular in the United States?

18. Name four American breeds of cattle developed in the United States, and describe their origin.

19. What is meant by *exotic* breeds of cattle? Name six exotic breeds that have been introduced into the United States in recent years.

20. Why were exotic breeds introduced into the United States?

21. What is probably the largest breed of cattle in the world? Where did it originate as a breed?

PART TWO

PHYSIOLOGY
OF
REPRODUCTION

CHAPTER 3

reproduction in the bull

The main role of the bull in reproduction is to produce viable spermatozoa capable of fertilizing the eggs produced by the cow. The bull must also be capable of delivering the spermatozoa to the reproductive tract of the cow when she is in heat and before the mature ovum is released from the ovary. This timing is important because the life of the spermatozoa in the female reproductive tract and the life of the ovum after ovulation are limited.

3.1 Anatomy of the Male Reproductive Tract

The reproductive tract of the bull includes the testicles; the accessory sex glands; the urethra, which leads from the testicles to the lumen of the penis; and the penis, through which spermatozoa pass during ejaculation. The various parts of the reproductive tract of the bull are shown in Figure 3.1.

The primary sex organs of the bull are the testicles, which are located in the scrotum. The bull possesses two testicles of about the same size, although in some bulls one is larger than the other. The

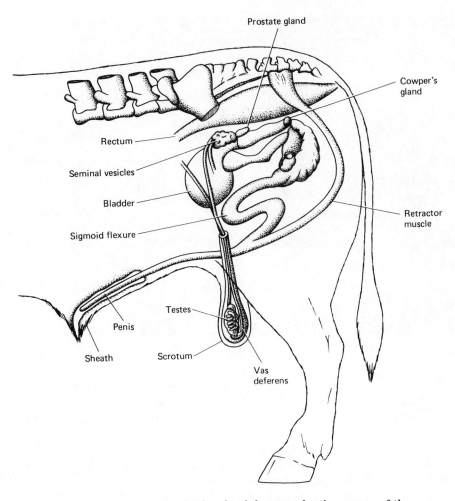

Figure 3.1.   Diagrammatic sketch of the reproductive organs of the bull.

testicles are oblong in shape, with the longer parts carried in a vertical position in the scrotum. The testicles of the mature bull have a firm consistency and lie free in the scrotum so they can be moved up or down to maintain the proper temperature for spermatozoa production and maintenance. Testicles of bull calves are in the scrotum at birth.

Within the testicles are long tubules called *seminiferous tubules* where the spermatozoa are produced when the bull calf reaches sexual maturity (called puberty) at 8 to 12 months of age. The seminiferous tubules are continuous with tubes in the *epididymis* and the *vas deferens.* Between the seminiferous tubules are cells called *interstitial cells* which are responsible for producing the male hormone, *testosterone.*

A connective tissue wall surrounds each seminiferous tubule. Within these walls is a layer of cells called spermatogonia, which are the primitive germ cells. These cells undergo a series of divisions in the process of spermatogenesis, as shown in Figure 3.2, to form first the primary spermatocytes, then the secondary spermatocytes, then the spermatids, and finally the spermatozoa. The spermatogonia and the primary spermatocytes possess pairs of chromosomes, but when the secondary spermatocytes are produced, they contain only half-pairs of chromosomes, as do the spermatozoa. Since the mature ovum also contains half-pairs of chromosomes, the union of the spermatozoa and egg at fertilization restores the normal pairs, or the total chromosome number of sixty in cattle. The spermatozoa produced in the seminiferous tubules eventually move to the epi-

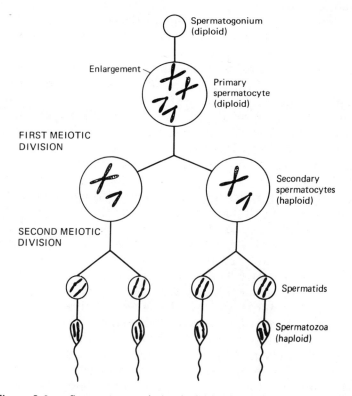

Figure 3.2.   Spermatogenesis (meiosis) in the testicles of the bull.

didymis where they are stored until they die and are resorbed or are ejaculated.

Billions of normal spermatozoa are produced within the testicles of the normal bull. The process of spermatogenesis in the bull has been estimated to require about 40 days from the beginning of sperm formation until the sperm reach the lumen of the seminiferous tubules. Another 14 to 21 days are required for them to pass to the tail of the epididymis. This length of time is variable, however. During the time the spermatozoa are in the seminiferous tubules they appear to be nourished by nurse, or Sertoli, cells.

The epididymis is a body of tissue located on the testis (the Greek *epi* means "upon," and *didymis* means "testis"). The seminiferous tubules of the testes are continuous with the tubules of the epididymis. It has been estimated that the length of the tubule in the epididymis of the bull is about 120 feet. The portion of the epididymis making direct contact with the testes and the seminiferous tubules is the head (*caput*) of the epididymis, the portion near the vas deferens is the tail (*cauda*) of the epididymis, and that portion between the head and tail of the epididymis is called the body (*corpus*) of the epididymis. The sperm mature and ripen in the epididymis and are stored there until ejaculated or resorbed. The tail of the epididymis is normally found on the ventral side of the testicle. In the mature bull the tail of the epididymis is usually distended with spermatozoa (Figure 3.3).

The tail of the epididymis on each testicle is continuous with the vas deferens, which connects this body with the urethra and penis. Sometimes bulls are vasectomized so they can be used for locating cows and heifers in estrus. A vasectomy is an operation in which a portion of each vas deferens is removed, preventing the passage of the sperm through the vas deferens so that no sperm can be eliminated when ejaculation occurs. Such an operation does not interfere with male hormone production or with the sex drive of the bull.

Both testicles of the bull are normally located in the scrotum, a pouch or bottlelike structure continuous with the skin of the body and located just anterior to the rear legs. The scrotum is covered with shorter and thinner hair than the rest of the body. The scrotum has a thermoregulatory function in that it keeps the testicles 4°C to 7°C cooler than the rest of the body. This cooler temperature is necessary for sperm production to occur. Two muscles function to raise and lower the testicles from the body. The *tunica dartos* muscle contracts when the bull is exposed to the cold and draws the testicles upward toward the body to keep them warm. When the temperature is hot, the *external cremaster* muscle relaxes, letting the testicles

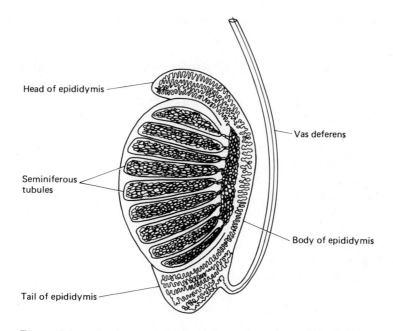

Figure 3.3.   Anatomy of the testicles and scrotum of the bull.

down from the body and causing a cooling effect. In cryptorchid males where both testicles are retained in the body cavity, no sperm are produced, and the male is sterile, although he produces testosterone and his sex drive (libido) is normal. *Monorchid* bulls, where one testicle is retained within the body cavity but the other is in the scrotum, are fertile.

Experiments in which the scrotum has been covered with material to raise its temperature have resulted in temporary sterility. When the covering is removed from the scrotum, however, sperm production resumes after a brief period required for recovery. Males of the first cross between the American bison and European cattle are sterile because the scrotum is carried so close to the body that the temperature of the testicles is too high for sperm production to occur.

Several hereditary defects affect the testicles of the bull. An autosomal recessive gene with incomplete penetrance has been reported to cause hypoplasia (underdevelopment) of the testicles in Swedish mountain cattle and the polled Finnoise breed in Finland. Both testicles may be affected, but the left is affected more often. Varying degrees of hypoplasia have been observed; in some males only the interstitial tissue of the testicles is affected, but libido may be present in some bulls. Fertility is usually low, but sterility sometimes occurs in affected bulls. The defect appears to be associated with the white coat color of the breeds. A similar condition has been reported in Ayrshires in Great Britain, but the mode of inheritance has not been reported. Cryptorchidism appears to be hereditary, but the actual mode of inheritance of this defect is not known.

The accessory sex glands include the seminal vesicles, the prostate, and the Cowper's glands. The seminal vesicles, erroneously named because they were once thought to be a storage place for the sperm, are paired and located near the point where the vas deferens (ampulla) joins the urethra. The secretions of the seminal vesicles make up about one half of the volume of the normal ejaculate of the bull. These secretions function in the transport of sperm through the reproductive tract during ejaculation.

The prostate gland surrounds the urethra near the opening of the bladder where the vas deferens enters the urethra. The paired Cowper's glands are located above the urethra at a point where it makes its exit from the pelvic cavity. Each Cowper's gland is about ½ to 1 inch in diameter in the mature bull. The secretions of the prostate and Cowper's glands are suddenly released at the time of ejaculation. Their main function appears to be the flushing of the tract and penis so the spermatozoa will not be damaged by accumulated urine or other detrimental materials as the sperm move through the tract at the time of ejaculation.

The penis is the copulatory organ of the bull and has two major functions. It serves as a passageway for urine when it is voided from the bladder and for sperm when they are ejaculated. It is also the means whereby semen is deposited in the reproductive tract of the cow during copulation. At the time of copulation the penis becomes erect and gorged with blood and is extended by the straightening of the sigmoid flexure due to the relaxation of the retractor muscle. Abnormalities of the penis sometimes occur in the bull. These include instances where the penis cannot be extended from the sheath because the sigmoid flexure does not straighten. This has been reported to be due to an autosomal recessive gene. The persistence of the penile frenulum has been reported to prevent normal protrusion of the penis in some Angus and Shorthorn bulls in the United States. Whether or not this is hereditary is not known, although heredity is indicated by the occurrence of this defect in herds where many individuals are related. Studies of Friesian bulls in West Germany detected several bulls in which the position of the penis was lower than normal when they mounted the cow and copulation could not take place. This was reported to be due to an autosomal dominant gene.

Bulls with the above-described defects are nonbreeders because even though they may produce a normal number of viable sperm, they cannot deliver them to the reproductive tract of the cow when she is in estrus. On rare occasions a bull may break his penis during attempts to copulate and cannot perform a normal service thereafter.

Semen is the ejaculate of the bull and contains spermatozoa and the secretions of the reproductive tract, especially the accessory sex glands. The volume of semen produced per ejaculate varies considerably with the age of the bull and the frequency of ejaculation. The average volume varies between 4 and 7 milliliters per ejaculate. The average number of sperm per milliliter of semen varies from 1 to 2 million, with a total of 5 to 15 billion sperm per ejaculate. Since only 1 sperm is needed to fertilize an egg, a large surplus of spermatozoa is introduced into the reproductive tract of the cow with a single ejaculation. Because of this large number, dilution of an ejaculate with appropriate extenders makes it possible to divide and use one semen sample on many females in the process of artificial insemination. Where the beef bull runs with the cows on pasture and may perform many services per day, the concentration of sperm and the volume of semen per ejaculate would be lower than the above-stated average figure.

Normally about 50 to 55 percent of the sperm in a fresh semen sample show a strong forward, progressive motility. Another 12 to 15 percent show weak motility without forward motion, and another 15 to 20 percent are not motile but alive. The remaining sperm are dead. This proportion of motile and nonmotile sperm was determined by using the hematocytometer for counting nonmotile sperm combined with the use of the live-dead stain (J. F. Lasley, *J. Animal Sci.* 10:211, 1951).

The shape of a normal bull sperm in semen is shown in Figure 3.4. The head of the sperm carries the genetic material, whereas the midpiece and tail supply the means of locomotion. The ability of the sperm to move forward much like a torpedo in water helps it come in contact with the egg in fertilization, although muscular movements of the female reproductive tract assist in transporting the semen to the site of fertilization in the upper two thirds of the fallopian tubes of the female.

Many abnormal forms of spermatozoa may be present in bull semen (Figure 3.5). These include tailless heads, headless tails, misshaped heads, coiled tails, two heads and one tail, etc. Abnormal sperm forms are due to upsets in spermatogenesis. Such upsets may be due to hereditary or environmental factors such as excessively high temperatures, disease, and possibly poor nutrition. Ten to 15 percent of abnormal sperm forms may occur in normal semen, but

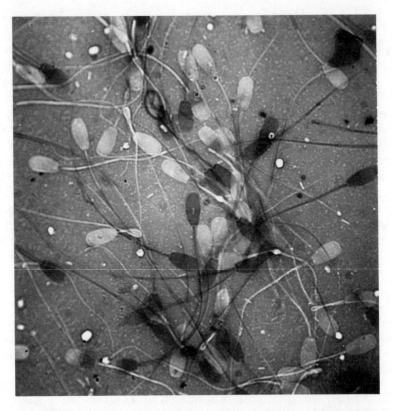

Figure 3.4.   Normal spermatozoa of the bull. The whitish sperm
were alive and the darker dead at the time slides were made of the
semen. (Courtesy Missouri Agricultural Experimental Station)

an excessively large percentage of abnormal sperm forms in the
semen indicates poor fertility or even sterility.

In Great Britain, an autosomal recessive trait in Guernsey and
Hereford bulls in which 40 to 95 percent of the heads were separated
from the tails has been reported. Affected bulls are subfertile or
sterile. A slight underdevelopment of the testicles was also noted in
bulls with this defect. In the United States an autosomal recessive
trait in which semen produced was low in concentration and high in
abnormal sperm forms has been reported. Affected bulls were of low
fertility. Knobbed sperm, or sperm with defective acrosomes, in
Friesians in Great Britain and the Netherlands have been reported.
An autosomal recessive mode of inheritance appears to be involved.
This condition results in a spot on the sperm head and is associated
with sterility.

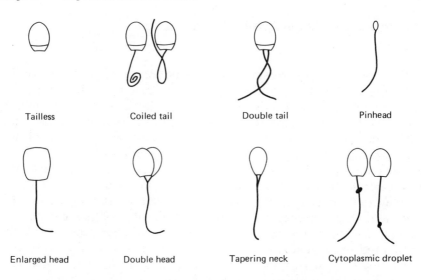

| Tailless | Coiled tail | Double tail | Pinhead |

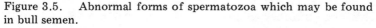

| Enlarged head | Double head | Tapering neck | Cytoplasmic droplet |

Figure 3.5.   Abnormal forms of spermatozoa which may be found in bull semen.

The sheath (or prepuce) of the bull is a long, narrow invagination of the skin and covers the extremity of the retracted penis. The opening of the prepuce is located slightly behind the navel and is surrounded by long preputial hairs, forming a tuft (Figure 3.6). In some bulls, especially Brahmans or those of partial Brahman breeding, the sheath is pendulous and hangs down from the body. Varying degrees of pendulous sheaths occur, and when the sheath hangs close to the ground the penis and/or surrounding tissues may be injured by contact with rocks and brush and may become infected. Less pendulous sheaths are desirable, and selection for such appears effective, indicating that inheritance is involved.

Figure 3.6.   Varying degrees of pendulous sheaths in bulls.

3.2   Role of the Bull in Artificial Insemination

Artificial insemination refers to the introduction of spermatozoa of the bull into the reproductive tract of the cow by artificial means. Since the normal bull produces 5 to 15 billion sperm per ejaculate, billions of them are wasted as only 1 sperm is required to fertilize each ovum. From 300 to 500 females can be inseminated with a single ejaculate if the semen is of good quality and properly extended. The idea was conceived many years ago to divide the ejaculate into many parts so that many females could be bred with a single ejaculate. Methods of storage have been perfected so that semen can be frozen and stored indefinitely and still retain its fertilizing capacity. Millions of cows in many countries of the world are now artificially inseminated each year. Genetically superior bulls can therefore sire several thousand offspring per year in countries all over the world, and some have sired up to 100,000 calves during their lifetime.

The large number of progeny per sire when artificial insemination is used makes it important to progeny-test and use only sires that are free of detrimental recessive genes for breeding. Otherwise, a bull that is a carrier of a detrimental recessive gene would transmit it to one-half of its offspring, and he could sire thousands of carriers of this defective gene in his lifetime.

Semen can be collected from the bull by the use of an artificial vagina. The bull may be stimulated into semen production by a cow in heat, other teaser animals, stimulation of the reproductive glands and organs by manual manipulation, or electrical stimulation. The artificial vagina (AV) was first developed by Russian scientists and is now widely used to collect semen for artificial insemination of cows throughout the world.

Semen harvested by means of the artificial vagina is usually clean and free of extraneous material. The internal temperature of the artificial vagina should be 43°C to 50°C and should have adequate length for the bull's penis. Artificial vaginas may be purchased from agricultural supply companies. When the bull mounts a cow restrained in a chute, the penis is directed into the artificial vagina where he ejaculates and the semen is collected in a glass container, usually a centrifuge tube, attached to a rubber directacone at the opposite end of the artificial vagina where the penis is directed. Care is taken not to expose the freshly collected semen to intense heat, cold, sunlight, or other substances that might damage or kill the sperm.

Freshly collected semen is usually checked for motility and normal morphology of the sperm before the extender is added to increase its volume and preserve the life of the sperm. Many different extenders such as those containing egg yolk, gelatin, milk, or other substances have been developed. An extender must not be harmful to the sperm because it is used to prolong sperm life during storage.

Semen from the bull may be extended and successfully stored at about 5°C for up to 4 days. If it is frozen, however, it can be stored indefinitely. When it is to be frozen, the extended semen is divided into quantities appropriate for inseminating one cow and then is frozen in liquid nitrogen at a temperature of –196°C (–320°F). Semen is frozen in 1-cubic-centimeter ampules or in plastic straws of various sizes ranging from ¼ to 1 cubic centimeter. Before introducing the semen into the cow it is quickly thawed, usually in an ice water bath. Many established bull studs collect, freeze, process, and distribute semen from bulls owned by private breeders.

STUDY QUESTIONS

1. What is the role of the bull in reproduction? How important is this role?

2. Where are the spermatozoa produced in the testicles of the bull?

3. Where is the male hormone, testosterone, produced?

4. Chromosome numbers must be kept constant from generation to generation in order for cattle species to survive. How is this done? What is the number of chromosomes in cattle?

5. What is the average amount of time required for the beginning of sperm formation until sperm have collected in the tail of the epididymis? Using this information, how long would one expect a bull to take to recover from an upset in spermatogenesis?

6. What are the parts of the epididymis? What is the function of the epididymis?

7. What muscles are required for raising or lowering the testicles in the scrotum? Why is this important in the male?

8. Why are males from the cross between the American bison and European cattle usually sterile?

9. What are two possible genetic defects in the testicles of the bull?

10. What are the accessory sex glands? What is their function?

11. What are some possible genetic defects affecting the penis of the bull that might interfere with his ability to reproduce? What is a nongenetic defect that could affect the penis of the bull?

12. How many spermatozoa are present in the normal ejaculate of the bull? How many sperm are needed to fertilize a single ovum? What use has been made of the surplus of spermatozoa in each ejaculate?

13. What percentage of sperm in a semen sample probably shows strong, forward motion when observed under the microscope?

14. What are some abnormal shapes of sperm in bull semen? Why may these abnormal forms be important, and what causes them?

15. What are some genetic defects of the sperm that may cause sterility in bulls?

16. What is a pendulous sheath? What problems may occur if the sheath is so pendulous that it touches the ground in some bulls?

17. What is artificial insemination, and how may it be used in cattle breeding?

18. How many calves may a bull sire in a lifetime through the use of artificial insemination? Why is it important to progeny-test bulls used for this purpose?

19. Name three methods that may be used for collecting semen from bulls. Which one appears to be the best and most practical?

20. What method may be used to store bull semen indefinitely? What is the practical significance of this method?

CHAPTER 4

reproduction in the cow

Reproduction in the cow is much more complex than in the bull. The bull merely supplies and delivers the spermatozoa that fertilizes the egg, but the cow must supply and deliver the ovum as well as nourish the developing young from conception to weaning. Besides this, the cow also contributes one half of the inheritance of each calf she produces.

4.1. Anatomy of the Female Reproductive Tract

The internal female reproductive organs of the cow include the ovaries, the oviducts or fallopian tubes, the uterine horns, the uterus, the cervix, and the vagina. The external portion of the female reproductive tract is the vulva. These are shown in Figure 4.1.

The reproductive tract of the cow is suspended in the abdominal cavity by means of the *broad ligament*. The broad ligament has three parts: the *mesovarium*, which supports the ovaries; the *mesosalpinx*, which supports the oviducts; and the *mesometrium*, which supports the uterus.

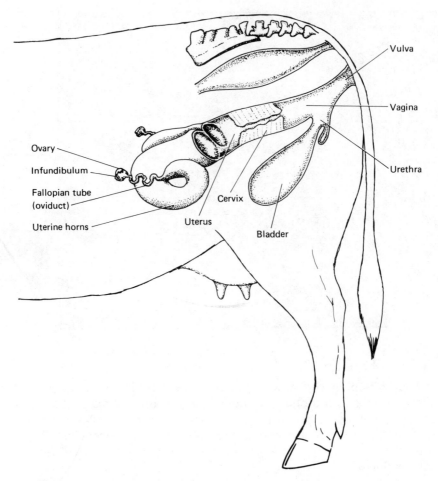

Figure 4.1.   Anatomy of the reproductive tract of the cow with the location of various parts within the body. (Courtesy National Association of Artificial Breeders)

4.1.1   The Ovaries and the Ova

The two ovaries of the cow are located in the abdominal cavity and unlike the testicles of the male produce mature ova at body temperature. The ovaries of the cow are shaped like an almond and weigh 12 to 20 grams in the mature female. A drawing of the female reproductive tract outside the body showing the location of the ovaries is shown in Figure 4.2.

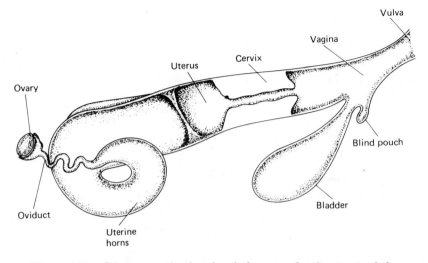

Figure 4.2.   Diagrammatic sketch of the reproductive tract of the cow. (Courtesy National Association of Artificial Breeders)

The mature ovum develops within a Graafian follicle which is filled with fluid. Before ovulation the mature follicle resembles a large blister on the skin. Follicles may develop at a young age in the heifer, but they usually do not mature until she reaches the age of puberty. Puberty varies between 7 and 15 months in the heifer, depending on the breed and nutrition of the individual as well as other factors. Ova usually do not mature and are not ovulated in pregnant cows or in cows after calving for a period of 20 to 40 days.

Ovulation is the process whereby the ovum is released from the ovary after the rupture of the Graafian follicle. A corpus luteum then develops at the point where the follicle ruptured (see Figure 4.3). At first the ruptured follicle is filled with a blood clot and is called the *corpus hemorrhagicum*. Connective and other tissue replaces the blood clot, and it becomes a *functional corpus luteum*. The functioning corpus luteum is sometimes referred to as the "yellow body" because of its yellowish-orange color. If pregnancy does not occur, the corpus luteum begins to regress and takes on a whitish color. It is then known as a *corpus albicans*.

4.1.2   The Fallopian Tubes

The fallopian tubes (oviducts) are two in number with each continuous with a uterine horn (Figure 4.1). The fallopian tubes in

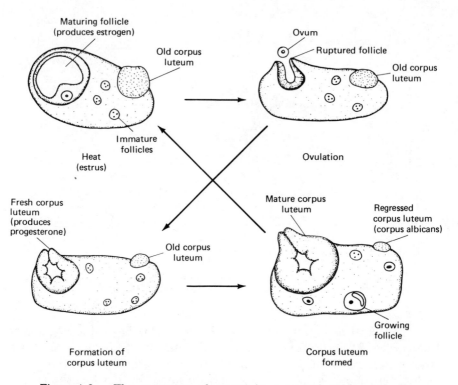

Figure 4.3.   The sequence of events in a typical 21-day estrual cycle in which pregnancy does not occur. (Courtesy National Association of Artificial Breeders)

the cow are 20 to 30 centimeters in length, although they would be longer if the coils were straightened. The end nearest the ovary expands to form a funnel-shaped structure called the *infundibulum*. At ovulation the egg finds its way into this funnel and is transported down the tubes toward the uterine horns. The transport of the sperm toward the egg and the egg toward the sperm is brought about by the beating of cilia on the outer layer of cells within the fallopian tubes and by muscular contraction of the tubes.

4.1.3   The Uterus and Uterine Horns

In the cow the body of the uterus is short and the horns relatively long. The uterine horns are well developed in cattle because it is here that the fetus grows and develops. The uterus receives its blood and nerve supply through the broad ligament within which the blood vessels are numerous. The inner surface of the uterus of the cow contains buttonlike projections called *caruncles*. The caruncles are found in the body of the uterus and the two uterine horns. In nonpregnant cows, the uterus and its horns have 70 to 120 caruncles, which are 12 to 15 millimeters in diameter. Each caruncle enlarges during pregnancy and attains a diameter of 8 to 11 centimeters. During pregnancy they appear spongy because of numerous crypts that receive the villi from the fetal chorion. The villi of the chorion in a localized area are called *cotyledons*. The cotyledons and the caruncles together are called the *placentome*.

4.1.4   The Cervix

The cervix is the opening of the uterus into the vagina and is sometimes referred to as the "mouth of the womb." The cervix is about 7 to 8 centimeters in length and contains characteristic annular folds or rings (about four in the cow). These often make it difficult to pass a tube or rod through the cervix in the process of artificial insemination. This is particularly true in heifers that have not produced a calf. In artificial insemination, however, the passage of an insemination tube is facilitated by applying movement of the cervix by means of the fingers of the inseminator working through the rectal wall. Then by moving the insemination tube from side to side it usually can be passed through the cervix and the semen deposited in the uterus.

During pregnancy the cervix is closed by a gelatinous plug which prevents the passage of bacteria or extraneous material into the uterus, thus protecting the fetus. The cervix becomes greatly enlarged (dilated) at calving time, forming a passageway for the calf into the vagina and through the vulva. The cervix usually has only one opening into the uterus, but a branched or double cervix has been reported to be inherited as an autosomal recessive trait with incomplete penetrance in Friesians and Herefords (K. Sittmann, *J. Hered.* 54:112, 1963) and as an autosomal dominant in the Pie Rouge breed of cows in the Netherlands. It is rare and does not appear to affect reproductive performance.

4.1.5   The Vagina and Vulva

The vagina is the portion of the female reproductive tract located between the cervix and the external opening of the reproductive tract called the vulva. Semen is deposited in the vagina by the bull during copulation. The vagina also serves as a passageway for the fetus at birth. The bladder opens into the ventral portion of the vagina so urine is also voided through this portion of the reproductive tract.

The vulva is the external opening of the female reproductive tract located just below the anus. It becomes greatly enlarged in most females as calving time approaches and is the final opening through which the calf passes during birth. It also becomes somewhat larger as the heifer reaches the age of maturity and is usually larger in older cows than in heifers.

4.1.6 Some Abnormalities of the Reproductive Tract

Underdevelopment (hypoplasia) of the Mullerian ducts has been reported in Friesians in Sweden and the United States (M. G. Fincher et al., *Cornell Vet.* 16:1–19, 1956). It is an autosomal recessive trait limited to the female. It is very similar to white heifer disease in Shorthorn heifers.

Underdevelopment of the Mullerian ducts has also been described in Shorthorn females. It appears to be due to the pleiotrophy (one gene affecting two or more traits) of an autosomal dominant gene with variable penetrance which determines the roan color (*RW* in the heterozygote) and the white color (*WW*) in the Shorthorn breed. The action of the gene appears to be limited to the Mullerian ducts, causing them to be underdeveloped and the females to be sterile. The action on the genital tract appears to be recessive (J. Rendel, *J. Genet.* 51:89–94, 1952) and appears in about 10 to 15 percent of white females.

Hypoplasia of the gonads has been reported at a high frequency in Swedish mountain cattle and the polled Finnoise breed in Finland. Both sexes are affected. It appears to be due to the pleiotrophic effect of the autosomal recessive gene (with incomplete penetrance) which determines the white coat color in these breeds. The hypoplasia can be partial or total. When it is total, no ovaries are present, and the female shows no estrus. When hypoplasia is partial, fertility is present in varying degrees (L. Lauvergne, *Bibliogr. Genet.* 20:1–60, 1966). The high incidence of the defect in the white Swedish breed has been reduced by selection against it, but it has not been completely eliminated.

About 90 to 95 percent of heifers born twin with a bull calf are sterile. Such heifers are known as *freemartins*. This condition appears to be due to common placental membranes of the twin pair, causing them to have a common blood supply during pregnancy. If the twins do not have a common blood supply during intrauterine life, the heifer is fertile. The fertility of the bull twin is affected little or not at all. The reproductive tract of the freemartin heifer is poorly developed and shows some masculinization. The condition may be identified by palpation of the female reproductive tract through the rectum in larger heifers or by culturing chromosomes in the lymphocytes in the blood of both twins. If the blood of the bull contains some dividing lymphocytes which carry two X chromosomes or if some of the dividing lymphocytes in the blood of the heifer twin contain XY chromosomes, this is indicative of a common intrauterine blood supply of the twins, and the heifer is sterile.

4.2 Physiology of Reproduction in the Cow

Physiology of reproduction in the cow involves an interrelationship between the anterior pituitary gland and the ovary. The ovaries of the heifer calf are inactive at birth but become active at the age of sexual maturity (puberty), which occurs between 4 and 14 months of age.

4.2.1 Puberty

Many factors affect the age of puberty in heifers. Puberty is reached at an older age in heifers maintained at high temperatures (above 80°F) than at lower temperatures (50°F). Nutrition also affects the time of onset of puberty. A low plane of nutrition usually delays puberty, whereas a high plane hastens its onset. Any factor responsible for slow growth may also delay puberty, perhaps because of the lower production of gonadotrophic hormones by the anterior pituitary gland or a lack of response of the ovaries to gonadotropins. On rare occasions fast-growing heifers and heifers large for their age may reach puberty while still nursing their mothers. Some such heifers exposed to a bull have been known to become pregnant while still nursing their mothers. This is a rare occurrence, however.

Differences among strains, breeds, sire groups, and crossbreds and inbreds indicate that puberty is affected by heredity. Jersey heifers are among those reaching puberty at a young age. Brahman heifers, on the other hand, usually reach puberty 6 to 12 months later than many heifers of the European breeds. Inbreeding delays the onset of puberty, whereas crossbreeding causes it to occur at a younger age. The age of puberty in different breeds and crosses is given in Table 4-1. In general, most heifers reach the age of puberty, with the exception of some Brahmans, in time to breed and produce their first calf at 6 years of age.

4.2.2   Estrus and the Estrous Cycle

Estrus is the occurrence of heat or the time the female will accept the male in the act of mating. The estrous cycle refers to the interval between successive estrous periods. The first estrous period in the heifer occurs at puberty, and it is due to the interaction of hormones produced by the anterior pituitary gland and the ovaries.

The anterior pituitary gland produces the follicle stimulating hormone (FSH) which causes the Graafian follicle to develop and mature in the ovary. As the follicle matures and produces a ripened ovum, it also produces estrogens which are responsible for the appearance of estrus and the initiation of the development of the uterus to receive the fertilized ovum if mating occurs. It also initiates the growth and development of the mammary gland tissue in the growing heifer.

Normal estrus, or heat, lasts from 10 to 26 hours in the cow, with an average of about 18 hours. The length of estrus varies quite widely, however, in different females. Some cows have a very short estrous period, whereas others have a long one. Some cows exhibit almost continuous estrus, which is due to a cystic follicle or one that

TABLE 4-1   Age and weight at puberty of heifers of different breeds and crosses[a]

Breed or cross	Number	Age at puberty (days)	Weight at puberty (pounds)
Hereford	62	415	603
Angus	64	366	561
HA & AH	132	371	585
Jersey crosses	117	322	482
S. Devon crosses	120	364	603
Limousin crosses	161	398	642
Charolais crosses	132	398	667
Simmental crosses	157	372	629
Hereford dams	504	389	612
Angus dams	441	363	592

[a]Data from *Progress Report No. 3*, ARS–NC–41, U.S. Meat Animal Research Center, Clay Center, Nebr., April 1976, p. 24.

does not rupture. A cow in almost constant estrus is known as a *nymphomaniac* (unusual sex desire). This condition may be successfully treated by inserting the hand into the rectum and picking up the ovary through the rectal wall and squeezing the follicle until it bursts. Sometimes hormones (gonadotropins) are used for treatment, but their use has not been as successful as the rupture of the follicle on the ovary by mechanical means. Nymphomania appears to be influenced somewhat by heredity, but environmental factors are also involved.

The estrous cycle varies between 18 and 24 days in the cow and averages about 21 days. One indication of pregnancy in the cow is the absence of estrus in the cow that has been bred. On rare occasions the interval between two estrous periods may be 8 to 10 days. This indicates that the corpus luteum was not functional as long as it normally would be. In other instances the interval between estrous periods may be 36 to 48 days, which indicates that either estrus was overlooked by the herder or the cow conceived when bred but shortly after the expected estrous period the fetus died and the female returned to estrus. Estrus sometimes occurs in pregnant cows, but a normal estrous cycle does not appear to be involved.

4.2.3   Ovulation

Ovulation is the term used to describe the release of the ripened ovum (or egg) from the Graafian follicle. Ovulation usually occurs 4 to 16 hours after the end of estrus and is triggered by the action of the luteinizing hormone (LH), secreted and released by the anterior pituitary gland.

Generally only one follicle ripens and is ovulated in the cow during an estrous period, but sometimes more than one is ovulated, producing twins, triplets, or even more young at birth. Twinning is rare, however, under normal conditions (see Table 4-2).

The ovary of the bovine female contains 50,000 to 70,000 primordial ova at birth, and she probably produces no new ones after that time. On the average only 5 to 10 of these ova ever mature and produce an offspring. Others that mature are wasted when mating and no pregnancy occurs. This represents a great waste of ova, especially in genetically superior females. The rate of development of these ova in the ovaries of cows and heifers may be greatly increased by proper treatment so that many follicles and ova may develop and many ova may be released from the ovary during one estrous period. This is known as *superovulation.*

TABLE 4-2   Incidence of multiple births in some breeds of cattle

| | | Estimated[b] | |
Breed	Twins per 100 births[a]	Triplets per 10,000 births	Quadruplets per 100,000 births
Angus	0.40–0.80	0.16–0.64	0.06–0.51
Charolais	1.50–2.00	2.25–4.00	3.38–8.00
Hereford	0.40–0.80	0.16–0.64	0.06–0.51
Holstein	2.00–3.00	4.00–9.00	8.00–27.00
Jersey	0.80–1.00	0.64–1.00	0.51–1.00
Simmental	4.50–5.00	20.25–25.00	91.13–125.00

[a]Incidence of twins represent averages of many reports in the literature. Twinning in individual herds or breeds may vary from these figures.

[b]Estimated from the method used in humans where the incidence of twins equals 1 out of n, triplets 1 out of n^2, and quadruplets 1 out of n^3.

Several treatments have been used to superovulate cows. One method is to stimulate the ovary to develop many follicles by injections of *pregnant mare serum* (PMS) which contains mostly FSH. Injections of PMS are given beginning the fifteenth day of the estrous cycle. The cow usually comes into estrus 5 days after treatment. When this estrus occurs, another hormone containing mostly luteinizing hormone (human chorionic gonadotropin) is given to induce ovulation. The ovary of the cow can be effectively stimulated between days 6 and 15 of the estrous cycle if prostaglandins are given 2 days later. About 10 percent of the treated cows do not respond to treatment to induce superovulation.

Superovulation is usually used in conjunction with the transfer of embryos. The transfer of embryos in cows refers to the transfer of a fertilized ovum from one cow into another cow that nourishes the foster calf from implantation until weaning. Superovulation is induced in the cow in order to produce several ova which can be transferred to several recipient cows. This makes it possible for one cow to produce many more offspring during her lifetime than under natural conditions.

Several commercial companies have been formed in recent years for the purpose of superovulating cows and transplanting embryos from one cow to another. These companies have met with varying degrees of success. The procedure is still quite expensive for each live calf produced.

A genetically superior cow is treated to produce several ova at one ovulation. The cow is then bred and fertilized eggs recovered at the proper time. One or more eggs can then be transferred to a recipient cow that provides the intrauterine nourishment for the fetus and actually gives birth to it and nurses it until weaning. Surgery is used to recover the embryos from the donor cows and to transfer them to the recipient cows. Recently, however, nonsurgical techniques have been developed for collecting embryos from donor cows.

Certain advantages have been proposed for embryo transplants. A genetically superior cow may produce up to 15 calves from a single treatment followed by surgery and transfer of these embryos to other cows. It may be possible to use immature heifer calves and infertile heifers for the production of embryos for transplant purposes. Transplants of embryos may make it possible to progeny-test cows at a relatively young age. This is not possible when the cow reproduces naturally. Embryo transplants may also be used to produce twins or other multiple births if this seems desirable.

Embryo transplants also have certain disadvantages. The procedure is very expensive and should be limited to the use of known genetically superior cows. It is possible that the number of treatments required for a cow to produce superovulations may be limited for a number of reasons. Approximately 10 percent of donor cows are subfertile after the first surgery because of adhesions within the reproductive tract and other causes. About 20 percent of donor cows are subfertile after the second surgery. Perhaps nonsurgical recovery of embryos from treated cows may prevent some of this difficulty. It is also possible that hormonal treatment to induce superovulations may impair future reproductive performance of donor cows, although the occurrence of this difficulty appears to be rare at this time. Some risk of death during the surgical process is also possible but not great. The skill of the person performing the surgery and the proper care and management of the animals help avoid some of these disadvantages.

4.2.4   The Corpus Luteum

After the egg is released from the follicle in the process of ovulation, a blood clot forms in the spot where ovulation has occurred. This blood clot is the initial stages of the formation of a *corpus luteum* (yellow body). The early corpus luteum is called a corpus hemorragicum because of its dark red color. The mature corpus luteum in the cow takes on a yellowish appearance — hence the name yellow body. A regressing corpus luteum becomes smaller and is whitish in color. It is called a corpus albican.

The mature corpus luteum produces the hormone progesterone. Progesterone prevents the growth of a new follicle and maintains pregnancy in the female. Progesterone also stimulates the growth of the uterine glands and the tissues of the mammary glands whose growth is initiated by estrogens secreted by the ovary.

Progesteronelike compounds may be used to synchronize estrus and ovulation in cows. Some of these compounds can be administered orally, whereas progesterone cannot because it is destroyed in the digestive tract.

In a group of mature, nonpregnant cows estrus occurs at random during a period of 18 to 24 days. Some of the cows may come into estrus on one day and others on another day. In a long breeding season cows may calve over a long period of time because they were bred at different times, and some of them may not have become pregnant when bred the first time. In addition, in a large herd of cows nursing calves estrus will occur at different times following parturition.

An understanding of the physiological factors involved in the appearance of estrus in cows has led to experiments in which estrus has been synchronized (caused to occur at the same time) in large groups of cows so they come into estrus and ovulate in a period of a few hours or a few days. This makes it possible to make maximum use of artificial insemination for breeding purposes and to provide adequate care for large numbers of females during a short calving period. Estrus can be synchronized only in sexually mature cows and heifers that are not pregnant and are cycling normally.

In estrus synchronization, progesteronelike compounds are given to a large number of cows in their feed. These compounds inactivate the ovary, preventing the growth and maturing of follicles and the occurrence of estrus. After being fed for several days, the compounds are withdrawn from the feed, allowing the ovaries of all of the females in the group to become active again and produce estrus and follicles in a synchronized manner.

Prostaglandins have also been used to synchronize estrus in nonpregnant and cycling cows and heifers. Injections of these compounds usually cause estrus to occur within 60 hours and ovulation in 90 hours after treatment when treated cows have a functional corpus luteum. Fertility in treated cows appears to be normal.

Plastic coils impregnated with progesteronelike compounds have also been used to synchronize estrus and ovulation in cows. The coils are inserted into the vagina of the cows and left for about 14 days, after which they are removed. Estrus and ovulation usually occur 56 to 72 hours after the coils are removed. Fertility in cows inseminated when the coils are removed appears to be normal.

A functional corpus luteum sometimes persists in the mature, nonpregnant cow. The progesterone it produces does not allow the ovary to develop normal follicles, and no ovulation or estrus occurs. Cows not exhibiting estrus for a long period of time should be examined to determine that they are not pregnant and whether a functional corpus luteum is present on the ovary. The manual removal of the corpus luteum by squeezing it from the ovary through the rectum usually results in estrus and ovulation in a few days, and the cow may be bred. The actual cause of a persistent corpus luteum is not definitely known, but several factors may be involved.

4.3   The Gestation Period

The period of pregnancy in the cow is also referred to as the gestation period. Defined in this way, pregnancy or gestation covers the period of time from mating until the calf is born. The fertilized eggs are usually located in the fallopian tubes at about 6 to 8 days after the cow has been in standing heat and has been bred. The developing embryo uses egg yolk and uterine secretions for nutritional purposes as it develops and forms the fetal membranes. The union of the developing fetus with the uterine tissue of the mother (called implantation) usually takes place from the eleventh to the fortieth day after ovulation. Implantation is a gradual process, with the union of the fetal and maternal membranes being rather loose at first. The fetal and maternal membranes become firmly attached by the beginning of the second month of pregnancy. Less than 5 percent of fetal death losses occur after that time. Most losses have occurred before that time and include those due to the failure of fertilization and losses of fertilized eggs and embryos before implantation.

In the cow the contact between the fetal membranes and the uterus is made at the caruncles to form the *placentome*, which includes both fetal and maternal tissues. Not all caruncles are functional during pregnancy, but the number becoming functional increases as pregnancy progresses. Cotyledons begin forming around the fetus about the fourth to fifth week of pregnancy and extend to other parts of the uterus and uterine horns as the fetus enlarges. Even the caruncles in the nongravid uterine horn may become functional. Each caruncle becomes enlarged to several times its original size as pregnancy progresses. Retained placentas after parturition are due to the failure of the maternal and fetal membranes to separate at the cotyledons.

4.4   The Embryo and Fetus

The developing individual is referred to as the embryo or the fetus. Basically, these two names refer to different periods in the life of the new individual. The new individual is an embryo until all of the body parts and organs are developed. Thereafter it is known as the fetus.

During pregnancy and after implantation the developing young receives its nourishment from the blood of the mother through the cotyledons. The placenta in the bovine is so impervious that even large molecules such as gamma globulins cannot pass from the blood of the mother to that of the young. For this reason the young must receive its supply of antibodies from the first milk, or colostrum, after birth. Urine excretion from the body of the fetus accumulates in the allantoic fluid. These excretions are relatively small in the fetus as compared to those after the calf is born and takes nourishment in the form of milk from its mother.

About 50 percent of the increase in fetal weight (growth) occurs during the last 2 months of gestation.Growth rate of the calf at birth, plus or minus 2 weeks, is about 1.0 to 1.5 pounds per day, but this growth rate varies with the genetic makeup and the sex of the calf. The weight of the calf at birth largely depends on the genetic constitution and nutrition of the developing fetus plus the length of gestation and the age and size of the cow. Although birth weights of calves from cows fed different levels of nutrition do not vary greatly, the nutrition of the dam does affect them slightly. The birth weights of calves from various beef breeds are shown in Table 4-3.

The heritability of birth weight in cattle varies between 35 and 40 percent. This indicates that birth weights are affected by heredity but that 60 to 65 percent of the variations in this trait are due to unknown environmental factors.

The length of gestation varies for individual calves, but it normally ranges between 270 and 290 days. The length of gestation for different breeds of beef cattle is shown in Table 4-4.

TABLE 4-3   Birth weights of calves from various beef breeds of cattle

Breed	Range in birth weight (lb)[a]
Angus	54–72
Hereford	63–82
Shorthorn	60–72
Charolais	87–108

[a]Summary of figures from many sources.

TABLE 4-4   Length of gestation in some breeds of beef cattle

Breed	Range in gestation (days)[a]
Angus	273-283
Hereford	279-287
Shorthorn	273-286
Charolais	278-288

[a]Summary of information from many different sources.

The heritability of gestation length in the beef breeds appears to be 45 to 50 percent, which suggests it is highly heritable. However, 50 to 55 percent of the variations in this trait are due to environmental factors. Heterosis appears to have little effect on gestation length because it averages about the same in crossbred calves as the average of the breeds making up the cross.

Prolonged gestation periods have been reported in several breeds of cattle in the world. Affected breeds include the Swedish Pie Rouge, the Finnish polled Finnoise, the Red Danish, Guernseys, Ayrshires, Friesians, native Japanese cattle, and the Pinzgauer of Austria. The length of gestation in some of these breeds is as long as 473 days and in others a few days shorter. In general, normal delivery in prolonged gestations is not possible, and calves have to be delivered by Caesarean section or by embryotomy. In most breeds the mode of inheritance of this trait has been reported as an autosomal recessive, with the homozygous recessive calf being carried an excessive length of time. Prolonged gestation has not been reported in American beef breeds.

Abortions, or the delivery of the fetus before full term, may be due to many causes such as injury to the cow, disease, and improper fetal development. It has been reported that European cows mated to American bison bulls often abort because of the production of an excessive amount of amniotic fluid. This suggests an incompatibility of the mother and fetus since the reciprocal cross does not appear to encounter this difficulty.

4.5   Parturition

Parturition is the act of giving birth to the young by the mother. Signs of approaching parturition in cows include an extended abdomen, an enlarged vulva, relaxation of muscles around the tail head,

an extended udder, the presence of milk, and a general restlessness exhibited by the cow. These signs vary within individual cows.
Cows near parturition will usually leave the herd and seek seclusion
in some isolated spot.

At parturition the unborn calf normally assumes a position in
which the nose and forelegs are directed toward the vaginal opening.
A position in which the calf is presented hind feet first with the
hocks up also occurs frequently. Any other mode of presentation is
considered abnormal and usually is accompanied by *dystocia* (difficult delivery). In abnormal presentations, assistance must often be
given by the herder or by a veterinarian or both the calf and mother
may be lost. (See Figure 5.3.)

Several stages of labor may be described in the cow. The first
stage is the preparatory stage, which includes the dilation of the
uterus and rhythmic contractions of the uterine muscles. These contractions force the fetus and membranes toward the cervix, causing it
to dilate. In the early preparatory stages, contractions of the uterus
occur every 15 to 20 minutes and last for about 20 to 25 seconds. In
the later preparatory stages, the uterine contractions increase in
strength and frequency so that they occur every few minutes. At the
end of this stage, the cervix is expanded so that the vagina and uterus
become a continuous canal. The placental membranes and the calf
are forced into the pelvis, and the fetal membranes burst, causing
fluid to flow from the vulva. A prolongation in the preparatory
stages indicates that there is some difficulty in presentation and that
assistance should be given.

The second stage of parturition in which the calf is delivered
normally quickly follows the first. The feet and head of the calf can
be seen extending from the vulva. One or both water bags rupture,
initiating contractions of the uterus. This normally forces the calf
through the birth canal. When expelled from the mother's body, the
calf is still attached to the fetal membranes, and it receives some
oxygen from the mother even if the birth process is prolonged. This
normally ensures an oxygen supply until the young begins to breathe
on its own. The second stage of parturition must be fairly rapid,
however, or the calf may suffocate. This is particularly true if the
calf is delivered hind feet first.

The third stage of parturition involves the expulsion of the
placental membranes, which is also associated with uterine contractions. The placenta is normally expelled within 24 hours after the
calf is delivered. The placental membranes may be retained for longer
periods in case of abortion, dystocia, and multiple births. Treatment
to remove the placental membranes often becomes necessary in such
cases.

Following the birth of the calf the reproductive tract of the cow is greatly enlarged, and mild uterine contractions may occur periodically. The uterus and the reproductive tract gradually return to their normal size. This is known as the *involution of the uterus* and usually is complete by 40 to 50 days following parturition.

The calf is born with little or no antibodies (gamma globulins) in the bloodstream because they are not able to pass through the placental membranes from the mother to her calf. For this reason it is imperative that the newborn calf receive the first milk (or colostrum) from its mother. Colostrum normally contains large amounts of antibodies which are absorbed into the bloodstream of the calf through the gut. The calf must obtain colostrum within a few hours after birth because the gut of the calf closes, and the antibodies cannot be absorbed into the blood stream.

Until recently little could be done to induce calving in pregnant cows. The discovery that a family of hormones secreted by the adrenal glands and known as corticosteroids could induce parturition has led to experiments to induce calving. Dexamethosone and Flumethosone are two corticoids commercially available for inducing parturition. When properly used they induce calving in 85 percent of the cases. Induced calvings should be limited to the last 2 or 3 weeks of pregnancy because calves are more likely to survive during that time than calves in earlier pregnancy. Cows usually calve within 34 to 60 hours after treatment with these compounds.

Induced parturitions may have several advantages. One of the main ones is that cows, and especially first-calf heifers, can be closely observed and assistance given to deliver her calf if it becomes necessary. Since induced parturitions shorten the period of pregnancy and lower birth weights, they may be useful in avoiding some calving difficulties. Survival of the calves and their growth rate after birth as well as milk production in treated cows appear to be normal in induced parturitions. Retained placentas frequently occur after induced parturitions with corticosteroids, and this may be one of the main reasons the procedure is not widely used on a practical basis.

Prostaglandins have also been used to induce parturitions in cows and heifers. They also increase the incidence of retained placentas, but they do not increase the incidence of abnormal presentations or calf mortality.

STUDY QUESTIONS

1. Why is reproduction in the cow much more complex than in the bull?
2. Name the internal female reproductive organs of the cow.

3. Define broad ligament, mesovarium, mesosalpinx, mesometrium, Graafian follicle, corpus hemorrhagicum, and corpus albicans.

4. What is puberty, and when does it usually occur in the heifer?

5. Where is the corpus luteum located in reference to the recently ruptured mature Graafian follicle?

6. By what means are ova directed into the infundibulum and transported down the fallopian tubes toward the uterus?

7. What are caruncles, cotyledons, and placentomes?

8. Why is it difficult to pass a probe through the cervix of the cow? Can it be done? Explain.

9. What is a branched or double cervix? What causes it? Does it interfere with reproductive performance?

10. What is white heifer disease? Is a similar condition found in heifers of other breeds? Explain.

11. What is a freemartin? Under what conditions does this occur? Is it due to heredity? Explain.

12. What methods may be used to determine if a heifer is a freemartin?

13. What breed of heifers reaches puberty at a young age? What breed of heifers reaches puberty at an older age than average?

14. Where are FSH and LH produced, and what is their relationship to estrus and the estrous cycle in cattle?

15. What is the condition in cattle known as nymphomania, and what is its cause? How can it be treated?

16. What two hormones are produced by the ovary? What are their functions?

17. What is probably wrong when a cow has two estrous periods within 8 to 10 days of each other?

18. What is ovulation, and when does it usually occur in the cow?

19. How many follicles usually ripen at one time in the cow?

20. How many primordial ova are present in the ovary at birth? How many more are produced after birth?

21. What is meant by superovulation? How can superovulation be induced in the cow?

22. Explain what is meant by embryo transplants. Why is superovulation used in conjunction with embryo transplants?

23. List some advantages that have been proposed for embryo transplants? List some disadvantages of this procedure.

24. Why can't the hormone progesterone be given orally and still be effective?

25. What is meant by the synchronization of estrus and ovulation? How is this done?

26. What is meant by a persistent corpus luteum? What are the results when this occurs?

27. How is the developing embryo nourished before implantation? How is the

fetus nourished after implantation? What is the difference between a fetus and an embryo?

28. Why is it necessary for a calf to obtain colostrum from its mother within a few hours after birth?

29. What is the heritability estimate for birth weight? For the gestation period? How much do environmental factors affect these two traits?

30. What is meant by prolonged gestation? What are the problems usually involved when it occurs, and is inheritance involved?

31. What are some of the signs of approaching parturition in the cow?

32. What is the normal position of the calf at the time of birth? If a calf is born hind legs first, could this cause some problems? Explain.

33. Describe the different stages of parturition in the cow.

34. What is meant by the involution of the uterus, and when does it usually occur?

35. What are some advantages and disadvantages of induced parturitions?

reproductive efficiency in beef cattle

CHAPTER 5

Reproductive efficiency may be defined as the ability of each mature cow in a herd to wean a calf each year. On a herd basis it may be defined as the percentage calf crop weaned for the total number of cows of breeding age in the herd.

Reproductive efficiency is the most important single trait in beef cattle production, because a calf must be born and survive if it is to produce a marketable product. A dry cow requires nearly as much feed as one that produces and nurses a calf each year. A dry cow will usually be fatter and heavier than a cow nursing a calf, which will add something to her value when she is sold for slaughter, but this additional weight is not equal in value to a good-quality, vigorous calf at weaning.

Reproductive efficiency in beef cattle is low and needs to be improved. Data from many agricultural experimental stations show that on the average about 83.3 percent of cows of breeding age in the herd give birth to a calf each year but that only about 71.6 percent wean a calf. This means that between 25 and 30 percent of the cows of breeding age in a herd fail to wean a calf each year. Some herds, however, wean a 90 percent or better calf crop year after year. These are well-managed herds that are exceptions rather than the rule. One herd in Missouri (H. Sewell et al., *Mo. Agric. Exp. Stn. Res. Bull.* 823, 1963) weaned a 92.1 percent calf crop over an 11-year period, illustrating that a high-percentage calf crop each year is possible.

5.1.   Influence of Heredity on Reproductive Efficiency

The heritability of reproductive efficiency is low, with most estimates between 0 and 10 percent. This low heritability estimate does not mean that this trait is not affected by heredity. It merely suggests that genes which act in an additive manner have little effect on this trait. Certainly many nonadditive genes such as those which are recessive in nature affect reproductive efficiency. The fact that this trait is improved by 5 to 10 percent when outbreeding or cross-breeding is practiced and declines when inbreeding is practiced suggests that other forms of nonadditive gene action such as over-dominance and possibly epistasis may affect this trait.

The repeatability of reproductive efficiency is also low (0 to 10 percent), which suggests that poor reproductive efficiency may not be a permanent characteristic of cows within a herd.

To improve reproductive efficiency from the hereditary stand-point, two general procedures are of importance. First, mating non-related parents gives hybrid vigor for this trait and improves it. Second, reproductive efficiency can be improved more rapidly by controlling or improving environmental conditions. Even though the heritability of reproductive efficiency is low, cattle producers should continue to cull infertile cows or cows of low fertility.

5.2   Factors Affecting Reproductive Efficiency in the Bull

Reproductive efficiency in beef cattle from the sire's standpoint may be low because it is vulnerable to environmental influences at several points of the reproductive cycle. These points include conception rate, embryonic death losses, and losses which occur between birth and weaning.

5.2.1   Conception Rate

Under natural breeding conditions reproductive efficiency of the bull can be influenced by his ability to produce viable sperm, the degree of his sex drive, and his ability to deposit enough viable sperm in the reproductive tract of the cow to fertilize the ovum. Failure in any one of these three phases can lower reproductive efficiency or result in a zero conception rate. Artificial insemination, of course, largely limits the effect on conception rate to the introduction of sufficient number of fertile, viable sperm into the female reproductive tract. It also adds the problem of finding cows in heat. Few herders are as good at this as the bull because the bull is with the cows day and night.

The close examination of the physical makeup of the bull together with an examination of the amount and quality of semen he produces helps find those of normal fertility.

Bulls vary considerably in many semen characteristics. Semen characteristics such as sperm concentration, sperm vigor, the percentage of live sperm, and the normal morphology of sperm appear to be low to moderately heritable. Physical defects related to breeding soundness such as defects of the prepuce, testicles, epididymis, vas deferens, and the feet and legs appear to be highly heritable (J. S. Brinks, *A.I. Dig.* 20:6, 1972). Selection for soundness in breeding bulls certainly appears to be important, but it also requires the best environment (feeding and management) for the best performance.

The consistency or firmness of the testicles as measured by means of a tonometer gave a correlation of 0.67 with fertility and showed a high correlation with semen quality and quantity (R. H. Foote and L. L. Larson, *Charolais Banner*, July 1973, pp. 188–189). A larger testicular circumference appears to be correlated with a greater sperm-producing ability. Body size is not a good indicator of testicular size, so testicular size should be measured in individual bulls. This measurement is highly repeatable, and bulls having smaller testicular circumference than average when young tend to have a smaller testicular circumference than average when they are mature. Testicular measurements should be taken and should be considered when selecting a potential herd bull prospect.

Exposure of a bull to too many cows during the breeding season together with poor nutrition can lower his reproductive efficiency. If several cows come into estrus on the same day when running with a single bull, the bull may not mate with one or more of them since some bulls seem to prefer certain cows to others. Any cow that is not bred will return to estrus 18 to 24 days later. Even if she conceives at the later estrous period, she will produce a lighter calf at weaning if all calves are weaned near the same time.

Under natural mating conditions, when a bull is exposed to a large number of cows, he may stay with cows in heat rather than graze, causing him to lose much weight. This is especially true under dry range or pasture conditions when feed is scarce. Poor nutrition resulting in a considerable weight loss can have an adverse effect on the bull's sex drive as well as his sperm-producing ability.

The number of cows that can be bred successfully by a single bull during the breeding season is quite variable, but some average figures may be given. A yearling bull can be pasture-mated to fifteen to twenty cows. If hand-mated, this number can be as many as twenty-five. A two-year-old bull (or older) may be pasture-mated with twenty-five to thirty cows, but if hand-mated, he may be bred to thirty to fifty cows in a breeding season. Hand-mating refers to bringing cows in heat to a corral where they are mated with the bull. Usually only one service is given per cow under such conditions.

For maximum use a sire should not be overfat. On the other hand, he should be kept in a thrifty condition throughout the year. It may become necessary to give bulls, especially young ones, some extra grain when pasture is dry and short or if the bull is in poor condition. Of course, salt and minerals should be available in the breeding pastures at all times.

Bulls may be run together in a pasture after the breeding season is over if a large number of bulls have been used. Some fighting will occur at first when bulls that are strangers to each other are brought together, and they should be given attention at this time to avoid injuries. After a day or two, the "peck order" will be established, and little or no fighting will occur. Under range conditions bulls sometimes congregate together in the fall of the year for a period of time while still exposed to the cows. This was noted by the author on an Arizona range several years ago. The bulls went their separate ways again, however, at a later date.

Freedom from internal and external parasites is necessary to keep bulls in good breeding condition. A poor condition from any cause would lower breeding efficiency.

Some beef cattle producers practice fertility testing of bulls shortly before the breeding season begins. Sterile bulls may be identified by this procedure, but it is sometimes difficult to identify bulls of lowered fertility. Semen tests may be of little value for identifying bulls that may experience a lowered fertility or become sterile later in the breeding season. Close observation of cows could determine whether or not the same cows are returning to estrus at regular intervals. If this happens, it is an indication that the bull they are exposed to is sterile or of lowered fertility and that he should be replaced by a fertile bull.

5.3 Factors Affecting Reproductive Efficiency in the Cow

Reproductive efficiency in the cow also depends on several reproductive functions. These include her ability to come into estrus, ovulate, and show normal estrous cycles when not pregnant and after puberty. In addition, the cow must conceive readily when bred, carry the fetus to a full term of pregnancy, deliver a live calf, and nourish and care for it from birth to weaning. She must also have the ability to rebreed regularly while nursing a calf so she can wean a calf each year. A long reproductive life span is also important (Figure 5.1). These different phases of reproduction will be discussed individually.

5.3.1 Estrus and the Estrous Cycle

Under normal conditions, when estrus does not occur, pregnancy is unlikely to occur. Thus, the occurrence of estrus is necessary for optimum reproductive efficiency of the cow.

A low plane of nutrition in which cows and heifers lose considerable weight has an adverse effect on estrus and the estrous cycle. A very low plane of nutrition in young heifers also delays the onset of puberty. In some experiments it has been observed that poor

Figure 5.1.   An Angus cow, now deceased, who produced 25 calves in 25 years. (Courtesy John H. Rush, Springfield, Missouri)

nutrition and weight losses in heifers that are already sexually mature cause estrus and normal estrous cycles to cease until an adequate plane of nutrition is again supplied. Once the estrous cycles begin again when the heifers are placed on an adequate plane of nutrition, normal fertility occurs. A low plane of nutrition may also increase the interval between calving and first estrus in mature cows but especially in first-calf heifers. In such cases this interval may be so long that the cow may fail to come into estrus and conceive during a short breeding season, causing her to be dry or open the next calving season (J. N. Wiltbank, *Proc. Annu. Conv. Natl. Assoc. Anim. Breeders*, 19:54, 1966). The energy required for lactation seems to take priority over ovarian activity to some extent, although the stimulation of suckling may also play an important part in increasing the interval from calving to first estrus.

High levels of nutrition for prepubertal heifers tend to hasten the onset of puberty and shorten the interval from calving to first estrus in cows during lactation. The effect is greater if cows have been on a medium to low level of nutrition prior to calving and then are placed on a high level of nutrition.

Research work disagrees as to the effects of a very high level of nutrition resulting in overfatness on reproduction in the cow. Early research work (O. A. Asdell, *Cornell Ext. Bull.* No. 305, 1934) indicated that fertility in grossly overfat animals is lowered, possibly because of the infiltration of fat into the ovary and the reproductive tract. This may interfere with ovarian activity and may set up a mechanical block so eggs cannot pass through the fallopian tubes to the uterine horns. Practical experience of purebred cattle producers with overfat show females also suggests that fat adversely affects reproductive performance. Some researchers (J. T. Reid, *J. Am. Vet. Med. Assoc.* 114:158, 1949) feel that sterility and overfatness in cows may have a common cause rather than one being the cause of the other.

5.3.2   Ovulation

Ovulation in the cow occurs 4 to 16 hours after the end of estrus. The life of the egg after ovulation is limited to 6 to 8 hours. The life of the sperm in the reproductive tract of the cow is between 24 and 48 hours. It is necessary for viable sperm and ova to be present in the reproductive tract at the proper time, or the fertilization rate will be reduced. In dairy cows the optimum conception rate is apparently obtained when inseminations or services are performed 13 to 18 hours before ovulation (G. W. Trimberger, *Nebr. Agric. Exp. Stm. Res. Bull.* 153, 1948). The same is probably true in beef cattle. When bulls run with the cows during the breeding season, a bull mates with the same cow many times during a single heat period, which may help ensure fertilization by introducing spermatozoa into the reproductive tract of the cow at the optimum time.

Between 30 and 40 percent of matings in cattle fail to produce pregnancies. Many factors may be responsible for these failures. One cause might be the failure of the Graafian follicle to rupture and release the ovum during estrus. This has been estimated to occur in between 5 and 10 percent of the cases in different experiments (D. A. Morros, *J. Dairy Sci.* 52:2, 224, 1969). Another cause of pregnancy failure is delayed ovulations; that is, ovulation occurs so late after normal service that either the sperm or ovum are dead or incapable of fertilization. In one study an incidence of 17.3 percent delayed ovulations was observed.

Estrous cycles of abnormal length (too short or too long) may be the cause of some failures of conception. These are due to abnormal ovarian function and more specifically to the improper function of the corpus luteum, probably due to a hormonal imbalance. A persistent corpus luteum will lengthen the period between estrous periods, whereas a short-lived corpus luteum will cause a very short estrous cycle.

The previous causes of poor conception combined with poor-quality semen, a poor environment in the uterus due to inflammations and infections, and conditions which may interfere with the normal transport of the sperm and ova are probably responsible for a low conception rate in cattle. Also, a few failures of conception may simply be due to mechanical failures of fertilization.

The factors discussed previously no doubt are important in repeat breeding cows. Culling such cows from the herd will tend to avoid some of these problems in the future and increase the overall herd conception rate. Close attention to proper nutrition, health of breeding animals, and the use of fertile bulls will help produce a high-percentage calf crop.

5.3.3 Embryonic and Fetal Death Losses

From the foregoing discussion it appears that most failures in reproductive efficiency occur before fertilization. No doubt other failures occur between fertilization and implantation. These failures are difficult to measure, however. An estimate of fetal death losses in cows after implantation may be obtained by conducting pregnancy tests after the end of the breeding season and comparing the number of cows tested pregnant with the number of cows that actually calve. The accuracy of this estimate depends on accurate pregnancy diagnosis and obtaining accurate records on those cows which calve. Data from the Missouri Agricultural Experimental Station Heterosis Project show that the percentage of cows diagnosed pregnant that failed to calve averaged 5.82 percent in 331 purebred cows producing purebred calves as compared to 4.87 percent in 673 purebred cows producing crossbred calves. These losses could be partially explained by abortions, some of which were unnoticed, and fetal deaths and resorptions. Both heredity and environment could be responsible for some of these losses. The fact that almost 1 percent fewer fetal deaths occurred in purebred cows producing crossbred calves than in purebred cows producing purebred calves indicates that heredity was involved.

The beef cattle producer has no sure way of controlling all of these losses. The percentage loss can be reduced, however, by supplying proper nutrition and management and by maintaining a high level of health among breeding animals in a herd.

5.3.4   Death Losses from Birth to Weaning

The majority of death losses from birth to weaning occurs within a few hours of birth. The movement of the calf from a uterine to an extrauterine environment is a critical period in its life, and there is a certain amount of death losses at this time. Calving difficulties are also important in determining whether or not the calf will be born dead or alive. Fortunately, the cattle producer can help prevent some of these losses by proper care and management. He or she probably has as much to do with controlling death losses at this time as in any other portion of the life cycle of the calf.

The introduction of exotic breeds of beef cattle of extremely large size and their use on smaller cows of popular breeds in the Americas has increased the incidence of calving difficulties and made livestock producers more aware of the causes of calving difficulties.

Death losses from birth to weaning average between 10 and 15 percent, with most losses occurring at or near parturition. A few losses occur later in lactation and may be due to illness, such as scours, and accidents. Losses near birth are mostly due to calving difficulties or to the failure of calves to obtain protection from their mothers through antibodies normally transmitted through the colostrum or first milk. These losses lower reproductive efficiency.

Calving difficulties occur at a much higher rate in first-calf heifers than in older, mature cows that have given birth to one or more calves. Two-year-old heifers are more likely to have calving difficulties than are three-year-old heifers which are calving for the first time.

5.4   Calving Heifers at Two Years of Age

Breeding heifers to calve at 2 rather than at 3 years of age offers the possibility of increasing lifetime production and decreasing costs per unit of production. Such a practice, however, has some disadvantages, including possible increased death losses among heifers and their calves, the possibility of stunting heifers when they calve too young, and the probability that two-year-old heifers nursing a calf may be slow in rebreeding or may not breed at all until the following year. Data from the Oklahoma Agricultural Experimental Station in which lifetime performance of 60 heifers calving at 2 years of age and 60 at 3 years of age showed that heifers calving at 2 years required more assistance at calving, but there was only a slight difference in mature weights among heifers in the two groups. The two-year-old heifers weaned about 0.9 more calves during their lifetime.

Research indicates that well-grown heifers can be bred to calve at 2 years of age without great difficulty (F. M. Peacock, *Florida Cattleman and Grower* 28:28, 1963) but that small and undersized heifers should be bred to calve at 3 years. Heifers bred to calve as two-year-olds should receive adequate nutrition to keep them thrifty, but overfatness should be avoided. Some researchers suggest that for heifers bred to calve at 2 years, their breeding season should begin 3 or 4 weeks before the regular breeding season for mature cows to allow a longer period between calving and exposure to the bull so they would be more likely to conceive. Certainly two-year-old heifers nursing a calf should be on a higher than average level of nutrition to increase the likelihood of conception.

5.5   Factors Affecting Calving Difficulties

Proper management can avoid some calving difficulties but not all of them. Some causes of calving difficulties and how to avoid them are important from the standpoint of increasing breeding efficiency.

5.5.1 Age of the Cow

The age of the cow has a very definite influence on the percentage of difficulty encountered at birth. One of the reasons for this is that two-year-old heifers have not reached their mature size potential, and therefore the size of the birth canal is smaller than in mature cows. Another reason for more difficulty in first-calf heifers is that their calves represent a higher percentage of their body weight (7.7) as compared to mature cows (6.8). This is shown in Table 5-1.

The effect of age of the cow on calving difficulties is demonstrated in Table 5-2. These data show that the percentage of dystocia was less in three-year-old cows than in two-year-olds and still less in four- to five-year-old cows. The percentage of Caesarean sections and the percentage of death losses at or near birth were also higher in the two-year-old heifers. A smaller percentage of calving difficulties and fewer Caesarean sections occurred in four- to five-year-old cows than

TABLE 5-1   Calf weight as a percentage of dam weight[a]

Breed of dam	First calf	Fourth calf
Angus	7.9	6.9
Hereford	7.6	7.0
Charolais	7.5	6.6
Average	7.7	6.8

[a]Calculated from data from the Missouri Agricultural Experimental Station Heterosis Project (unpublished).

TABLE 5-2   Calving difficulties in cows of different ages[a]

	2-year-old heifers	3-year-old cows	4- & 5-year-old cows
Percentage of difficulties	54.6	27.1	7.5
Percentage of calves pulled	46.8	18.9	5.3
Percentage of Caesareans	5.0	2.4	0.5
Percentage of abnormal presentations	3.7	5.8	1.7
Percentage dead within 72 hr of birth	8.4	4.5	5.3
Percentage dead from 72 hr of birth to weaning	3.9	4.5	2.5

[a]Data from *Progress Report No. 4*, ARS-NC-48, Germ Plasm Evaluation Program, U.S. Meat Animal Research Center, Clay Center, Nebr., June 1976.

in younger ones. Some death losses of calves at or following birth
occurred even in older cows.

5.5.2   Breed of Cow

The breed of cow has a definite influence on the incidence of
calving difficulties. Much of the difficulty, but not all, is related to
the size of the females within a breed. Even within a breed, however,
cows vary in size. Cows within a breed or in different breeds which
are about the same size may differ in the amount of calving difficulties encountered.

Data from the Missouri Agricultural Experimental Station
Heterosis Project indicate that there is a difference in dystocia for
different breeds of cows (J. J. Sagebiel et al., *J. Anim. Sci.* 29:235,
1969). This is illustrated in Table 5-3. Cows of the Angus, Charolais,
and Hereford breeds produced both purebred and crossbred calves.
Dystocia is usually increased when cows of a small breed are mated
with bulls of a larger breed, especially first-calf heifers.

5.5.3   Breed of Bull

Data from many experiments show clearly that the breed of
bull to which cows are bred can influence the amount of dystocia
encountered. This is so well known that many commercial cattle
producers breed first-calf heifers to a small bull. Data from the U.S.
Meat Animal Research Center as well as from other experimental
stations show that using Jersey bulls on first-calf heifers reduces
calving difficulties. Although the calves produced from such matings
are more likely to survive than when bulls from large breeds are used,

TABLE 5-3   Differences in dystocia in cows of different breeds

Breed of Cow[a]	Dystocia[b] score		Calving diffi- culties (%)		Severe diffi- culties (%)	
	M	F	M	F	M	F
Angus	2.38	1.92	32.4	30.5	20.0	18.2
Hereford	1.61	1.28	18.7	11.1	15.2	9.6
Charolais	1.10	1.10	2.6	1.6	0.0	0.0

[a]From *J. Anim. Sci.* 29:235, 1969.
[b]The higher the dystocia score, the more difficult the birth. Significant
cow breed effects were noted for both sexes for the three traits studied.

the calves may make slower gains during their lifetime than calves sired by larger bulls. All aspects of the life cycle must be considered when breeding for greater efficiency.

Bulls within the same breed differ in the amount of dystocia occurring in cows bred to them. A part of the increased difficulty is due to large bulls siring larger calves. Bulls with broad, rough shoulders and hips and with heavy muscling often produce calves with similar characteristics, which tends to increase the incidence of calving difficulties. Bulls with these characteristics should be avoided as much as possible.

5.5.4   Sex of the Calf

Bull calves average 3 to 6 pounds heavier at birth than heifer calves. As a result somewhat more calving difficulties are encountered with bull calves than with heifer calves. As yet, however, it is not possible to control the sex of calves produced so there is no way to reduce calving difficulties due to the sex of the calf.

5.5.5   Gestation Length

Calves are increasing in weight from 1 to 1.5 pounds per day, and sometimes more, during the latter stages of gestation. Longer gestation periods tend to be correlated with heavier birth weights, and these conditions do favor more calving difficulties. Induced parturitions may be helpful in producing lighter birth weights and less calving difficulties, but this has not been proved experimentally.

5.5.6   Pelvic Area

The birth canal, or pelvic area, increases as the size of the heifer or cow increases. This may be partially responsible for increased calving difficulties in two-year-old heifers, although some heifers with large pelvic areas may need assistance at calving. This is especially true if the calf is larger than average. Devices have been developed for measuring the pelvic area in heifers, as shown in Figure 5.2.

5.5.7   Position of the Calf at Birth

About 4 to 5 percent of calves assume an abnormal position at delivery. These positions include the head turned back, one or both

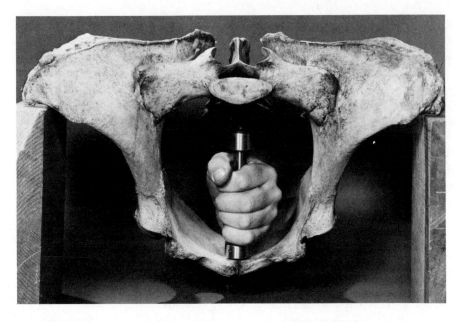

Figure 5.2.   The pelvimeter used for measuring the pelvic area in
beef heifers. (Courtesy Rainbow End Ranch, Douglas, Arizona)

forelegs turned back, a breech position, the calf turned sidewise or
on its back, etc. (see Figure 5.3). These abnormal presentations
usually have to be corrected before the calf can be delivered. They
require the assistance of a veterinarian or an experienced herder to
correct them. If it is not possible to correct the position, it may be
necessary to perform a Caesarean section for delivery of the calf. In
such cases the services of a veterinarian are needed.

5.5.8   Cow Too Fat

Extremely fat cows often experience more difficulty at birth
than those moderate in flesh. This may be due to reduction in size of
the birth canal because of the accumulated fat deposits. Something
more than this may be involved, however, since calves from extremely
fat cows appear to be weaker and to suffer higher death losses between birth and weaning. The condition of the cow does not appear
to have a great influence on the birth weight of her calf except when
degrees of finish are extreme.

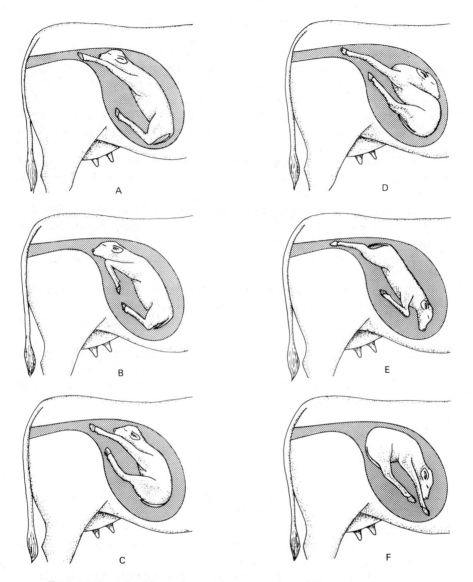

Figure 5.3. A. Normal anterior position and presentation. B. Anterior position with front legs bent at the knees. C. Anterior position with hind feet in the pelvis. D. Anterior position with head and neck turned back. E. Posterior presentation. F. Posterior presentation (breech).

5.5.9 Procedures for Reducing Calving Difficulties

Because 5 to 10 percent of the calves born die at birth or soon afterwards and about one-half of these losses are due to dystocia, procedures for preventing such death losses are important. Methods of preventing some of these losses will now be described.

Replacement heifers should be properly grown out and developed during their first winter and should weigh 650 pounds or more at the beginning of the breeding season.

Extremely large bulls with rough shoulders and bulging hips and with heavy muscling should not be used on yearling heifers or small cows.

During gestation, pregnant cows and heifers should be fed enough to keep them in thrifty condition, but they should not become very thin or very fat. It is important to feed them a ration containing the proper amount of protein, vitamin A, minerals, and salt.

Cows and heifers should be closely observed during the calving season so that those needing assistance at calving can be cared for. Proper equipment and facilities for assistance should be provided in case help is needed.

Records of calving ease for all calves born will help identify sires and dams responsible for calving difficulties. Affected cows can be culled, and in some cases sires may be identified and culled or at least not used on first-calf heifers or small cows.

5.6 Death Losses from Birth to Weaning

Some losses occur in calves which are born alive and appear to be normal and vigorous. A few calves are weak from birth and gradually grow weaker until they die. Such calves may have been injured in the birth process, or they may possess some genetic or nongenetic defect that eventually causes their death.

Even some healthy and vigorous calves may die from accidents, disease, or other causes. One important cause of death losses in calves which are alive and vigorous at birth may be due to their failure to obtain sufficient colostrum which contains antibodies from their mother, or they may be unable to absorb these antibodies from their intestines into their bloodstream.

Antibodies in the bloodstream are necessary for a calf to resist infections. The calf is born without antibodies in its blood because these large molecules encounter difficulty in passing from the blood of the mother to that of the fetus through the placental membranes. The placental membranes in the cow are impervious to the large antibody molecules.

The main source of antibodies to the newborn calf is the colostrum or first milk. The level of antibodies at this time is about 100 times the level normally found in the blood. Antibody levels in the colostrum are high but begin to decline with each milking after the cow gives birth to her calf. By the third milking the antibody concentration has been reduced by about one third.

High levels of antibodies in the blood of the calf are necessary for its ability to resist infections. The ability of the calf to absorb antibodies is maximal at birth. By 20 hours after birth the absorption rate is only 10 to 12 percent efficient, and by 36 hours the absorption rate is zero in most calves. In some calves, however, the efficiency of absorption is very low, even within 3 or 4 hours after birth. Thus, the more colostrum the calf gets the first 5 hours, the better. The herder should be certain that the newborn calf nurses its mother as soon as it is able to stand. Some cows, especially first-calf heifers, are reluctant to let their calves nurse, and it is necessary to restrain them so the calf can nurse. This problem is not encountered in the majority of cows and heifers, but when it does occur, the failure of the calf to obtain colostrum at the proper time may result in its death later.

Some first-calf heifers produce little or no milk the first day or two after calving. Some herders collect colostrum from older cows that freshen and freeze it so as to have an adequate supply for newborn calves as the need arises. Inadequate amounts of antibodies in the newborn calf may not result in its death but may make it more susceptible to infections such as pneumonia and scours.

The half-life of antibodies received by the calf from its mother is about 1 to 2 weeks. The calf, however, normally begins to produce antibodies of its own so that a normal level of antibodies is present in the blood after those received from the mother have disappeared. Reports indicate that in a few cases calves are not able to produce their own antibodies and may later succumb to infections (P. Nansen, *Acta Pathol. Microbiol. Scand.* 80:49, 1970).

STUDY QUESTIONS

1. Define reproductive efficiency.
2. Why is reproductive efficiency one of the most important single economic traits in beef cattle production?

3. Why is the average percentage calf crop weaned about 70 to 75 percent in beef cattle in the United States but in some herds a 90 to 95 percent calf crop is always weaned?

4. The heritability of reproductive efficiency in beef cattle appears to be 0 to 10 percent. What does this indicate should be done to improve this trait?

5. What is the repeatability of reproductive efficiency in beef cattle? What does this mean from a practical standpoint?

6. From the bull's standpoint, what stages of the reproductive cycle may lower his breeding efficiency?

7. Is the size of the testicles in the bull related to his reproductive efficiency? Explain.

8. What may be some adverse effects when a bull is exposed to a very large number of cows during a single breeding season?

9. How many cows should be run with a bull of different ages?

10. What effect can extremely poor condition have on the reproductive efficiency of a bull?

11. What is accomplished by semen-checking a bull before the beginning of the breeding season? What are the limitations of such tests?

12. What are some of the reproductive functions in the female that are subject to adverse effects on her fertility?

13. What happens to estrus and the estrous cycle when heifers are placed on a starvation diet for 3 or 4 months? What happens when they are again placed on an adequate good-quality ration?

14. What are some of the effects of a high level of nutrition in cows and heifers?

15. Why do grossly overfat animals usually possess low fertility?

16. Define ovulation. When is the optimum time to breed or artificially inseminate cows and heifers?

17. It has been reported that 30 to 40 percent of matings in cattle do not end in pregnancy. Why?

18. What may be the underlying cause, or causes, when estrous cycles in cows are abnormally long or abnormally short?

19. When do most failures of reproduction occur? What percentage of death losses probably occur in fetuses after the end of the breeding season? Are these genetic?

20. Why do so many death losses occur at birth or shortly afterward in calves?

21. Discuss the advantages and disadvantages of calving heifers at 2 rather than 3 years of age.

22. If heifers are bred to calve at 2 years of age, what precautions should be taken?

23. What is the relationship of the age of the cow to the amount of dystocia?

24. Does the breed of the cow affect the amount of dystocia? Explain.

25. In what way does the breed of bull affect dystocia?

26. Why does the sex of the calf affect dystocia in cows, and what can be done about it?

27. The length of the gestation period is related to dystocia. How? What can be done about it, if anything?

28. What is the normal position of delivery of a calf at birth? What percentage of abnormal positions are encountered?

29. Why do extremely fat cows have more difficulty calving than cows of medium fatness?

30. Outline procedures that might be followed for reducing calving difficulties.

31. What is colostrum? Of what importance is colostrum to the newborn calf?

artificial insemination of beef cattle

Artificial insemination refers to the placement of spermatozoa of the male into the reproductive tract of the female by artificial means. This is in contrast to natural insemination in which the male deposits spermatozoa in the reproductive tract of the female during copulation. Many more females can be bred per male by artificial than by natural service.

A normal, healthy bull produces from 4 to 6 billion spermatozoa per ejaculate. Many viable sperm are required per service, but only one viable sperm is required to fertilize the ovum which is released from the ovary shortly after the end of estrus. Because of the great surplus of sperm introduced into the reproductive tract of the female during natural mating, people conceived the idea many years ago of increasing the volume of an ejaculate of the male by adding an appropriate extender (or diluter). The extended semen could then be divided into many parts so large numbers of females could be artificially inseminated with a single ejaculate. After years of research it is now possible to use a single extended ejaculate of the bull containing a high proportion of viable sperm for the insemination of

300 to 500 cows. This makes it possible to obtain thousands of off-spring in a year from a genetically superior sire.

6.1   History of Artificial Insemination

The first use of artificial insemination in animals goes back several centuries in time. Documentary evidence dating to 1322 states that an Arab chieftain placed a wad of cotton in the reproductive tract of his prize mare and left it there for 24 hours. At night he removed the wad of cotton, crept to the camp of an enemy chieftain who owned an excellent stallion, and let the stallion smell the wad of cotton, causing him to ejaculate into a piece of cotton held in readiness at the proper location. The owner of the mare then returned to his camp and placed the cotton containing the semen into the reproductive tract of the mare. She conceived and produced a foal.

In 1782, Spallanzani, a professor of physiology at the University of Pavia, used artificial insemination to successfully breed a bitch that had produced one previous litter. The semen was obtained from a young dog through spontaneous ejaculation. Slightly over 1.0 cubic centimeter was injected into the uterus of the bitch by means of a syringe. Sixty-two days later the bitch whelped three living pups.

Ivanoff, a Russian researcher, pioneered in artificial insemination in several species of farm animals including horses, cattle, and sheep. As a result of his work, 19,800 cows were successfully inseminated in Russia in 1931. An average of 100 cows were inseminated per bull.

The first cooperative artificial breeding association for cattle was organized in Denmark in 1936. E. J. Perry of the United States later visited Denmark and saw the possibilities of artificial insemination in livestock breeding. In 1938, researchers at many U.S. colleges were already using artificial insemination on a small scale for breeding ewes, mares, sows, and cows. Research was being conducted on the collection, dilution, and storage of spermatozoa for artificial insemination purposes. Methods for evaluating the fertilizing capacity of sperm were also being investigated.

Artificial insemination was first widely used for the breeding of dairy cattle. Gradually, however, more and more beef cows were bred by this method. Millions of cattle in many countries of the world are now being bred by this method. Today, 50 percent or more of the dairy cattle in the United States are bred artificially. Research is continuing on methods of improving the efficiency of this method of breeding.

6.2 Some Advantages of Artificial Insemination

One of the main advantages of artificial insemination is the greater and more widespread use of sires. In the United States approximately 3,500 cows are now bred per sire per year. Some bulls are used for breeding 50,000 or more cows per sire per year. This number includes both beef and dairy cattle. When natural mating is practiced, only 30 to 50 calves can be bred per sire per year.

Artificial insemination makes it possible to speed up the genetic improvement of cattle through the greater use of superior sires. It has been estimated that through the use of artificial insemination and superior proved sires, genetic improvement in milk production is two to three times more rapid than if natural service alone were used. No doubt gains made by slaughter steers sired by superior beef bulls used artificially are faster and more efficient than gains made by steers from average commercial bulls used in natural service.

Artificial insemination helps preserve the health of superior sires because they do not have direct contact with cows in mating as is true in natural mating. Thus, the bull used artificially is not exposed to diseases such as leptospirosis, brucellosis, etc.

Through the use of artificial insemination the time required to prove a young sire by progeny testing is greatly reduced. In addition, a young bull can be used on cows in many herds, which makes the progeny test results more meaningful.

Bulls which are unable to serve cows naturally because of accidents and injuries can easily be used by artificial insemination.

Artificial insemination requires that only fertile semen be used for breeding purposes. This is determined by laboratory tests for viability and actual artificial insemination results. Thus, bulls of low fertility are readily recognized and not used.

Artificial insemination makes it possible for a small breeder to use genetically superior sires that could not be used by natural mating because of the expense involved. Semen from even the most genetically superior bulls may be obtained for $5 to $25 per vial. The average cost per vial is about $10.

Through the use of frozen semen, a genetically superior sire can be used for many years after his death. Frozen semen stored for as long as 25 years has been used to produce calves in one instance.

The use of frozen semen makes it possible to breed cows to a bull located thousands of miles away. Semen from some of the exotic breeds of bulls has been used in this way, and it is of considerable value in increasing the number of cattle from a new breed in a country in a short period of time.

Frozen semen and artificial insemination can be used as a means of estimating the amount of genetic progress made through selection over a period of time. Semen from bulls in the beginning of the selection experiment can be frozen and stored for long periods. It can then be used periodically by artificial insemination along with semen from bulls produced later in the experiment so the performance of the progeny of the two groups can be compared, and the amount of genetic improvement that has been made can be estimated.

In the future it may be possible to partially control the sex of the calf by treating and processing semen. If this could be done, it would have many advantages, and the use of artificial insemination would probably be required.

6.3   Some Limitations of Artificial Insemination

Not all sires used for artificial insemination are genetically superior for quantitative traits such as rate and efficiency of gains, conformation, and carcass desirability. U.S. Department of Agriculture studies indicate that only about one third of the proven dairy sires actually improve the performance of their progeny in herds where they are used. Since more attention is paid to proving dairy than beef sires, possibly a smaller percentage of beef bulls is actually genetically superior. Milk production cannot be measured directly in dairy bulls, whereas rate and efficiency of gain can be measured in individual beef bulls. Since most performance traits in beef cattle are highly heritable, measuring such traits in the individual beef bulls should be effective and would require less time and expense than progeny tests.

Many beef sires used artificially have not been tested free of detrimental recessive genes. If a bull carries a detrimental recessive gene, he will transmit it to approximately one half of his offspring. Thus, the use of such a bull to produce thousands of offspring will cause a widespread distribution of any detrimental recessive gene he carries. Before being used on large numbers of cows, beef sires should be progeny-tested to determine that they are not carrying detrimental recessive genes. Using a sire on at least thirty-five of his own daughters with no defective offspring produced proves him free of any detrimental recessive gene at the 99 percent level of probability.

Artificial insemination requires skill, training, and experience on the part of the inseminator in handling semen and in the actual process of insemination. The inseminator must also have a good knowledge of reproduction in the cow and in the collection and evaluation of semen. The processing of semen is now usually handled by commercial companies who hire trained workers for this purpose. Therefore, the inseminator is not required to be an expert in these areas but should be aware of the procedures involved.

One of the main problems of successfully using artificial insemination in beef cattle is the difficulty of finding each individual cow when she is in heat. Unlike dairy cows which are brought to the milking parlor twice daily, beef cows often run in pastures of large acreages and cannot be observed so closely. Close observation and the use of heat detectors and/or teaser animals for finding cows in heat are quite helpful in this respect.

6.4   Collection of Semen

For best results with artificial insemination it is necessary to obtain semen from the bull that contains large numbers of viable spermatozoa and is free of harmful contaminants. Exposure to extremely high or low temperatures, direct sunlight, or water and other contaminants is harmful to sperm. Thus, any possible detrimental effects on sperm must be avoided.

Sexual preparation of bulls before actual semen collections are made often increases the semen output per ejaculate. The preparation appears to be less effective in beef than in dairy bulls.

A bull may be sexually prepared in a number of ways. One is to allow a bull to have two or three false mounts prior to ejaculation. Keeping a teaser animal close to the bull or changing its location often increases semen output. Some beef bulls will not mount a cow that is not in estrus or that is in a chute but will mount one that is in standing heat and confined to a lot or corral.

The injection of certain hormones prior to ejaculation often increases sperm output. As the frequency of ejaculations increases, the semen volume, sperm concentration per ejaculate, and total number of sperm per ejaculate decrease. The average ejaculate in bulls ranges in volume from 4 to 6 cubic centimeters with 4 to 6 billion sperm per ejaculate.

6.4.1   The Artificial Vagina

Russian scientists first developed the artificial vagina about 1930. It has proved to be the most practical and satisfactory way of collecting semen from the bull. Through its proper use a normal ejaculate free of extraneous material is usually obtained.

An artificial vagina for collecting semen from the bull is shown in Figure 6.1. An artificial vagina consists of a thick rubber hose about 3 to 3.5 inches in diameter and about 18 to 22 inches long. A thin rubber liner is run inside this hose and folded back on either end so that warm water can be introduced into the cavity between the two layers. A thin rubber cone (called a directacone) which tapers to

Figure 6.1.   An artificial vagina for collecting semen from the bull. A. Outer rubber hose. B. Inner rubber lining. C. Thin rubber directacone. D. Centrifuge tube. E. Assembled artificial vagina.

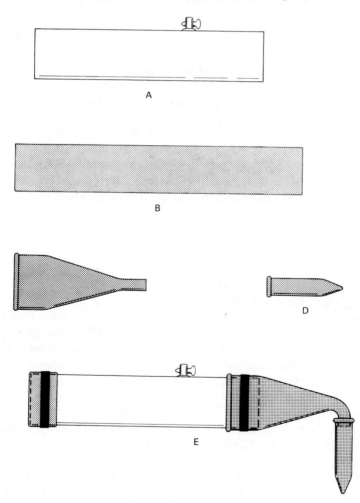

a small size at one end is then attached to the assembled rubber hoses with the small end attached to a graduated centrifuge tube. The temperature of the artificial vagina should be 43°C to 50°C. The inner liner of the artificial vagina is usually lubricated with a lubricant that is not toxic to the sperm. When the bull mounts the cow and extends his penis, it is guided into the open end of the artificial vagina. When the bull makes the copulatory thrust, semen is ejaculated into the opposite end of the artificial vagina and is collected in the centrifuge tube. If the centrifuge tube is graduated, the semen volume can be read directly. Freshly collected semen should be protected against cold or heat shock to prevent damage to the sperm. All equipment should be clean and free of materials that might also be detrimental.

6.4.2   Manual Manipulation

To collect semen by manual manipulation, the arm of the person performing the stimulation is inserted into the rectum of the bull and a gentle, stroking pressure is applied to the ampulla and vas deferens. A second person is required to collect the semen when emission occurs. Attempts to collect semen from bulls by this method often fail but are more successful when attempted by an experienced operator. Many samples are contaminated by bacteria and foreign matter, and the motility of the sperm may be low. The method may be used when semen is collected from a bull that cannot mount a cow because of an injury.

6.4.3   Electrical Ejaculation

A special apparatus may be used to cause ejaculation in the bull. The apparatus (Figure 6.2) includes a voltmeter, a transformer, a milliammeter, and one or more electrodes. A specially built electrode is inserted into the bull's rectum, and an alternating current of 3 to 30 volts is turned on. This electrical current excites the ejaculatory nerve centers of the bull, resulting in a partial to complete ejaculation. Periods of electrical stimulation are usually alternated between short rest periods to produce ejaculation.

Electrical ejaculation of the bull often produces extra amounts of secretions from the accessory sex glands, which lowers the sperm concentration per unit of semen. When properly used, electrical ejaculation produces a clean sample of semen, however. One advantage of this method of collection is that semen can be obtained from bulls that have been injured and cannot be collected from by using the artificial vagina or that lack libido and will not mount and ejaculate.

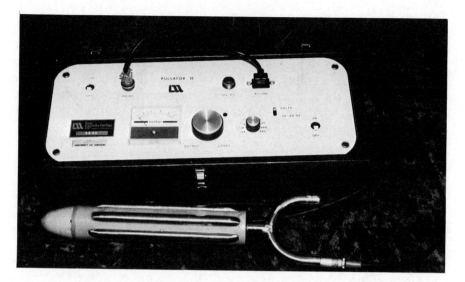

Figure 6.2.   Apparatus for the electrical ejaculation of bulls. (Courtesy Dr. C. J. Bierschwal, University of Missouri, Columbia)

6.4.4   Vaginal Collection

One of the first methods used in semen collection was to allow the bull to serve a cow and then collect semen from the anterior portion of the cow's vagina. A syringe attached to a glass or plastic tube, a sponge, or vaginal spoon may be used to collect the semen in this manner.

Vaginal collection is probably the least desirable method of semen collection because a sample collected in this manner may be contaminated with urine and mucus secretions, and there is danger of spreading diseases of the genital organs. Another disadvantage of this method is that usually only a part of the semen deposited in the vagina after copulation can be recovered.

6.5   Semen Evaluation

One of the advantages of artificial insemination is that semen samples can be examined prior to use for quantity and quality of the spermatozoa it contains. Those samples not meeting the required standards can be discarded. Laboratory tests for good-quality semen indicate its fertilizing capacity but do not guarantee it. A live calf born from the use of semen is the best indication of fertility.

6.5.1   Semen Appearance

The appearance of semen at the time of collection gives some idea of its quality. A good-quality semen sample should be viscous and have a white, milky color. The whiter and more viscous the semen, usually the higher the sperm concentration. A close visual examination of some samples with a high concentration of sperm will often reveal swirling movement within the sample.

6.5.2   Semen Concentration

Several methods are available for estimating the number of spermatozoa per milliliter of bull semen. The hemocytometer, normally used for counting the number of leukocytes or red blood cells, can also be used to count sperm concentration. A second method involves the determination of opacity rating of a semen sample. The semen is diluted and the opacity of the sample compared with the opacity standards of a previously determined sperm concentration. This is a rapid means of estimating sperm concentration. The packed-cell method, similar to the hematocrit method used for determining the volume of red blood cells in blood, may also be used for estimating sperm concentration. Photoelectric colorimetery in which the amount of light passing through a semen sample is compared to that passing through a standard solution of semen is a fourth method of determining sperm concentration. This method is rapid and highly repeatable and is used extensively.

6.5.3   Estimation of Sperm Quality

Many methods have been developed for estimating sperm quality. These include estimating the percentage of motility, percentage of live sperm, reaction rates, methylene blue reduction time, morphology, and hydrogen-ion content. An estimation of motility by placing a drop of semen on a clean, clear glass slide that has been warmed to 38°C is widely used. Observing the droplet under the low-power objective of the microscope allows the observer to estimate the percentage of spermatozoa which possesses strong, progressive motion. Semen samples which possess strong motility are usually considered of good quality. By making a dry smear of semen on a clean glass slide and then drying and staining the smear it is possible to observe the morphology of the sperm under the oil emersion of the microscope. Low-quality semen may contain up to 30 to 35 percent abnormal forms. A high percentage of abnormal spermatozoa indicates an upset in spermatogenesis in the testicles.

6.5.4   Extension of Semen

A semen extender is a solution added to semen to increase its volume. In addition, a good extender will be favorable for the survival of spermatozoa which retain their fertilizing capacity over a period of time. A good extender maintains a normal pH of 6.5 to 6.7 for an extended period, protects sperm from temperature shock, and supplies needed nutrients and other substances which prolong sperm survival. When mixed, the semen and extender should be at the same temperature. Usually a semen sample is extended to a volume necessary to provide the desired number of viable sperm necessary for conception. Glass vials (ampules) usually contain 0.7 to 1.0 milliliter of extended semen, which is enough for one insemination. If semen is processed in plastic straws, 0.25 to 0.50 milliliter is used for a single insemination.

Extenders containing substances such as egg yolk, gelatin, and milk, fruit, and vegetable juices as well as many other substances have been developed. The extenders most widely used contain egg yolk or milk. An egg yolk citrate extender is the one most widely used in the United States today. An extender using boiled whole milk or skim milk also has been widely used for commercial artificial insemination. The antibiotics penicillin and streptomycin have been used in extenders to improve the keeping qualities of bull semen and to improve the conception rate by 5 to 10 percent.

6.5.5   Storage of Semen

If only fresh semen could be used for artificial insemination, much extra semen would go to waste. For this reason much research work was done in the early days of artificial insemination to perfect methods of storage of semen so it could be stored for use over a period of several days.

Early efforts were directed toward the storage of liquid semen at temperatures above freezing and lower than body temperature. Research showed that semen stored at 5°C would retain its fertilizing capacity up to 4 days. When commercial artificial insemination was first practiced, semen was shipped in thermos bottles containing cracked ice. Sometimes semen stored in this manner was dropped from light aircraft by parachute near the premises of the local inseminator.

British scientists discovered in 1949 that extenders containing glycerol could be used for freezing fowl semen. The discovery was made when a laboratory technician accidentally mistook a bottle containing glycerol for a bottle containing a diluter medium. Later it was found that the addition of glycerol to a diluter increased the resistance of sperm to freezing. Later this procedure was adapted to the freezing of bull semen and greatly increased the use of frozen semen for artificial insemination all over the world.

Successful freezing of semen requires that ice crystals not be formed in the freezing process because this would kill the sperm. The addition of glycerol to a good extender prior to freezing dehydrates the sperm, and they can be frozen without crystals being formed. Semen from some bulls does not freeze as well as it does from others, but the cause or causes have not been fully determined.

Once semen is frozen, it is stored in liquid nitrogen at a temperature of $-196°C$ ($-320°F$). Fertility declines only a little, if at all, for 2 or 3 years when semen is stored in this manner. Calves have been produced from frozen semen stored for 20 to 25 years.

Before use, frozen semen should be rapidly rewarmed to melting temperature because this gives a better sperm survival than slow rewarming. Many technicians thaw the frozen semen for 8 to 10 minutes in ice water before it is used for artificial insemination. Frozen semen is now used all over the world, with very little liquid semen being used. Semen is frozen in several different kinds of containers. Glass ampules containing 0.5 to 1.00 cubic centimeter or enough for one insemination are used extensively in the United States. Some semen is frozen as pellets, but this involves problems with the identification of each sample. Plastic straws about 5 inches in length and containing 0.25 to 0.5 milliliter of semen are commonly used for storage and insemination purposes. The straws require less storage space and yield a high recovery rate of spermatozoa when thawed, and fewer sperm are lost in the insemination process. Shell freezing has also been used as well as lyophilization of spermatozoa. Lyophilization involves the removal of 90 to 95 percent of the moisture from the semen. Only a small percentage of sperm is motile when prepared in this way and placed in an extender for use. Only one successful pregnancy using such semen has been reported.

Many established bull studs collect, freeze, process, and store bull semen for private breeders for a fee. As a general rule, semen processed by these companies is used by the owner, although some owners sell such semen to other breeders. A breeder's certificate to use and register calves from such semen is sometimes required. Breeder's certificates vary in price but may be $50 or more.

Purebred beef cattle associations have strict regulations on the use of artificial insemination for breeding purposes. Regulations are not the same for all registry associations, and they are revised from time to time. Blood typing of the bull is required by most associations, and a limited number of breeders can be co-owners of a bull. Most associations allow use of frozen semen after the death of a bull, but in some cases there is a time limit after which it can no longer be used for insemination purposes.

6.5.6   Insemination of the Cow

Cows usually stay in heat for 12 to 36 hours, although in warm climates and on the range standing heat may be much less than this. The optimum time to inseminate a cow is usually thought to be 6 to 10 hours after the first signs of estrus or when the cow is first in standing heat.

When artificial insemination was first practiced, liquid semen was used for insemination purposes. A speculum (glass or steel) was inserted into the vagina of the cow and the cervix located by means of a spotlight worn on the head of the inseminator. Semen was placed into the cervix by means of a glass pipette attached to a syringe. This procedure was later replaced by intrauterine insemination using a plastic tube attached to a syringe or disposable plastic bulb. The inseminator wore a thin plastic glove and inserted his or her hand into the rectum of the cow, picking up the uterine horns through the rectal wall. A plastic tube containing the semen was then inserted into the cow's vagina to the cervix and by careful manipulation was passed through the cervix, and semen was deposited in the uterus or uterine horn. In some cows, especially heifers, difficulty is encountered in passing the tube through the cervix so cervical inseminations are made. Many inseminators today, however, prefer deep cervical to intrauterine inseminations.

6.5.7   Heat Detection in Cows

Finding cows when they are in heat is probably the most difficult problem encountered in the practice of artificial insemination. When running with the cows, the bull is on the job night and day. When people attempt to find cows in heat, they usually limit their observation to two and maybe three times per day. This will miss many cows when estrous periods are shorter than normal. Missing an estrous period will cause the cow to calve 18 to 24 days later, and her calf will weigh less and bring up to $20 less when sold at weaning. It has been estimated that 60 to 68 percent of beef females exhibit estrous symptoms between 6 P.M. and 6 A.M. (R. L. Godke and J. L. Krider, *Charolais Bull-O-Gram*, Feb.–March 1977, p. 24). Checking cows three times per day could miss 8 to 10 percent of cows in heat. Obviously checking cows more often or using apparatuses to mark them will detect more cows in heat.

The female bovine is homosexual in the expression of estrus. When a cow comes into heat, other cows not in heat will take turns mounting her, and in some instances she will try and mount them, often without success. Occasionally a cow in the latter stages of pregnancy will be persistent in mounting a cow in heat, and in 7 to 15 percent of pregnant cows estrus will be exhibited. The length of the estrous period is quite variable in different cows and under different conditions. The estrous period can last only a few hours in some females and 1 to 2 days in others. Cows with short estrous periods initiated in the nighttime hours will not be detected by herders.

A herd of cows may be handled in several different ways to increase the efficiency of finding those in heat. In range cattle grazing pastures of thousands of acres, bunching all cows early in the morning and again late in the afternoon often helps locate more cows in heat. The bunching together seems to have a psychological effect on the herd, with cows in estrus being sought out and mounted by others. This works very well with heifers and dry cows, but if care is not exercised, such a practice places undue stress on the calves, especially those which are very young. Range cattle usually gather around a water source in the middle of the day, and inspection at this time will locate many cows in heat. Large bull calves in the herd will often follow cows in heat or those in the initial stages of heat.

Some breeders of purebred cattle separate calves from the cows and bring them together again in the morning and again in the evening for the calves to nurse. While separated from the cows, the calves have access to a grain ration. When handled in this manner, calves consume enough grain to make rapid gains. Also, since the cows can be observed twice each day, those in heat may be detected. Perhaps also limiting the nursing to twice each day may allow cows to show estrus in a shorter time after calving than if their calves nursed at several intervals during the day. The stimulus of repeated nursing seems to prevent the appearance of estrus in some cows.

Several different schemes have been devised for detecting cows in heat. One commercial product consists of a plastic bag filled with a dye that is cemented on top of the midline of the back just in front of the tail head. When mounted by other cows, the pressure will change the color of the detector patch from white to red. If properly used, these patches can increase the number of cows found in estrus during the breeding season. The patches should be supplemented with other means of detecting estrus, however.

Altered or treated males are often used for detecting cows in heat. These are sometimes referred to as teaser or gomer bulls. A chin ball marking halter is widely used to hold a marking device on such bulls. The harness holding the marker in place is attached to a container filled with brightly colored ink or paint. When a bull wearing this apparatus mounts a cow in heat, a bright-colored mark is left on her back. The chin ball marker should be refilled every 5 to 7 days during the breeding season. Changing the color of the ink or paint in the chin ball marker every 16 to 18 days will help identify recycling females. Some herders apply colored grease to the brisket of the teaser bull, but the disadvantage of this system is that the grease has to be replaced almost daily in a heavy breeding season.

Teaser or gomer bulls are those which run with the cow herd but have been altered so they cannot serve a cow and induce pregnancy. Usually males used for this purpose are selected because of their vigor and sex drive. Some cattle producers prefer certain dairy or crossbred bulls for this purpose because they appear to be more aggressive sexually than purebred bulls. Some altered bulls can be used for more than 1 year, but others have to be replaced by younger bulls when they become sluggish and lose some of their sex drive.

Several techniques are used to alter teaser bulls. A *vasectomy* is one type of operation that can be used. This operation involves tying off, severing, and removing a portion of the vas deferens leading to each testicle. The operation is performed by making an incision through the scrotum near the abdomen. Vasectomized bulls are still able to serve a cow, but no sperm are deposited in the reproductive tract of the cow because their passageway from the testicles is blocked.

Another method of altering teaser bulls is a *penectomy*. This technique involves removing various lengths of penis so the bull cannot copulate with the cow. When a portion of the penis is removed, the urethra is split and sutured to the surface of the skin surrounding the urethral opening so the bull can urinate. Usually the operation works best when performed in young bull calves, but it can be performed in older bulls.

Still another operation performed on teaser bulls is called *penile displacement*. This operation involves relocating the penis so that at erection it is extended at a 35- to 45-degree angle from the abdomen. Although the penis is fully extended, it is extended at such an angle that copulation usually cannot occur.

Teaser bulls may be operated on and the penis fixed or tied to the tough connective tissues of the abdominal wall. This does not allow the teaser bull to extend his penis when he mounts a cow. Still another operation is to close the prepucial opening around a bloat trocar cannula near its opening by purse string sutures. This prevents the extension of the penis when the bull mounts the cow. Recently a device called a Pen-O-Block has been used to prepare teaser bulls. This apparatus consists of a rigid plastic tube and cannula which is inserted into the bull's sheath and secured in place by washers and cotter pins according to the manufacturer's directions. This apparatus also prevents the extension of the penis so that copulation cannot occur. Most of the techniques used to prevent the extension of the penis have the advantage that bulls with such operations can be restored to normal breeding operations again through simple surgical techniques. However, most of them have to relearn the sex act, and many bulls have severely reduced libido after the use of the Pen-O-Block.

Nymphomaniac cows can also be used as teaser animals. Nymphomaniac cows show an abnormally strong sex desire because of the presence of large cystic follicles on their ovaries. The abnormal sexual behavior is probably due to a hormonal imbalance. Some efforts have been made to prepare teaser animals by treating open cows with injections of the male hormone testosterone and fitting them with a chin ball marker. Some success has been reported with this procedure.

Hormonal-treated steers have also been used as teaser animals. Treatment of steers with testosterone and estrogens has stimulated them to seek out females in heat with considerable success. Heifers treated in the same way have also been successfully used as teaser animals.

6.6 Pregnancy Testing of Cows and Heifers

It is desirable to check cows for pregnancy at the end of the breeding season. Open females can be culled without feeding them for another year. Pregnancy tests can be conducted on several hundred cows in a single day if sufficient help and good facilities are available. The pregnancy tests can be conducted by a veterinarian or a herder who has been trained and possesses skill in this particular area. Ultrasonic devices can also be used for detecting pregnancy in cows, but they do not appear to be as useful in cows as in sows and ewes.

STUDY QUESTIONS

1. Define artificial insemination, and contrast it with natural insemination.

2. Why was the idea of artificial insemination first conceived? How many cows can be bred by artificial insemination with a single ejaculate of the bull?

3. By whom and in what species was artificial insemination first used? Does this story seem plausible?

4. Who was the first man to practice artificial insemination on a practical basis?

5. When was the first cooperative artificial breeding association established and in what country?

6. When and by whom was the first cooperative artificial breeding association established in the United States?

7. List some of the advantages and disadvantages of artificial insemination.

8. Of what importance was the discovery that semen from male animals could be frozen, thawed, and still retain its fertilizing capacity? How was this discovery first made?

9. List a number of ways that frozen semen can be used to advantage.

10. Discuss the advantages of collecting semen by means of manual manipulation.

11. What is probably the best method of collecting semen from the bull? Why?

12. What should be the appearance of a sample of bull semen when first collected?

13. Discuss some methods of evaluating sperm quality.

14. What is a semen extender, and what is its main purpose?

15. How long can semen be stored in the liquid form? In the frozen state?

16. Why are some antibiotics added to stored semen?

17. List and describe various methods used in attempting to store semen.

18. Describe the method now used for artificially inseminating the cow. What method was first used for this purpose?

19. What is probably the most important problem in using artificial insemination on a practical basis? Why?

20. Why is the bull more efficient at finding cows in heat than herders?

21. Do pregnant cows ever come into heat? Explain.

22. What is a nymphomaniac cow? A gomer bull? A penectomy? A vasectomy?

23. Discuss several practical ways that cows can be managed so that greater efficiency can be obtained in finding cows in heat.

24. Describe some operations now used to alter bulls so they can be used for finding cows in heat but cannot impregnate cows.

25. Why is it important to pregnancy-test cows and heifers at the end of the breeding season?

PART THREE

GENETICS OF BEEF CATTLE BREEDING

CHAPTER 7

principles of beef cattle genetics

Beef cattle breeders have become acquainted with many principles of genetics in recent years. One step in this direction was the occurrence of snorter dwarfism in purebred beef cattle which became of great economic importance in the 1960s. To control the appearance of this defect, breeders had to learn about progeny testing for a recessive gene and how to conduct such tests. More recently, more attention has been directed to finding genetically superior breeding animals through performance and progeny testing. Hybrid vigor from cross-breeding has also been of great importance. Most breeders are familiar with the meaning of hybrid vigor or heterosis. In this chapter we shall discuss some of the fundamental principles of genetics as they apply to the improvement of beef cattle through breeding.

7.1 The Cell

The bodies of beef cattle consist of trillions of microscopic building blocks called cells. Many different kinds and shapes of cells are present in the animal's body. Different cells perform different functions. Even though they are extremely small, their structure is very complex, as are their functions.

The animal's body possesses two major kinds of cells. Body cells, or *somatic cells*, are responsible for the function and structure of the body. Many cells such as the blood cells and cells of the tissues and organs are of this kind. A second group of cells is the sex cells (or gametes), which are the ovum of the female and the spermatozoa of the male. Gametes function in transmitting inheritance (genes) from one generation to another. Body cells do not function in this manner.

A diagram of a cell is shown in Figure 7.1. One can get an idea from this drawing how complex a cell may be because it contains many parts and inclusions or bodies called *organelles*. Most cells contain a cell wall, a nucleus, and cytoplasm, but there are exceptions. For example, the red blood cells of beef cattle and many other mammals do not possess a nucleus. Spermatozoa, on the other hand, possess a nucleus but very little cytoplasm. All parts of the body cells are important because each function is necessary for the proper growth, development, and functioning of the body. In this chapter we shall direct most of our attention to the sex cells because through

Figure 7.1.   Diagrammatic sketch of a cell.

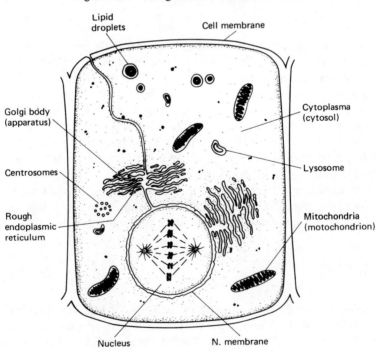

them the genetic makeup of individuals is determined in each suc-ceeding generation.

7.1.1   The Chromosomes

The nucleus of the cell carries the genetic material in beef cattle, with little or none carried in the cytoplasm. Within the nucleus are the chromosomes, the carriers of the genetic material. Chromo-somes (*chromo* = color; *somes* = body) can be stained and observed at the proper stage of cell division known as the *metaphase*. At this stage of cell division each chromosome appears doubled, with the two equal halves connected at a point known as the *centromere*. This is shown in Figure 7.2. Some chromosomes are longer than others, with the longer ones presumably carrying more genes be-cause the genes (hereditary material) are carried lengthwise on the chromosomes.

Beef cattle have a total of sixty chromosomes or thirty pairs because chromosomes occur in pairs in the body cells and primitive sex cells. Members of each pair are similar in length and form, but the thirty pairs differ in these respects. One pair of the thirty pairs of chromosomes in cattle is called the *sex chromosomes* because they

Figure 7.2.   Chromosomes of beef cattle. Note that the autosomes at the metaphase appear "V"-shaped, whereas the X chromosome resembles a large "X." (Courtesy University of Missouri, Columbia)

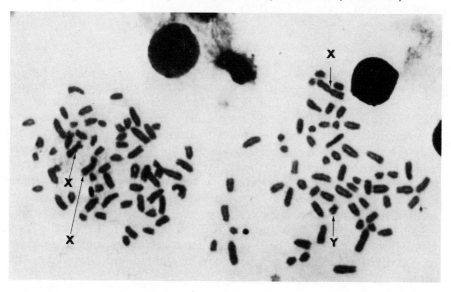

are responsible for determining the sex of the individual. The other twenty-nine pairs of chromosomes are called *autosomes.* In cattle the sex chromosomes in females are both X chromosomes (XX), whereas in the male one of the pair is an X and the other is a Y (XY). As shown in Figure 7.2, the X chromosome is larger than the Y chromosome. Therefore, the female possesses a few more genes than the male. A rough measurement of the relative length of the X and Y chromosomes in cattle suggests that a cow carries about 95 percent of her genetic material on her autosomes and 5 percent on her sex chromosomes. These measurements also suggest that the genes carried on the sex chromosomes of the cow exceed those of the bull by about 1.5 percent.

Of the pairs of chromosomes in body cells and primitive sex cells, one of each pair came from the sire and one from the dam. The sperm and the ovum, however, possess only half-pairs of chromosomes, so the parents transmit only a sample one half of their inheritance to each offspring. The union of the sperm and egg restores the pair number of chromosomes in the new individual. This is illustrated in Figure 7.3. If the new individual receives an X chromosome from its sire, it will be a female because any calf always receives an X chromosome (never a Y) from its mother. If it gets an X from the mother and a Y from the father, it is a male or XY. Since chromosomes in the primitive germ cells of the male occur in pairs (XY), the male produces an X-bearing and a Y-bearing sperm. Because of this the male controls the sex by transmitting either an X- or Y-bearing sperm to each offspring.

7.1.2   Chromosome Abnormalities

Chromosome abnormalities sometimes occur in cattle, humans, and other animals. These differ from the normal chromosome complement in several ways, including abnormalities in numbers, structure, and mixtures of cells in different individuals. More chromosome abnormalities have been described in humans than in cattle, probably because they have been studied more thoroughly in humans. Some chromosome abnormalities have been described in beef cattle, however. Sometimes the occurrence of chromosome abnormalities is associated with defects. Some of these occur and have been described in cattle. Most have been observed by culturing white blood cells (lymphocytes) and staining a dried smear of these cells on a glass slide. Some of the cells in the smear are in the metaphase stage of cell division, and chromosomes can be stained and observed at this stage.

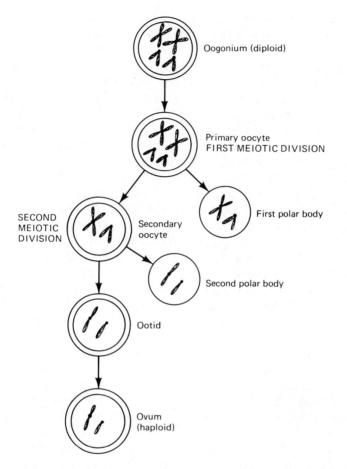

Figure 7.3.   Oogenesis or ovum production in the cow. An illustration of how the paired chromosomes separate into half pairs in the gametes. When the sperm fertilizes the egg, normal pairs are restored.

A chromosome analysis of the blood of nine cattle with lymphatic leukemia showed that aneuploidy in varying degrees was observed in seven of them. Aneuploid cells are those that have one or more chromosomes in excess or one or more chromosomes missing from the cell nucleus.

Structural abnormalities of chromosomes also occur. They are probably caused by breaks in one or more chromosomes that are followed by chromosomal rearrangements within a pair or between different pairs. Thus, a part of a chromosome may be missing (a deletion), or it may be duplicated (a translocation or duplication).

The chromosome rearrangement most often reported in cattle is the presence of a large metacentric chromosome (a translocation) or one less chromosome than the usual sixty, which is known as a deletion. A metacentric chromosome is one in which a doubled chromosome at the metaphase is connected at the center and resembles the letter X. Both conditions have been noted in the male cells of all calves in a set of quintuplets and in females in a set of triplets. The same condition was found in association with bovine lymphatic leukemia. Evidence suggests that this abnormality is not specifically associated with the lymphatic leukemia but that it is an inherited condition.

An abnormality of chromosome structure has been reported in several breeds of cattle. It is called a Robertsonian translocation because it was first described by a man named Robertson. It is called a translocation because it is the result of the union (at the ends) of two autosomes from different pairs (or two different pairs of homologous chromosomes). As shown in Figure 7.2, the dividing autosomes of cattle at the metaphase are connected at the ends, which causes them to appear V-shaped. The union of these two V-shaped chromosomes to form a single large X-shaped chromosome at the metaphase is a translocation (Figure 7.4). Carriers of this abnormality cannot be determined without studying the chromosome complement, but females carrying it are of lowered fertility. The translocation abnormality appears to be widely distributed in cattle of European origin.

A chromosome abnormality involving mixed cell populations occurs in cattle. In other words, some of the cells carry two X chromosomes, and some carry an X and a Y. A true Holstein hermaphrodite about 5 months of age possessed some cells which were XX and some XY. This individual possessed a scrotum without testicles, a penis, and a uterus with an ovary on the right side.

A mixture of cells, some of which contain XX and some XY chromosomes, is quite common in the freemartin heifer born twin with a bull calf. The XX cells are from the heifer's own cell line, whereas the XY cells are from the cell line of her twin brother. Since the placental membranes of a freemartin heifer are fused with those of her twin brother, they have a common blood supply during intrauterine life which results in the mixed blood cell population. In such cases the female is sterile because of the improper development of her reproductive tract. The fertility of the bull twin does not appear to be affected. About 90 to 95 percent of bull-heifer twins have a common intrauterine blood supply resulting in the sterility of the heifer. When they do not have a common intrauterine blood supply, the heifer is fertile. Culturing the chromosomes of cells in the blood of bull-heifer twins will determine whether or not they had a com-

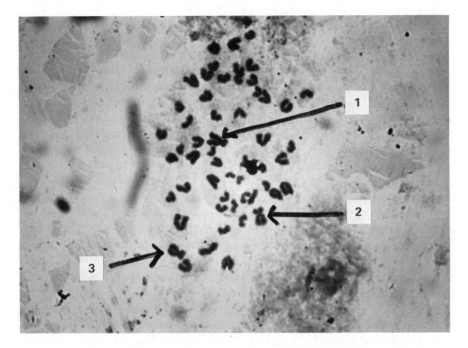

Figure 7.4.   Chromosome abnormality in cattle in which there is an extra chromosome resembling the large X chromosomes. Numbers with arrows point to these three chromosomes. (Courtesy University of Missouri, Columbia)

mon intrauterine blood supply. If both XX and XY cells are found in either the heifer or bull twin or both, the heifer is a freemartin.

7.2   The Gene

The gene is the smallest unit of inheritance. The gene is a portion of a long, helical DNA molecule carried on the chromosomes. Genes are responsible for triggering all biochemical reactions in the animal body. Much is known about the structure and functions of genes, but we shall not discuss them in detail here. Our attention will be largely directed to pairs of genes and how they express themselves phenotypically in the individual. We shall also discuss ways of identifying individuals which possess superior genes and therefore should transmit them to their offspring. In other words, we shall emphasize beef cattle genetics and how it may be applied to the improvement of beef cattle through breeding.

7.3 Segregation of Genes

Genes, because they are carried on the chromosomes, normally occur in pairs in body cells and in half-pairs in the gametes. A single pair of genes is carried at the same location (called a locus) on a pair of chromosomes referred to as *homologous chromosomes* because they are alike (homo). One gene of each pair comes from the mother and the other from the father. If an individual becomes a parent, it will transmit one gene of each pair, but not both, to each of its offspring.

It is customary in genetics to represent each gene by a letter of the alphabet. Two genes carried at the same locus on homologous chromosomes are called allelomorphs (alleles for short) and may be identified as gene *A* or gene *a*. Sometimes more than two alleles are found at the same locus in some members of a population. If three occur, they could be identified as genes A, A^1, and a. We shall limit our discussion here to just two alleles because this is a sufficient number to illustrate the principles to be discussed. Different pairs of alleles can be assigned different letters of the alphabet. For example, genes *B* and *b* would belong to a different set of alleles than genes *A* and *a*, and the two sets would either be on different chromosomes or on different parts (loci) of the same chromosome. When the latter occurs, the two pairs of genes (*Aa* and *Bb*) are said to be linked.

To illustrate how genes in the parents segregate in the sex cells and recombine in the offspring at fertilization, let us first use the single pair of alleles *A* and *a*. With these two genes we can have three different pairings: *AA*, *Aa*, and *aa*. An individual that is *AA* is said to be of genotype *AA* and is homozygous *AA*. If a mutation in one of these genes does not occur, a parent of this genotype will transmit an *A* gene to each of its offspring. A parent of genotype *aa* is also homozygous and barring a mutation will transmit only an *a* gene to each of its offspring. An individual of *Aa* genotype is heterozygous (because it possesses unlike genes within a pair) and will transmit the *A* gene to 50 percent of its offspring and the *a* gene to the other 50 percent.

One pair of genes is not the only pair of genes that segregates into the sex cells of a parent. Thousands of pairs segregate at the same time, and it would be impossible to illustrate here how all of these segregate. Illustrations using two pairs of genes, or a few with more than two, will give an idea of how this occurs. To illustrate the segregation of pairs of genes into the sex cells, we must use a rule of probability which states the following: The probability of two or more *independent events* occurring together is the product of each separate probability. Two events are independent if the occurrence of one has no influence on the occurrence of the other. Gene pairs *Aa* and *Bb* are independent if they are carried on two different chromosome pairs, and for this illustration we shall assume this is true.

Let us assume that an individual is of genotype *AaBb* and that the two pairs of genes are carried on two different pairs of homologous chromosomes. What would be the probability of genes *AB* being transmitted to a progeny through its sex cells? The probability that this individual will transmit the *A* gene to an offspring is one-half. The probability that it will transmit the *B* gene to an offspring is also one half. Thus, the probability that it will transmit both an *A* and a *B* gene to each offspring is one half times one half or one fourth. Similarly, the probability of such a parent transmitting a combination of either genes *Ab*, *aB*, or *ab* to an offspring would also be one fourth. The possibility of combinations of genes transmitted to their offspring for various genotypes involving two pairs of genes is shown in Table 7-1.

For further illustration, let us use an individual that is of genotype *AaBbCcDd* with each one of the four different pairs of genes carried on a different pair of homologous chromosomes so that they are not linked. The probability that such an individual would transmit genes *ABCD* to an offspring through its sex cells would be one half to the fourth power or one sixteenth. Furthermore, an individual of this genotype could transmit only one of sixteen possible combi-

TABLE 7-1   Proportion of different combinations of genes transmitted by a parent to its offspring when that parent is of a certain genotype

Genotype of parent	Probability of this parent transmitting this combination of genes to an offspring
AABB	*AB* 1.00
AABb	*AB* 0.50; *Ab* 0.50
AAbb	*Ab* 1.00
AaBB	*AB* 0.50; *aB* 0.50
AaBb	*AB* 0.25; *Ab* 0.25; *aB* 0.25 *ab* 0.25
Aabb	*Ab* 0.50; *ab* 0.50
aaBB	*aB* 1.00
aaBb	*aB* 0.50; *ab* 0.50
aabb	*ab* 1.00

Note: The genotype is the combination of genes an individual possesses in its gene pairs. The combinations of genes transmitted through the sex cells are in half-pairs.

nations of genes to an offspring. Since thousands of pairs of genes are carried by a parent, this explains why even different offspring of the same two parents may not closely resemble each other or why some two or more offspring resemble each other more than others.

The number of different combinations of genes that can be transmitted by a heterozygous parent to its offspring through the gametes is shown in Table 7-2.

7.4 Recombination of Genes in the Offspring

The same law of probability used for predicting possible combinations of genes in the gametes of a parent can also be used for predicting the probability of certain genes pairing in the offspring of two parents.

Assume we have two parents both of whom are of genotype *Aa*. What is the probability (P) that their offspring will be *AA*, *Aa*, or *aa*? This is illustrated as follows:

	Father	*Mother*
	Aa ×	*Aa*

P of *AA* offspring	½ × ½ or ¼
P of *Aa* offspring	½ × ½ or ¼
P of *aA* offspring	½ × ½ or ¼ ²⁄₄ or ½
P of *aa* offspring	½ × ½ or ¼

TABLE 7-2 Number of different combinations of genes a parent can transmit to its offspring if these different pairs of genes are carried on different pairs of chromosomes

Genotype of parent	Number of different combinations of genes transmitted through the sex cells
Aa	2
AaBb	4
AaBbCc	8
AaBbCcDd	16
AaBbCcDdEe	32
AaBbCcDdEeFf	64

Note: The formula to use is 2^n, where *n* is the number of different pairs of heterozygous gene pairs (heterozygous means unlike pairs of genes such as *AaBb*).

Thus, these probabilities give us the familiar 1 *AA*:2 *Aa*:1 *aa* ratio. This ratio holds only when large numbers of offspring are produced. If only four offspring are produced by such parents, they could be any one or any combination of all three genotypes. For example, the probability of these parents (*Aa* × *Aa*) producing four *aa* offspring would be one fourth to the fourth power or 1/256.

The probability of offspring being produced by parents carrying two different pairs of genes follows the same law of probability. For example, the probability of two parents that are *AaBb* producing an *AABB* offspring is one fourth times one fourth or one sixteenth. Probabilities of these same parents producing other genotypes in their offspring may be calculated in the same manner. This is shown in Table 7-3.

The transmission of genes and combinations of genes by a parent becomes even more complex as the number of pairs of genes considered increases. This is illustrated in Table 7-2. Note that in this table an individual of genotype *AaBbCcDdEeFf* can transmit sixty-four different combinations of genes to its offspring. An individual of genotype *AABBCCDDEEFF* which is homozygous for this same six pairs of genes could transmit only one combination of genes

TABLE 7-3   Different combinations of genes in the offspring of parents of different genotypes (genetic makeup)

Genotype		Probability of genotype of offspring
Father	*Mother*	
AA	*AA*	*AA* 1.00
Aa	*AA*	*Aa* 0.50; *AA* 0.50
aa	*AA*	*Aa* 1.00
Aa	*Aa*	*AA* 0.25; *Aa* 0.50; *aa* 0.25
aa	*aa*	*aa* 1.00
AABB	*AABB*	*AABB* 1.00
aabb	*AABB*	*AaBb* 1.00
aabb	*aabb*	*aabb* 1.00
AaBb	*AaBb*	*AABB* 0.0625
AaBb	*AaBb*	*AABb* 0.1250
AaBb	*AaBb*	*AAbb* 0.0625
AaBb	*AaBb*	*AaBB* 0.1250
AaBb	*AaBb*	*AaBb* 0.2500
AaBb	*AaBb*	*Aabb* 0.1250
AaBb	*AaBb*	*aaBB* 0.0625
AaBb	*AaBb*	*aaBb* 0.1250
AaBb	*AaBb*	*aabb* 0.0625

(*ABCDEF*) to its offspring. Which parent, or genotype, do you think
would be the best breeder, and why?

7.5 Phenotypic Expression of Genes

In the discussion of the segregation and recombination of genes we
purposely avoided mentioning dominance, recessiveness, overdominance, and other ways genes may express themselves phenotypically.
We did this because gene pairs segregate and recombine in the gametes
in the ways mentioned regardless of their phenotypic expression. The
phenotypic expression of genes, however, probably has more meaning
to beef cattle breeders because they select their breeding animals
mostly by phenotype.

Phenotype refers to differences among individuals in a population that we can measure by means of our senses. Phenotype includes
rate of gain, carcass quality, coat color, and many other traits. The
phenotype of beef cattle is determined by both heredity and environment. When a trait such as coat color is affected by one gene or a few
pairs of genes, environmental effects are usually small. When a trait
such as rate of gain is affected by many genes, environmental effects
are usually larger. For traits affected by both heredity and environment, it is difficult for the breeder to determine if an animal is superior or inferior because of heredity or environment or both.

In this discussion we shall direct our attention to how genes
express themselves phenotypically and disregard environmental
effects. In a later chapter we shall discuss both hereditary and
environmental effects on economic traits in beef cattle.

Genes express themselves phenotypically in two general ways
known as additive and nonadditive. The differences between the two
are such that they require different kinds of mating and selection
systems.

7.5.1 Nonadditive Gene Action

Nonadditive gene action includes dominance-recessiveness, overdominance, and epistasis.

A dominant gene is usually represented by a capital letter and a
recessive gene by a lowercase letter. For example, in beef cattle
polledness (*P*) is dominant to horns (*p*). When *P* and *p* occur together,
the individual is polled because the polled gene is dominant to the
horned one.

With two genes such as *P* and *p*, three different pairings are possible. These pairings are *PP*, *Pp*, and *pp*. A bull that is *PP* is pure for polledness (homozygous), and he will produce all polled calves regardless of whether or not he is mated to polled or horned cows. The *Pp* individual is not pure polled (he is heterozygous) and will produce some polled and some horned calves unless he is mated with a pure polled (*PP*) cow. The horned bull (*pp*), of course, does not possess the polled gene and will transmit only a horned gene to his progeny. What we have said about the dominant polled gene (*P*) and the recessive horned gene (*p*) applies to all other dominant and recessive genes regardless of the traits affected.

Six possible matings can be made when a single pair of alleles such as *P* and *p* are involved. These matings and the progeny they would be expected to produce are shown in Table 7-4.

Polled individuals can be pure polled (*PP*) or heterozygous polled (*Pp*), as stated previously. The pure polled individual should produce all polled offspring. The heterozygous polled individual will produce offspring which are both horned and polled. The problem is to determine if an individual is pure or heterozygous polled. This often requires a progeny test, which will be discussed later, but if a polled individual has produced one horned offspring or one of its parents was horned, it is heterozygous.

A gene may be *partially dominant* rather than completely dominant. A good example has been described in comprest Herefords. The gene which causes a Hereford to be comprest (small and compact) is partially dominant to the gene in Herefords for conventional size (*c*). The normal, conventional-sized Hereford is of genotype

TABLE 7-4   Six possible matings with possible genotypes and phenotypes from such matings

Genotype of parents		Genotypes and phenotypes of the progeny
Sire	Dam	
PP	PP	All PP or pure polled
PP	Pp	All progeny polled but 50% PP and 50% Pp
PP	pp	All progeny polled but heterozygous polled (Pp)
Pp	PP	All progeny polled but 50% PP and 50% Pp
Pp	Pp	25% of progeny pure polled, 50% heterozygous polled, and 25% horned
pp	pp	All pp and horned

cc. The comprest individual is heterozygous or *Cc.* One comprest gene (*C*) makes the individual a comprest, but two comprest genes (*CC*) cause the individual to be a dwarf. It usually dies early in life.

Several partially dominant genes are known in animals. Only a few have been described in cattle, however. The effects of partially dominant genes which have been described are usually lethal (deadly) or harmful in the homozygous pairings as was true with the comprest Hereford. Dexter cattle carry only one gene (*D*) for small size or are of genotype *Dd.* An individual of genotype *DD* is known as a bulldog calf and is dead at birth. A *dd* individual is of normal size and is called a Kerry.

Some genes for coat color in species other than cattle are not harmful in the heterozygote as far as is known but are lethal in the homozygote. The dominant white gene (*W*) in horses is not harmful in the heterozygous *Ww* individual but is lethal in the homozygous *WW.* A similar condition is found in yellow mice where the heterozygote *Yy* lives, but the homozygous *YY* dies. The Blue Merle Collie is normal and carries one dilution gene (*D*). The White Merle Collie carries two dilution genes (*DD*), which is harmful but not lethal, causing deafness and other defects.

The white coat color in Charolais appears to be partially dominant to the black of the Angus and the red of the Hereford. A crossbred Charolais X Hereford calf is pink in coat color, but the face is white. A Charolais X Angus crossbred calf is often a dirty white color. It is possible that purebred Charolais bred up from a Hereford base might carry the gene for white face but not show it because the body color is white.

More partially dominant genes with a harmful effect may be present in cattle breeds, but they have not yet been identified and reported.

Overdominance is a type of gene action in which the heterozygous individual is superior to either homozygote. In other words the A^1A^2 individual is superior to those that are A^1A^1 or A^2A^2.

Although snorter dwarfism in beef cattle is usually referred to as a recessive trait, under some conditions it illustrates the principle of overdominance. A few years ago when short-legged, compact, and early maturing cattle were preferred, selection often seemed to favor the carrier of the recessive dwarf gene (genotype *Dd*) because one dwarf gene had an effect on the appearance of some individuals. When the long, rangy, fast-growing individuals were preferred, as is true at the present time, preference for the dwarf gene was much less.

Overdominance appears to have an important effect on traits related to physical fitness. It is difficult, however, to distinguish between the overdominant and the dominant phenotypic effect. From the practical standpoint distinguishing between the two may not be important.

Since overdominance is expressed in the heterozygous individual, it brings up an important problem in beef cattle breeding, because when heterozygous individuals are mated together, only about 50 percent of their offspring are heterozygous. Thus, the parents look or perform better than they breed. For example,

$$P_1 \qquad A^1A^2 \qquad\qquad X \qquad\qquad A^1A^2$$

$$F_1 \qquad\qquad \begin{array}{l} 1\ A^1A^1 \\ 2\ A^1A^2 : 50\ \% \\ 1\ A^2A^2 \end{array}$$

On the other hand, mating individuals that are A^1A^1 with those that are A^2A^2 will give offspring that are 100 percent A^1A^2 or heterozygous. Thus, the parents tend to breed better than they look or perform.

These examples show that selecting the most vigorous and better-performing individuals when overdominance is important can be disappointing.

Most breeders avoid inbreeding, which tends to make individuals more homozygous for many gene pairs. This suggests that heterozygous individuals which do not breed true are the ones breeders often select for. This may not be as undesirable as one might think because heterozygous females are more efficient mothers, as a general rule, than those that are homozygous.

Epistasis is still another form of nonadditive gene action. Epistasis is due to the interaction of one pair of genes at one locus with other pairs of genes at different loci. It differs from overdominance in that overdominance involves the interaction between alleles of a single pair of genes at a single locus.

Several examples of epistasis have been reported in beef cattle. Coat color in Angus X Hereford crossbreds is an example. In Herefords the white face gene (*W*) is dominant to the solid colored face (*w*). The black coat color gene in the Angus (*B*) is dominant to the red coat color gene (*b*). Crossing Herefords with Angus usually gives a black white-faced calf, although there usually is not as much white on the face of the crossbreds as on the face of purebred Herefords. Since the white face of the Hereford covers up the black solid face of the Angus, the white face gene (*W*) is said to be epistatic to the black gene (B). Sometimes the Angus X Hereford crossbred produces a red calf with a white face. This suggests that the Black Angus parent carried the red gene (genotype Bb) and that the red white-faced calf received a red gene from each of its parents (genotype of calf, bb). Since this red calf has a white face, this indicates that the white face gene (W) is also epistatic to the red gene (b) for coat color in these two breeds.

Epistasis is also involved in the presence or absence of horns and scurs in beef cattle. Experimental results indicate that one pair of genes (P for polledness and p for horns) is responsible for the presence or absence of horns and another pair for scurs (S_c for scurs and S_n for the lack of scurs). The gene for scurs (S_c) is dominant in males and recessive in females. Horned individuals (genotype pp) can carry the gene for scurs (S_c) because the horns cover up the appearance of scurs (S_c). On the other hand, scurs appear in polled individuals in some instances so that the gene for scurs (S_c) is epistatic to the polled gene (P). How these two genes interact to produce horns, polledness, and scurs is illustrated in Table 7-5.

The inheritance of horns and scurs is not this simple in some breeds of cattle such as the Charolais, Brahman, and breeds from a Brahman base. Some of the exotic breeds appear to have a more complex mode of inheritance of horns and polledness. At least in breeds derived from a Brahman cross, the gene for scurs (S_c) appears to be dominant in both males and some females, and in some individuals one or more pairs of additional genes may be involved.

One pair of genes in a body cell is surrounded by thousands of other pairs of genes which can influence how this pair of genes expresses itself phenotypically. These surrounding genes are called

TABLE 7-5　Combinations of genes responsible for horns, scurs, and polledness in beef cattle

	Phenotype	
Genotype	Male	Female
PPS_nS_n	Polled	Polled
PpS_nS_n	Polled	Polled
ppS_nS_n	Horned	Horned
PPS_cS_n	Scurred	Polled
PpS_cS_n	Scurred	Polled
ppS_cS_n	Horned[a]	Horned[a]
PPS_cS_c	Scurred	Scurred
PpS_cS_c	Scurred	Scurred
ppS_cS_c	Horned[a]	Horned[a]

[a]Horned but are carrying the scurred gene.

modifying genes, and their effect appears to be epistatic. For example, the Hereford color pattern is probably due to one pair of genes, but the amount of white on Herefords varies from a large to a small amount. The variation in amount of white is probably due to modifying genes.

Epistasis may be responsible for producing a trait in crossbred calves not seen in calves of the pure breeds making up the cross. For illustration, let us assume that a combination of genes *A* and *B* is necessary for a trait to appear phenotypically. Breed 1, however, is *AAbb*, and breed 2 is *aaBB*, with the trait not appearing in either of the two pure breeds. When they are crossed, all offspring would be *AaBb* and would show the trait. Some of the animals in each of the two breeds probably would not be homozygous *AA* or *BB* so that not all calves would show the trait.

We cannot be certain at this time how important epistasis might be in beef cattle breeding. If it affects coat color, it probably also affects performance and other traits. If it is important, its effects can be utilized largely by crossing unrelated lines within a breed or by crossbreeding.

Additive gene action also affects many traits in beef cattle. In additive gene action the phenotypic effects of one gene adds to the phenotypic effects of another. Genes which contribute or add to the phenotype are usually denoted by a capital letter such as *A*, *B*, or *C* and are called *contributing genes*. In this kind of gene action genes that contribute nothing to the phenotype are called *neutral genes* and are represented by lowercase letters such as *a*, *b*, or *c*. Dominance, overdominance, and epistasis are separate and apart from additive gene action.

The following genotypes and phenotypes illustrate additive gene action. This, of course, is a hypothetical example using two pairs of genes. We shall assume the trait involved is postweaning rate of gain and that animals of genotype *aabb* (the residual genotype) gain 2 pounds per day, whereas genes *A* and *B* each add 0.20 pound to daily gain. In this example we shall disregard environmental effects. The genotypes and their corresponding phenotypes would be as follows:

Genotype	*Phenotype, rate of gain (lb/day)*
aabb	2.00
Aabb	2.20
AAbb or *AaBb* or *aaBB*	2.40
AABb or *AaBB*	2.60
AABB	2.80

This shows that as each contributing A or B is added to the genotype it increases the rate of gain. Actually there may be dozens of pairs of genes (contributing or neutral) affecting a trait, and each contributing gene may not add the same amount to the phenotype as others. The same general principle in our illustration will apply, however. Actually, postweaning rate of gain appears to be affected by additive gene action in a way similar to that shown in our example except many pairs of genes are probably involved. Several other traits in beef cattle appear to be affected by this kind of gene action.

7.6   Utilization of Additive and Nonadditive Gene Action

Since many of the traits in beef cattle are determined by several pairs of genes, it is possible that some of these genes act in an additive and some in a nonadditive manner. Therefore, some traits are mostly affected by additive genes, some by nonadditive genes, and some by both.

For traits affected largely by additive gene action, selection should be based on the phenotype of the individual. The mating system of choice would be to find the best and mate the best to the best. Additive gene action is indicated when a trait is medium to highly heritable, it is not improved by crossbreeding, and it is affected little or not at all by inbreeding.

For traits affected largely by nonadditive gene action, improvement may be brought about by crossing unrelated lines within a breed or by crossing different breeds. Nonadditive gene action is indicated when a trait is lowly heritable, it is improved by crossbreeding, but it is accompanied by a decline in the trait when inbreeding is practiced.

Some traits may be affected by both additive and nonadditive genes. For such traits, selection for the best and mating the best to the best would make some improvement, as would crossbreeding. Such traits are indicated when they are medium to lowly heritable, they are improved by crossbreeding, but they show some decline when inbreeding is practiced.

Breeders usually select for more than one trait at a time in their beef herd. Some of these traits may also be determined by either additive or nonadditive gene action or both. In such cases traits affected largely by additive gene action should be improved in the pure breeds by finding the best and mating the best to the best. The superior breeds can then be crossed to improve nonadditive traits which show hybrid vigor.

7.7 Some Other Facts about Genes

When we were discussing dominance and recessiveness, we probably gave the impression that when a trait was affected by a dominant gene, the presence of a dominant gene always caused the trait to appear. We also suggested that when a trait is due to a recessive gene, an individual possessing two recessive genes within a pair always showed the recessive trait. This is not always true because some genes lack penetrance and do not express themselves phenotypically.

Penetrance refers to the percentage of times a trait is expected because of an animal's genotype and actually appears in the phenotype. If it is expected to show up 100 percent of the time and does, this is *complete penetrance*. If it shows up only 50 percent of the time when expected, it has 50 percent penetrance. Genes which have less than 100 percent penetrance show *incomplete penetrance*. Genes with incomplete penetrance may be dominant or recessive in their phenotypic expression.

One kind of shaking in chickens has been reported to be due to a recessive gene with incomplete penetrance. All chickens which shake would be recessive, but when shakers were mated, only 34 percent of their offspring shook. The penetrance of this gene would then be 34 percent.

A condition in humans known as *manic depressive psychosis* has been reported to be due to the dominant gene (*P*). Homozygous individuals (*PP*) always show the trait so the penetrance is 100 percent. Heterozygous individuals (*Pp*) show the trait only 20 percent of the time so the penetrance in the heterozygote is 20 percent.

From the practical standpoint, if penetrance is incomplete, the breeder might cull individuals showing the trait, but others not showing it would still be carrying the gene or genes.

Varied expressivity refers to the degree to which a trait is expressed phenotypically. This means that the penetrance of the gene is 100 percent but that individuals with the same trait vary from one extreme to another. Scurs in cattle vary from a small scab to a large horny projection on the head, or a scur may appear on one side of the head but not on the other. This is an example of variable expressivity.

Several factors may be responsible for incomplete penetrance and varied expressivity of genes. One includes the external environment such as temperature and sunlight. Another is the internal environment, which involves the hormones within the body and modifying genes.

A gene-determined trait may appear phenotypically anytime in the life of the individual. Some traits appear before birth, others shortly afterward, whereas others may not appear until later in life. For example, many of the economic traits are not fully expressed until animals are more than 1 or 2 years of age. They are not expressed at birth. Therefore, selection should be delayed until each individual has the opportunity to express its phenotype for these traits.

Pleiotropy may be defined as the situation in which one gene may affect two or more traits. Many examples of pleiotropy are known in animals, but few have been reported in beef cattle. A white (albino) or partially albino (gray) coat color in some animals is associated with defects or even the death of the individual. We mentioned earlier that the dominant white gene (W) in horses is lethal when it is homozygous (WW). Most blue-eyed white cats are deaf, and partially albino Herefords have defects in some of their white blood cells that make them more susceptible to certain infectious diseases. These are probably instances where the same gene affects two or more traits, although it is possible that two genes affecting two different traits are so closely located on the same chromosome (linked) that they would very seldom separate in the sex cells by crossing over. The traits such genes affect would usually be inherited together, making it appear that both traits were determined by the same gene.

Several genes with small or minor phenotypic effects may affect two or more traits. For example, fast gains in beef cattle are usually associated with more efficient gains. Geneticists usually say that two traits such as this are genetically correlated; that is, some of the same genes affect both traits.

STUDY QUESTIONS

1. What are some of the reasons that beef cattle breeders have become more familiar with breeding principles in recent years?

2. What are the two major kinds of cells the animal body contains?

3. What are organelles?

4. At what stage of cell division may cells be stained and observed? How do the chromosomes appear at the time they can be stained?

5. How many pairs of chromosomes does the normal body cell of beef cattle contain? What are autosomes?

6. Does the bull or the cow control the sex of the offspring? How is the sex controlled?

7. What kinds of chromosome abnormalities have been observed in cattle? What is meant by a Robertsonian translocation?

8. What is a freemartin? What sort of test can be used to detect such an animal?

9. How is the number of chromosomes in cattle kept constant from one generation to another?

10. What are homologous chromosomes? What are allelomorphs?

11. State an important law of probabilities that may be used to estimate the probability of a gene occurring in a gamete?

12. What is the probability that an individual of genotype *AaBbCcDd* mated with another individual that is of genotype *AaBbCcDd* will produce an offspring of genotype *AaBbCcDd*?

13. How many different genotypes can be produced by two parents that are of genotype *AABBCCDD*? How many can be produced by two parents that are *AaBbCcDd*?

14. Distinguish between the genotype and the phenotype of an individual.

15. What kinds of phenotypic expression of genes may be classed as nonadditive?

16. Distinguish between overdominance and epistasis.

17. How can it be determined if a polled individual is homozygous polled (*PP*) or heterozygous polled (*Pp*)?

18. How many different kinds of matings are possible if two alleles such as *A* and *a* are involved? What are they?

19. Distinguish between a partially dominant and a dominant gene.

20. Describe a partially dominant gene in cattle.

21. Give an example of epistasis in beef cattle.

22. Explain how it is possible to have a defect in a crossbred calf that is not present in either of the two parental breeds used to produce the crossbred calf.

23. What is polledness? What are scurs? How are they inherited in the Angus, Hereford, and Shorthorn?

24. Assume you have a herd of superior horned Hereford cows. How would you develop a herd of pure polled Herefords starting with these original cows?

25. What are modifying genes? Give an example in cattle.

26. Define additive gene action, contributing genes, neutral genes, and residual genotype.

27. Since many traits in beef cattle are affected by additive gene action, nonadditive gene action, or both, how can the commercial cattle producer design a program to best utilize both?

28. How do we know if a trait affected by many pairs of genes is due mostly to additive or nonadditive gene action?

29. Define penetrance, complete penetrance, incomplete penetrance, and variable expressivity.
30. Distinguish between penetrance and variable expressivity.
31. What is meant by pleiotropy? By genetic correlations among traits?

inherited traits in beef cattle

Two kinds of inheritance affect traits in beef cattle. One is known as qualitative inheritance, characterized by a sharp distinction among phenotypes; these traits are little affected by environment and are determined by one pair, or a limited pair, of genes. Quantitative inheritance, on the other hand, involves traits in which there is no sharp distinction among phenotypes; they are usually greatly affected by environment and are determined by many pairs of genes. Our attention in this chapter will be directed mostly to qualitative traits that have a simple mode of inheritance, although we shall mention a few affected by many genes. Quantitative traits will be discussed in more detail in a later chapter.

8.1 Mutations

Mutations are changes in genes which produce a new phenotype when the right combination of genes is present in the individual. More specifically, a mutation is a change in the code sent by the gene (a portion of a DNA molecule) by means of messenger RNA to the ribosomes in the cytoplasm to build a particular protein. Proteins consist of chains of amino acids, and differences in proteins are due to differences in numbers, kinds, and arrangements of amino acids in the protein molecule. A new mutation is a change in the code sent to the cytoplasm by means of messenger RNA for building a certain protein. A new mutation therefore produces a new or different kind of protein in place of the one produced by the original gene. Enzymes, which trigger many biochemical reactions in the body, are proteins subject to new mutations. When a mutation affecting a particular enzyme occurs, that enzyme is not produced, and this may be detrimental or even fatal to the individual.

Mutations may occur in body (somatic) cells or in the sex cells. Only those occurring in the sex cells are transmitted from one generation to another. Body cell mutations end with the death of the individual within whose body cells they occur. A black spot on the coat of a red Hereford is probably a body cell mutation in which red has mutated to black. Since black is dominant to red, only one gene mutation would be necessary to have its effect.

Many mutations have occurred in beef cattle in the past and will continue to occur in the future. Genetic differences among individuals are due basically to the occurrence and accumulation of mutations in a population. New mutations may be induced by exposing the individual to radiation, certain drugs, etc., but this is not practical because the induction of a mutation cannot be controlled. It is not possible to produce a specific mutation when desired. In general, the induced mutations are the same as those which occur naturally.

Most new mutations are recessive and are detrimental. A few are not harmful and may be preferred by man in selection. For example, primitive cattle possessed horns probably because they helped the individual to survive in the wild state. No doubt mutations of the genes for horns to polledness occurred in the ancient ancestors of domestic cattle, but the polled condition probably had no selective advantage and may even have been detrimental to survival at that time. Under domestication, however, man has selected for the polled condition because polled animals are more suitable to feedlot conditions and do not cause injury to each other, as sometimes occurs when cattle are horned.

Breeding animals no doubt carry many recessive genes for genetic defects in the heterozygous state. Many of these never appear in a herd where no inbreeding or linebreeding is practiced. Mating relatives tends to pair recessive genes so they are expressed phenotypically. Perhaps this is the reason that most breeders prefer to out-breed and avoid inbreeding.

When recessive defects do appear in a herd, a breeder attempts to eliminate them by discarding defective animals and their close relatives. This may prevent the occurrence of the recessive defect, but it seldom eliminates the gene from the herd or flock.

8.2   Inheritance of Some Desirable Traits

The mode of inheritance of many traits in beef cattle is known, but it is not known for others. The trait first must be observed, and then proper matings must be made to determine if inheritance is involved, and if it is what mode of inheritance is responsible for the inheritance of the trait. Some inherited traits have been reported on the basis of scanty data, and the inheritance reported may not be accurate. This is due to the absence of enough animals to complete accurate breeding tests.

8.2.1   Horns-Polledness

In the Angus, Hereford, and Shorthorn, polledness is dominant and the presence of horns recessive. Scurs are due to another gene (S_c) which is dominant in males and recessive in females. The symbol S_n is assigned to the gene for no scurs. The scurred gene in Charolais, some of the exotic breeds, and new breeds developed from Brahman crosses sometimes shows dominance in both the male and female.

Another gene known as the African horn gene (A_f) is dominant to polledness in the male and recessive to it in the female. It may be present in varying frequencies in many breeds. Other pairs of genes still not reported may also affect horns and polledness.

8.2.2   Fertility

Fertility is determined by many pairs of genes, probably because many physiological pathways are involved. Apparently few genes with an additive effect are involved in the production of this trait. On the other hand, nonadditive genes (dominance-recessiveness, over-dominance, and epistasis) have an important effect. Many inherited defects of fertility are known to be due to dominance-recessiveness.

8.2.3 Weaning Weight

The weaning weight trait is also determined by many (poly) genes. Some genes are additive since the trait is low to medium in heritability[1] and can be improved by selection. It is also affected by nonadditive genes because it shows some hybrid vigor when crossbreeding is practiced.

8.2.4 Postweaning Rate of Gain

The postweaning rate of gain trait is determined by many pairs of genes many of which have an additive effect, as shown by a heritability estimate of 40 to 45 percent. It is also affected to a certain extent by nonadditive genes because it shows some hybrid vigor when crossbreeding is practiced.

8.2.5 Postweaning Efficiency of Gain

The postweaning efficiency of gain trait is also determined by many pairs of genes. The genes appear to act in an additive manner, as shown by a heritability estimate of 35 to 40 percent and a very small effect of heterosis. This indicates that feed efficiency is little affected by nonadditive gene action.

8.2.6 Carcass Traits

Most carcass traits are also determined by many genes, most of which have an additive effect, as shown by a heritability estimate of 30 to 70 percent. Nonadditive gene action has little effect on these traits because they show little or no hybrid vigor when crossbreeding is practiced.

8.2.7 Conformation

Conformation at different ages appears to be due to many pairs of genes with an additive effect. Conformation score shows only a small amount of hybrid vigor, which indicates it is little affected by nonadditive genes.

---

[1] A heritability estimate is an estimate of the portion of the total variation (phenotypic) due to heredity.

8.2.8   Mature Size

The mature size trait is also affected by polygenes with an additive effect, as shown by a heritability estimate of 50 to 70 percent. It shows little hybrid vigor in crossbred animals.

8.2.9   Temperament

Temperament and disposition in beef cattle appear to be affected by polygenes, although the phenotypic expression of the genes involved is not fully known. Selection for a better temperament and disposition appears to be effective, but no information is available on the effects of hybrid vigor. Certainly, environment and how animals are handled can affect these traits.

8.2.10   Genetic Resistance to Disease and Parasites

Little has been done to determine the mode of inheritance of genetic resistance to disease and parasites in beef cattle. Several lines of evidence suggest that there is genetic resistance and susceptibility in some individuals in this species. Breed differences exist in the susceptibility to pinkeye and cancer eye in cattle. Brahmans appear to be more resistant to certain parasites than the European breeds, and a new breed called the Droughmaster has been developed in Australia that is resistant to tick infestations. Furthermore, some individuals within a breed appear to be more resistant and some more susceptible to diseases and parasites.

Possibly some forms of genetic resistance and susceptibility are due to a few genes with a nonadditive effect, whereas others may be due to the effects of additive genes. The inheritance of such traits has not been fully determined.

8.3   Inheritance of Coat Color

Coat color in beef cattle is of little economic importance unless it is related to a defect or associated with poorer performance in a particular environment. Sometimes, however, a desired color is found in a particular breed or crossbred which cattle feeders prefer. For example, the "black baldies" resulting from the cross of the Angus and Hereford breeds usually give a black calf with a white face.

Little is known about the inheritance of coat colors in some breeds of cattle. The phenotypic expression of many genes has been proposed but may not be entirely accurate, and the knowledge of them may be incomplete. Genes affecting coat color which have been described are as follows:

A and a.   Glass-eyed albinism due to a dominant gene (A) has been described in Herefords in the United States. The coat is usually white except for some slight blemishes. The iris is a pale blue surrounded by a whitish circle, giving a "glassy" appearance. These albinos are not tolerant to light.

A_t and a_t.   True albinism in which the cattle lack pigment in the hair, skin, and eyes has been described in Holsteins and some other breeds. The eyes are pink or red in color.

A_p and a_p.   Another type of albinism (pseudo) has been described in Guernseys and Holsteins. Pigment is lacking in the hair and skin, but the eyes are colored. The calves are born white but later turn a creamy shade due to some pigment produced in the hair, giving a "ghost" pattern.

B and b.   Black pigment (B) in the hair is dominant to red (b) and covers up (is epistatic to) the blackish gene (B_s). Black and red appear to be the basic colors in cattle.

B_s and b_s.   The blackish gene B_s is dominant to its own allele, called nonblackish (b_s). This is a different pair of genes than the B and b. A blackish individual shows black spotting, as found in many Jerseys. Two red genes bb must be present in the individual in order for the blackish gene to express itself phenotypically. Blackish is not expressed in the presence of the black gene B.

B_r and b_r.   This is still another pair of alleles and is responsible for brindling B_r and nonbrindling b_r. A brindle pattern is one in which red and black hairs alternate, giving alternate vertical stripes. The blackish gene (B_s) must be present in order for the brindling gene (B_r) to express itself. The red genotype (bb) must also be present because the black gene (B) covers up the expression of the blackish gene (B_s).

D and d.   The dominant dilution gene (D) appears to dilute both black and red colors to a lighter shade when present in the DD or D_d genotypes. Animals of genotype d_d would show no dilution of red or black.

D_c and d_c.   Another pair of dilution genes, apparently different from D and d, is present in the Charolais breed. The D_c gene appears to act as a partially dominant gene in that in the homozygous (D_cD_c) form it dilutes colored pigment to a white. In the heterozygote (D_cd_c) it dilutes red to a light red or yellow color and black to a dirty white.

D_i and d_i. This is another pair of dilution genes, but the gene causing the dilution is recessive (d_id_i). The d_i gene appears to be present in breeds such as Jerseys, which carry the blackish gene (B_s).

I_n and i_n. This pair of alleles has been suggested as modifiers of the solid-colored gene (S), producing white in the inguinal region. Solid-colored individuals of I_nI_n or I_ni_n would have no white but those of genotype i_ni_n would.

L_w and l_w. This pair of alleles has been suggested as modifiers of the white spotting gene (ss). The L_w gene appears to be partially dominant, which spotted individuals of ssL_wL_w almost colored, those of genotype ssL_wL_w intermediate, and those of genotype ssl_wl_w almost white. The L_w and l_w genes may also affect the expression of the Hereford pattern (S_h).

M and L. This pair of genes appears to modify the blackish color, with M being dominant in males and recessive in females. Both MM males and females are dark blackish in color, while LL males and females are light blackish. Males of genotype ML are dark, but the females are light.

P_l and p_l. This pair of genes affects pigment in the posterior positions of the legs. The p_lp_l genotype appears to modify the expression of the spotting genes (ss) and those for the Hereford pattern (S_h). Individuals of genotype P_LP_L or P_lp_l will have pigmented legs, whereas those of genotype p_lp_l will not. Individuals possessing the Hereford pattern (S_h) will have pigmented legs if they are of genotype P_lP_l or P_lp_l, and they will have a brockled face since the P_l gene appears to affect both the legs and face.

P_s and p_s. The P_s gene causes pigmented skin spots, whereas the p_s gene does not. P_sP_s and P_sp_s individuals will also probably have skin spots and smutty noses.

R and W. The roaning gene R appears to be an incomplete dominant which can act on any color. A roan color is one in which white and colored hairs are mingled in the coat of the individual. A white individual would be WW, a roan RW, and a colored individual RR.

R_e and r_e. The R_e gene which causes red pigment around the eye in Herefords appears to be completely dominant to the r_e gene.

Thus, $R_e R_e$ and $R_e r_e$ individuals have red pigment around the eyes, whereas those of genotype $r_e r_e$ do not.

R_m and r_m.   It has been postulated that the recessive gene r_m in the homozygous genotype $r_m r_m$ causes the roan individual to be red. Additional unreported genes which modify the roan color may also be involved since roan individuals vary from almost white to red.

S_c, S_d, S_h, S, and s.   This is a multiple allelic series which involves genes occupying the same locus on a chromosome in a population. Only one of these alleles can occupy a locus in the sex cells, whereas two can occupy the same locus in an individual. The S gene causes the individual to be a solid (self) color all over the body. The recessive (s) gene causes the individual to be white spotted. Alleles S_c, S_d, and S_h appear to be dominant or partially dominant to S and s. The S_h gene is responsible for producing the Hereford pattern. The S_d gene causes Dutch belting, a color in which a white belt surrounds the middle of the animal. The color-sided gene (S_c) can be expressed as a white stripe around the upper part of the tail or as a broad, white stripe on both the ventral and dorsal surface of the body. The gene is found in Longhorns, Pinzgauers, and Charolais as well as other breeds.

W and w.   The recessive ww individual shows whitening along the underline and between the rear legs up to the tail. The WW and Ww individuals do not show this coloring.

W_n and w_n.   A recessive white gene $w_n w_n$ has been postulated for the White Nellore breed. Individuals of genotypes $W_n W_n$ and $W_n w_n$ would be colored, whereas those of genotype $w_n w_n$ would be white.

W_p and w_p.   A dominant white coat color has also been postulated for White English Park cattle. Thus, $W_p W_p$ and $W_p w_p$ individuals would be white, while the $w_p w_p$ individuals would be colored. The dominant W_p gene appears to vary in expressivity and restricts pigmentation to the muzzle, ears, and feet. The $W_p w_p$ individuals may show pigmentation on the entire head and may show spotting and speckling on the necks and sides. The W_p gene is also found in the Longhorn and Galloway breeds.

W_r and w_r.   A gene which restricts the amount of white in white-spotted individuals (ss) has also been postulated. If it does exist, the $W_r W_r$ and $W_r w_r$ individuals would have the amount of white they possess restricted to only a white spot on the forehead and to the switch of the tail. The w_r gene in the $w_r w_r$ genotype would have no effect on these white spots.

Black Angus in the United States carry the black gene B at a high frequency (about 0.93) and the red gene b at a low frequency (0.07). It has been estimated that about 1 calf in each 200 born to Black Angus parents in the United States are red (genotype bb). If true, this suggests that the red gene b persists in the Black Angus breed even after many generations of selection against it. The frequency of the red gene would be expected to vary in different herds. Angus may also carry the recessive inguinal white gene i_n. Thus, Black Angus with white in the flanks and udder region should be of genotype BBi_ni_n or Bbi_ni_n. Angus cattle do not appear to carry the recessive dilution gene ii or the blackish gene B_s. If the breed does possess these genes, they are present at a very low frequency.

Herefords are homozygous for the red gene bb and the Hereford pattern gene S_h. Some Herefords may carry the brindling gene B_r, but it does not express itself in the pure breed because they lack the blackish gene B_s. Some Herefords appear to carry the skin spotting gene P_s, which can also cause smutty noses. Some Herefords carry the R_n gene, which limits the amount of white on the neck, and the R_e gene, which causes red pigment to be present around the eyes. Other Herefords appear to carry the partially dominant L_w gene, which limits the amount of white spotting on the coat. The gene for pigmented legs (P_l) is also carried by some individuals of the Hereford breed.

Shorthorns are characterized by three different coat colors, white WW, roan WR, and red RR. It has been postulated that some Shorthorns carry the recessive gene r_n which in the r_nr_n genotype causes roan individuals to be red. The gene for pigmented skin spots P_s is also present in some Shorthorns. A few Shorthorns also carry the white spotting gene in the homozygous form (ss) because a few red and roan individuals show white spotting. Other pairs of genes carried by at least some Shorthorns include P_s and p_s, L_w and l_w, R_n and r_n, and P_l and p_l.

Charolais cattle are nearly white in color because they possess two dominant dilution genes DD. In the Dd or heterozygous form one dilution gene dilutes red to a light red or yellow and black to a dirty white. This breed also carries the color-sided gene S_c, and in crosses with some of the colored breeds this gene is expressed in the crossbreds as a white stripe around the upper part of the tail or as a broad white stripe on both the ventral and dorsal surfaces of the body. Occasionally Charolais graded up from a Hereford base may carry the Hereford pattern gene S_h.

The *Simmental* breed carries the red gene *b* in the homozygous *bb* genotype. The white spotting gene *ss* appears to be the same in Simmentals as in other spotted breeds. Some Simmentals appear to possess the S_h gene and possibly the dominant R_e gene, which causes red around the eye. As in other white-spotted cattle, the Simmental breed varies from mostly pigmented to varying degrees of white, indicating that genes are present in some individuals which modify the expression of the spotting genes *ss*. The Simmental breed appears to possess one or more pairs of dilution genes that may be different from the dilution gene carried by the Charolais. In the Simmental, the dilution genes do not dilute red and black as much as those in the Charolais, but they dilute black more than red.

Coat color in many of the *exotic beef breeds* has not been studied as thoroughly as it has in some of the older breeds, such as the Angus, Hereford, and Shorthorn. It seems very likely that many of the genes mentioned for coat color here are present in the exotic breeds. It is also possible that other pairs of genes not reported here affect coat color in cattle.

8.4   Detrimental and Lethal Genes

Many years ago reports on cattle genetics dealt with inherited detrimental and lethal (deadly) traits. More recently, attention has been directed toward the inheritance of economic traits determined by many pairs of genes (polygenes) and how these traits may be improved through selection and the mating of nonrelatives.

Detrimental and lethal defects cause trouble for beef breeders only when they appear in his or her herd. When they do occur, they are usually of great concern to the breeder because of the bad name which might be given to breeding animals in the herd. The breeder should be on the lookout for inherited defects at all times and should be able to recognize defects when they do appear.

Most defects in beef cattle, as in other species of animals, are either recessive or partially dominant and have their phenotypic effect in the homozygous genotype. Such defects become more evident when inbreeding and linebreeding are practiced, because these mating systems tend to make more pairs of genes homozygous in the inbred individual regardless of their phenotypic expression. Thus, inbreeding and linebreeding tend to pair recessive and partially dominant genes so their effects appear phenotypically. Outbreeding and crossbreeding have the opposite genetic effect because they tend to make more pairs of genes heterozygous. This tends to cover up recessive genes and does not allow them to express themselves phenotypically. Perhaps this is one of the main reasons livestock breeders prefer outbreeding to any form of inbreeding.

Hundreds of genetic and possible genetic defects have been reported in cattle. A complete description of each trait would require considerable space. We shall limit our discussion here to defects which have been reported as genetic and whose mode of inheritance is known.

8.4.1   Recessive Defects

Defects reported here are all classed as recessives, although in a few cases penetrance may not be complete and the genes may vary in their expression.

Achondroplasia.   This is the failure of the normal development of the bone cartilage, resulting in dwarfism, described in Angus cattle in 1941. A similar condition has been described in Holstein, Guernsey, Jersey, and Telemark cattle. A form of achondroplasia called *snorter dwarfism* has been described mainly in Angus and Hereford cattle. It apparently has been found at a low frequency in some other breeds.

Agnathia (shortened lower jaw).   This condition was reported in four calves in a single herd in Kentucky. It has not been described in other breeds, so it may be a rare mutation.

Alopecia (loss of hair).   Several recessive forms have been described. Sometimes calves are born with a normal hair coat, but later a progressive loss of hair occurs, with the skin becoming completely bald. Alopecia has been described in Holsteins and Herefords.

Ankylosis (stiffening or fixation of joints).   The condition is lethal and has been described in Holsteins and Herefords.

Anopthalmia (absence of one or both eyes).   Found in Friesians in Egypt.

Arthrogryposis (retention of joints in a flexed position).   It has been described in numerous breeds including the Chianina, Simmental, Shorthorn, Hereford, and Charolais. In Charolais, it is often accompanied by a cleft palate.

Ataxia (muscular incoordination).   The condition is detrimental because some calves cannot stand and nurse. It has been described in Herefords, Shorthorns, Jerseys, and Holsteins.

Brachygnathy (parrot beak, lower jaw shortened). Affected calves die shortly after birth. Described in milking Shorthorns.

Brachymely (short limbs). Described in European Friesians in Egypt. Other defects such as the lack of a rectum often accompany the defect.

Cardiac ventricles (abnormal hearts). Abnormalities of the arteries, heart, and veins are involved. Calves may live for 9 months in some instances. Described in Herefords in the United States.

Cataract (loss of transparency of the crystalline lens of the eye). Affected calves are born blind. Described in Holsteins and Jerseys.

Continuous spasms. Affected calves are called *doddlers* because they are seized by spasms which cause their movements to be incoordinated. Described in Herefords in the United States.

Dermatosporaxia (brittle skin). Described in Belgian cattle breeds.

Ectromelia (amputated limbs). Described in European Friesians. Affected calves usually die.

Edema (dropsy). An accumulation of fluids under the skin. Described in Ayrshires and Friesians.

Encephalomyopathy (internal hydrocephalus). Described in Herefords in Nebraska. Affected calves were also blind at birth because of the defect. Two forms may be involved, with one acting as a dominant.

Epilepsy (paralytic seizures). Described in the Brown Swiss and the Pie Rouge of Sweden. A similar condition has been observed in the Angus in the United States.

Epitheliogenesis (defective skin). A lethal condition described in Holsteins, Jerseys, and Ayrshires. It is probably present in some other breeds.

Fused teats. The fore and rear teats are grown together. Described in Guernseys and Herefords.

Harelip and cleft palate. Affected calves usually die at birth or soon afterward. Found in the European Pie Rouge.

Hernia — umbilical. Some reports indicate it is a dominant and others a recessive. Present in many breeds and limited to males only.

Hoofless (lack of hooves on one or more legs). Calves born with this defect died or were eliminated shortly after birth. Nine calves in a purebred Angus herd where inbreeding was practiced were identified.

Hydrocephalus (water on the brain). Reported in Ayrshires, Friesians, Jerseys, and Herefords.

Hypotrichosis (poorly developed hair).   Reported in Herefords, Swedish Friesians, and British Friesians.

Ichthyosis (fish scales disease).   Both sexes have a small amount of hair, and the skin is keratinized. Affected calves die shortly after birth. Described in Red Danish cattle, Norwegian Reds, American Friesians, Pinzgauers, and a native Japanese breed.

Impacted molars (impacted teeth).   A lethal condition described in Jerseys and milking Shorthorns.

Impotentia coeundia (inability to copulate).   Affected bulls could not serve a cow because the sigmoid flexure of the penis could not straighten. Reported in Swedish Reds and Whites, Swedish Polls, Swedish Lowlands, and Dutch Friesians.

Longheaded dwarf.   Affected calves have compact, thick, low-set bodies. As the calves mature, the head becomes longer and narrower. Described in the Angus breed.

Mandibular pragnathism.   Improper fitting of the upper and lower incisors or lower incisors and the dental pads. Results in inefficient grazing and malnutrition. Found in Herefords, milking Shorthorns, Jerseys, Holsteins, and Ayrshires.

Microencephalitis (small brain).   The size of the cerebral hemisphere is considerably reduced, as is the cerebellum. Calves are usually dead at birth. Described in Herefords in New Zealand.

Multiple eye defects (blindness).   Many defects of the eye occurring in late fetal life and causing blindness. Found in the Jersey breed.

Neuraxial edema (fluid in the nerves).   Affected calves were unable to stand and showed signs of spasms. Edema of the terminal portions of the myelinated fiber bundles was observed. Reported in Herefords.

Osteoarthritis (degeneration of joints and bones).   The joints become enlarged, and the muscles of the limbs may degenerate with age in both sexes. Described in Jerseys and Holsteins.

Phalanges reduced (creeper calves).   The metacarpal and metatarsal bones are shorter than normal. Calves cannot stand but crawl about on their knees and hocks. Described in Swedish cattle.

Polycythemia (increased number of red blood cells).   This condition developed early in life and persisted throughout the first year of life. Many affected calves died in calfhood, but at maturity the condition was less intense. Described in inbred Jerseys.

Porphyria.   The accumulation of portions of hemoglobin in the bones and teeth, causing them to turn a reddish brown. Affected animals are sensitive to sunlight. Described in Holsteins, Herefords, and Shorthorns.

Probatocephaly (swollen forehead). Reproduction and performance of affected individuals were greatly reduced. Other defects also appeared. Penetrance of the gene in the heterozygote was about 20 percent. Described in Limousins.

Screw tail (kinky tail). Characterized by the fusion of two or more pairs of coccygeal vertebrae. The tail is shortened and twisted like a corkscrew. Described in the Red Poll breed.

Short spine. Characterized by an extreme shortening of the vertebral column and the thorax. Calves are usually born dead. Described in Norwegian Mountain breeds.

Spasms. Affected calves appeared normal at birth but showed signs of spasmodic incoordination and muscular contractions 2 to 5 days later. Affected calves died in a few weeks. Described in Jerseys and Herefords.

Spastic paresis (lameness of hind limbs). When affected animals stand, the back is arched, and there is an imbalance of the hind limbs. Described in many breeds, including the Angus, Friesian, Shorthorn, Charolais, and Hereford. The penetrance of the gene is low (incomplete).

Strabism (squinting of the eyes). Described in Jerseys in the United States and Shorthorns in Great Britain.

Strabismus (crossed eyes). The eyes were normal at birth but became crossed at 6 to 12 months of age The eyes of mature animals became severely crossed. Described in Jerseys.

Syndactyly (mule foot). The hooves are not split as in normal cattle. The condition is more common and complete in the forelegs. Described in Holsteins but present in many breeds.

Tendon contracture. Tendons of the legs were rigid and caused their contracture, making delivery difficult. Many calves were born dead. Described in milking Shorthorns.

Tongue defect. Affected animals of both sexes continually twist their tongues outside the mouth. Some animals are more severely affected than others. Described in Jerseys and Pie Rouge cattle in Europe.

White heifer disease. This condition has long been recognized in the Shorthorn breed, but females in other breeds may be affected even though not white in color. It appears to be more prevalent in white animals, however. It is characterized by a persistent hymen and incomplete cervix. Two or more pairs of genes may be involved.

Wry tail (crooked tail).   The base of the tail is set at an angle to the backbone due to the sacral vertebrae being set off center. Described in Jerseys.

8.4.2   Partially Dominant

A form of achondroplasia has been described in the Dexter breed of cattle originating in Ireland. In the heterozygote, bone growth is retarded, producing a short, thick compact individual. In the homozygote, a dwarf monster is produced which has a short, bulging head resembling a bulldog. These calves are often referred to as bulldogs and are aborted early or are dead at birth. The same gene is present at a low frequency in some other breeds.

A form of achondroplasia in Herefords known as *comprest* and another in Shorthorns known as *compact* have been reported. These may be due to the same gene as the bulldog gene or to its alleles.

8.4.3   Dominant

Bouclure (curly hair).   The viability of the calves is usually normal. The condition is found in the Ayrshire and some Swedish breeds of cattle.

Divided ears.   The ears of affected animals are divided in two. Described in Brahmans in the United States. The penetrance of the gene is incomplete.

Duck-legged cattle (very short legs).   Reported in Herefords in Texas.

Hypertrichosis (long hair).   Reported in European Friesians.

Hypertrophy — muscular   (double muscling). This condition is found in many breeds of cattle, especially those of European origin. The mode of inheritance is not clear, but many think it is probably an autosomal dominant with incomplete penetrance.

Incisors (poorly developed teeth).   Described in American Friesians. Affected animals cannot graze properly. Penetrance of the gene appears to be incomplete.

Multiple lipomatosis.   This condition is characterized by large growths of fatty tumors developing in the peritoneal region. They do not become malignant. Found in Holsteins.

Notched ears.   Affected individuals have a peculiar notch in the margin of both ears. The gene shows incomplete penetrance and is found in Jerseys and Ayrshires.

Ocular colobomata (eye defect).   Defect of the eye occurring early in pregnancy. At birth a portion of the eye is missing. Found in Charolais.

Polydactyly (extra toes).   Affected animals have an extra toe on each front foot with accompanying lameness. Found in Herefords.

Smooth tongue.   The tongue is smooth, and the skin covering is fragile and bleeds easily in affected animals. The affected animals also have velvety hair and small horns and show extra salivation. The penetrance of the gene is incomplete. The condition has been described in Holsteins and Brown Swiss.

8.4.4   Sex-Linked Recessives

Hypotrichosis (poorly developed hair).   Affected calves were completely hairless at birth and were toothless. All died within 6 months of birth. Found in the Maine-Anjou breed.

Polydactyly (extra front feet).   Described in Herefords in the United States. Affected animals possessed extra digits on both front legs.

Sex-linked lethal.   A family of Holstein cows was found which transmitted a sex-linked recessive lethal gene responsible for the death of the male offspring.

8.4.5   Sex-Linked Semidominant

Hypotrichosis (streaked hairlessness).   Seventeen females and no males descended from one registered Holstein cow showed this condition. Affected animals were very susceptible to extreme cold.

8.4.6   Polygenes

Cancer eye.   The eyes or eyelids of affected animals became cancerous. This condition was found mainly in Herefords in the United States. Susceptibility to cancer eye was about 30 percent heritable.

Parrot mouth (short lower jaw).   The degree of shortness was quite variable. When it was extremely short, grazing was difficult. Described in Jerseys, Holsteins, and Guernseys. It is also present in other breeds.

8.5 Avoiding the Appearance of Inherited Defects

Most inherited defects are recessive, as mentioned earlier and further illustrated by a large proportion of the defects described in Section 8.2. Because of their recessive mode of inheritance, genes for such defects may be present in a line, family, or herd of cattle for many generations without appearing the homozygous genotype. Undoubtedly, many hidden recessive genes which are detrimental in nature are present in each individual.

The practice of inbreeding tends to pair recessive genes so they appear phenotypically. Because many recessive traits are undesirable, the breeder usually tries to rid a herd of these recessive genes. This is much harder to do than breeding to produce recessive traits. For example, it is much easier to breed Red (*bb*) Angus than Black (*BB* or *Bb*) Angus, because mating red *bb* with red *bb* should give all red individuals *bb*. Mating black with black, if the parents are heterozygous *Bb*, will produce some black and some red offspring. To get rid of the black gene *B*, one merely discards all animals showing the black color.

8.5.1 Controlling the Eliminating Recessive Genes

It is much easier to control the appearance of a recessive gene in a herd than it is to eliminate it. The appearance of a recessive gene may be prevented simply by making sure that one parent is homozygous dominant *DD*. To eliminate a recessive gene, both parents must be homozygous dominant. Mating parents that are unrelated helps prevent the occurrence of recessive defects because unrelated parents are less likely to be carrying the same recessive gene.

Certain recessive genes have been eliminated in some breeds. For example, the recessive gene for horns is apparently not present in purebred Angus, nor is the solid red face (recessive) present in purebred Herefords. Part of the reason these recessive genes have been eliminated from these two breeds is due to intense selection against them. Angus breeders have always culled any purebred calves with horns, whereas purebred Hereford breeders have kept only breeding animals with a white face. Why Black Angus breeders have eliminated the recessive gene for horns from their breed but have failed to eliminate the recessive red gene is an interesting question. Possibly black carriers of the red gene *Bb* are preferred in selection.

The following is the recommended procedure for the elimination of any recessive gene from a herd. (1) Cull all recessive individuals and their parents because such parents are carrying the recessive gene (genotype *Dd*). If practical, the other close relatives of the recessive individual should also be culled, although some may be carriers of the recessive gene and others may not. (2) Since carriers of the recessive gene *Dd* usually cannot be distinguished phenotypically from the homozygous dominant *DD* individuals, a progeny test must be conducted to determine if the individuals are homozygous dominant *DD* or heterozygous dominant *Dd*. When heterozygous individuals are determined in any way, they must also be eliminated.

Mating individuals to be tested with homozygous recessive individuals can be used as a progeny test. If one homozygous recessive progeny is produced from such matings, the individual being tested is carrying the recessive gene and is of genotype *Dd*. If no recessive offspring is produced from five such matings, the individual being tested is homozygous dominant *DD* at the 95 percent level of probability. Seven offspring none of which are recessive indicates the individual being tested is homozygous dominant *DD* at the 99 percent level of probability.

Known carriers *Dd* of a recessive gene may be used as tester animals if homozygous recessive individuals are not available. Known carriers are those individuals which have produced at least one homozygous recessive offspring or one of whose parents was homozygous recessive. Again, one homozygous recessive offspring from such matings proves the parent being tested a carrier of the recessive gene (genotype *Dd*). Eleven offspring with no recessive offspring produced indicates the animal tested is homozygous dominant *DD* at the 95 percent level of probability, whereas sixteen offspring none of which are recessive indicates the parent tested is homozygous dominant *DD* at the 99 percent level of probability.

Mating a parent to its own unselected offspring tests for any recessive gene the parent might be carrying, not just a specific one. A large number of matings is required, which makes it impossible to progeny-test a cow. Bulls can be progeny-tested in this manner, especially if they have been bred to many cows by artificial insemination. Twenty-three nonrecessive offspring from matings with different unselected daughters proves the bull free of any recessive gene at the 95 percent level of probability. A total of thirty-five nonrecessive offspring from matings with different unselected daughters proves a bull free of any recessive gene at the 99 percent level of probability. Only one recessive offspring, however, proves him and the daughter producing the recessive calf both carriers of a recessive gene. Such a progeny test requires added time and expense and results in at least 25 percent inbreeding in the offspring.

8.6   How to Determine if a Defect Is Inherited

Many defects are due to heredity, and many are due to environment. Environmental factors such as infections of the young during intrauterine life, detrimental substances consumed by the pregnant mother, and accidents of development can cause defects. These are not genetic traits, although resistance or susceptibility to adverse conditions caused by these environmental factors may have a genetic base.

Several methods may be used to determine whether or not a trait is inherited. When a defect occurs, the literature should be searched to determine if it has been previously described as a genetic trait. If it has not been reported previously, a pedigree study can be made, if pedigrees are available, to determine if defective calves are

Figure 8.1.   Pedigrees of six young bulls with a persistent penile frenulum. A recessive trait is indicated because all pedigrees trace back to ancestors by two different pathways.

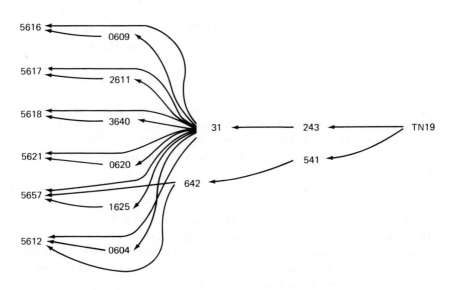

related or inbred. Inbreeding pairs recessive genes and causes such traits to appear phenotypically. An arrow diagram showing common ancestors in the pedigrees of six Angus bull calves with a penal defect is given in Figure 8.1. The trait appears to be inherited because all six calves trace to the same ancestor. The trait appears to be recessive because the pathways to the common ancestor go through the sire and dam of each defective calf. If all defective calves trace to a common ancestor only through one parent (but sometimes both), a dominant mode of inheritance is indicated.

The final proof of whether or not a trait is inherited involves the remating of parents that have produced the same defective type of offspring to determine if defective individuals are again produced. If they are, inheritance is definitely indicated. If they are not, inheritance probably is not involved. The ratio of normal to defective offspring from parents previously producing defective offspring would also suggest the mode of inheritance of the trait.

STUDY QUESTIONS

1. What are the two general kinds of inheritance that affect economic traits in beef cattle? How do they differ?

2. Fundamentally, what is a mutation?

3. Is a body cell (somatic cell) mutation ever transmitted from parents to their offspring? Explain.

4. Is it practical to cause mutations to occur in beef cattle? Explain.

5. What is the mode of inheritance of most new mutations? Are most new mutations desirable or undesirable?

6. Which is hardest to do, eliminate a recessive gene from a herd or prevent its appearance? Explain.

7. Explain the inheritance of horns and scurs in Angus, Herefords, and Shorthorns. Does this explanation of the mode of inheritance hold true in some of the exotic breeds and those with some Brahman ancestry? Why?

8. What is meant by polygenes? What traits in beef cattle appear to be determined by polygenes?

9. Does inheritance play a part in the resistance or susceptibility of beef cattle to disease and parasites? Explain.

10. What is meant by a *true albino*? Does this occur in cattle?

11. How many different pairs of genes have been reported to be involved in coat color inheritance in cattle? Is this a complete list? Is the list absolutely correct?

12. Explain how one could cross a Jersey cow on a Hereford bull and produce a brindle calf when neither parent was brindle.

13. What color is a calf produced by the cross of a Charolais on a Hereford? Why is it this color?

14. Assume you want to breed purebred Herefords that are less susceptible to cancer eye. What is one possible way to do this?

15. Some purebred Simmentals are red and white, and some are yellow and white. How can this be explained?

16. If you cross a Hereford bull on an Angus cow, what are two expected characteristics one would expect in the crossbred offspring?

17. How many recessive defects are listed in this chapter? How many dominant defects are listed? Why are there more recessive than dominant defects?

18. Why is it important to have a list of such defects available?

19. Assume a recessive defect appears in your herd of beef cattle. How would you set about eliminating it?

20. What advantages and disadvantages might be encountered in progeny-testing a bull by mating him to thirty-five of his own unselected daughters?

21. If a defect has not been reported previously as being inherited, what are some ways one can prove whether or not it is inherited?

22. How often will the purebred breeder have occasion to use information summarized in this chapter?

identifying superior breeding stock

Phenotypic variation in economic traits is the raw material with which the beef cattle breeder must work to improve his livestock through breeding. The tools he has available to mold this raw material include selection, inbreeding, linebreeding, outbreeding, and crossbreeding. We shall discuss the practical importance of variation and selection in this chapter. We shall discuss inbreeding, linebreeding, outbreeding, and crossbreeding in Chapter 10.

9.1 Variation

Economic traits in beef cattle are determined by many pairs of genes. The phenotypes for such traits are not all or none as is true of many traits such as black or red or dwarf and normal where one pair or a few pairs of genes is involved. The phenotypes of individuals for economic traits affected by many pairs of genes (quantitative traits) are not clear-cut but show continuous variation from one extreme to another. Furthermore, most economic traits are influenced a great deal by environment. For these reasons care must be taken to accurately measure the phenotypes of individuals within a herd and to minimize mistaking hereditary for environmental effects and vice versa. What the breeder attempts to do is increase as much as possible the correlation between an individual's phenotype and its genotype.

9.1.1   Heritability Estimates

Heritability estimates may be defined as the proportion of the total phenotypic variation that is due to heredity. Heritability estimates are calculated on the basis of how much more relatives resemble each other for a particular trait than nonrelatives. Relatives should resemble each other more than nonrelatives because they should have more genes in common. Many kinds of relatives have been used for the calculation of heritability estimates all over the world. Average heritability estimates for various traits in beef cattle are given in Table 9-1. It should be kept in mind that these are aver-

TABLE 9-1   Heritability estimates for certain economic traits in beef cattle

Economic trait	*Percentage heritable*
Conception rate	0–10
Calving interval	0–10
Length of gestation	30–40
Birth weight	35–40
Weaning weight	25–30
Mothering ability	35–40
Weaning conformation score	20–25
Postweaning daily gains	40–45
Postweaning daily gains—pasture	30–35
Postweaning efficiency of gains	35–40
Postweaning daily feed consumption	50–55
Final feedlot weight	50–55
Yearling weight	50–55
Conformation score at slaughter	35–40
Slaughter grade	40–45
Carcass grade	35–40
Dressing percentage	35–40
Ribeye area	60–65
Tenderness score	50–60
Marbling score	40–45
Fat thickness	40–45
Percentage retail product	25–30
Susceptibility to cancer eye	25–30

age figures from many reports in the literature and that in a particular herd, they may be above or below these average figures.

The degree of heritability of a trait may indicate several things
to the beef cattle breeder. For example, a heritability estimate of 30
percent means that approximately 30 percent of the variation in a
group of individuals is due to heredity. The remaining 70 percent is
probably due to environment. A heritability estimate of 30 percent
indicates that some progress can be made in improving this trait by
finding the best and mating the best to the best. A low heritability
estimate (0 to 15 percent) shows that little progress can be expected
by selecting and mating superior individuals. A heritability estimate
as low as this also suggests that the trait may be improved by outbreeding and crossbreeding and by improving the environment.

Environment is standardized when heritability estimates are calculated to reduce environmental variations as much as possible.
Realized heritability estimates are those resulting from actual selection experiments. Basically, such a heritability estimate shows how
much of the superiority of individuals selected for parents was
actually transmitted to (or reflected in) their offspring.

9.1.2   Some Problems of Finding Superior Breeding Stock

Two major problems are encountered by beef cattle breeders
when they try to find and use superior breeding stock. These problems are related to each other to a certain extent. One problem is
that the breeder tends to mistake superiority or inferiority that is
due to environment for that due to heredity and vice versa. A second
problem is that even when the breeder selects the best individuals for
parents, there is a tendency for the average of the offspring to move
back (regress) toward the average of the population from which the
parents were selected. In other words, the parents tend to look better
than they breed.

Beef animals may be superior because of heredity or environment or both. Superiority due to heredity is what is desired. Superiority due to environment will not be transmitted by the parents to
their offspring.

The heredity of an individual is fixed at conception when the
sperm and egg unite at fertilization. It remains the same thereafter
for all practical purposes and cannot be changed by the breeder. For
traits affected by many genes, the superiority of an individual is
estimated by comparing the record of the individual for a trait with
the average of all its contemporaries for that trait. The proportion

of the variation due to environment can be reduced and the proportion due to heredity can be increased by comparing individuals in an environment as standard as possible. When this is done, a superior individual is more likely to be superior because of heredity. It is impossible, however, to completely remove variations due to environment even though they can be reduced.

The average offspring of superior parents tend to regress toward the mean of the population from which the parents were selected for two reasons. The parents may be superior because they are heterozygous (overdominance) or because they had a superior environment. Heterozygous parents do not breed true, and they produce some homozygous and some heterozygous offspring. For example, let us assume that heterozygous *Aa* individuals gain 3.0 pounds per day, whereas the homozygous *AA* and *aa* individuals gain 2.0 pounds per day. The mating of two heterozygous parents (*Aa*) would give the following results:

Father		*Mother*
Aa		*Aa*
3.0 lb/day	×	*3.0 lb/day*

Avg. of parents = 3.00 lb

Average of their offspring	$\frac{1}{4}$ *AA* or 2.00 lb/day $\frac{2}{4}$ *Aa* or 3.00 lb/day $\frac{1}{4}$ *aa* or 2.00 lb/day

The average for the offspring would be 2.00 + 3.00 + 3.00 + 2.00 divided by four (10 divided by 4) or 2.50 pounds. Thus, the average of the offspring was less (3.00 versus 2.50 pounds) than that of the parents.

The average of the offspring can be less than that of the parents because the parents were superior because of environment. For example, let us assume that parents with postweaning daily gains of 3.00 pounds were selected from a group where the average was 2.00 pounds. The superiority of the parents would be 1.00 pound per day (3.00 − 2.00). If the heritability of daily gain is 40 percent, this suggests that only about 40 percent of the superiority of the parents (1.00 × 0.40 = 0.40) or a gain of 0.40 pound per day would be transmitted to their offspring, which would have an expected average daily gain of about 2.40 pounds. Thus, the average of the offspring was less (3.00 − 2.40 or 0.60) than the parents because the parents were superior because of environment. Superiority due to environment cannot be transmitted to offspring.

Several things can be done by the beef cattle breeder to minimize these problems. They cannot be eliminated entirely, however. Some of the mistakes made in selecting heterozygous parents may be overcome by selecting individuals from families or lines that have a history of being superior. In other words, selecting superior individuals from superior parents would be recommended. Also, if the individuals are inbred, they should tend to be less heterozygous than noninbreds. Some mistakes of confusing superiority due to environment for that due to heredity may be avoided by comparing individuals in the same herd under the same conditions and by adjusting records of all contemporary individuals to a standard basis for all individuals being compared. Reductions in variations due to environment leave a larger proportion of the remaining variation that is due to heredity.

Each individual probably has a genetic limit to its type and performance. For example, even with the best care and feed a Hereford cow will not give as much milk as a Holstein. She just does not have the proper genetic makeup to produce that much milk. Her genetic limit may be due to too few milk-secreting cells in the udder, a lack of udder capacity, a digestive tract that is too small to digest enough feed for a high level of milk production, a less efficient physiological system for milk production, and an inherent tendency to store excess nutrients digested as fat rather than to convert them to milk production. Effective selection for improved performance tends to raise the genetic limit for each new generation.

9.2   Selection

Selection may be defined as the causing or allowing of certain individuals to produce the next generation. Selection implies that those selected as parents are superior to the average of the group from which they are selected for the traits considered.

Two general kinds of selection are practiced. Selection by nature is called *natural selection* and favors the survival of individuals best suited to a particular environment. Selection practiced by people is called *artificial selection* and is not the same as that practiced by nature, although natural selection assists artificial selection by culling those individuals which are weaker and less fertile. Our attention here will be directed mostly to artificial selection practiced by people.

People have the opportunity to select breeding animals on the basis of records on the individual and/or its relatives. Information on relatives may be used because the relatives should contain many of the same genes as the individual, depending on how closely they are related to the individual. In general, full sibs are related to the individual by 50 percent, half-sibs by 25 percent, and progeny by 50 percent. They are related to their parents by 50 percent when inbreeding is not practiced. These percentages refer to the percentages of genes relatives have in common over and above the average of the population, because they are related. Inbreeding tends to increase the relationship of individuals within the same inbred family or line.

9.2.1   Selection Based on Individuality

Selection for quantitative traits involves rearing and feeding a group of animals of near the same age at the same time and under as nearly the same conditions as possible. Pretest conditions should also be as nearly the same as it is possible to make them. The record of each individual should then be compared with the average of its contemporaries. This means that one half of the individuals in the group will be average or below and that one half will be average or above. The higher the individual ranks above the average for a certain trait, the more superior it should be.

The *trait ratio* has been used to describe by a single figure where an individual ranks in comparison to its contemporaries. The trait ratio may be calculated from the following formula:

$$\text{trait ratio} = \frac{\text{record of the individual}}{\text{average of records of its contemporaries}} \times 100$$

This ratio may be used to compare individuals for a single trait or for a number of traits in an index. When the ratio is above 100, the individual is superior to the average. When it is below 100, the individual is below average. A trait ratio for an individual of 120 means that individual is 20 percent above the average. A trait ratio of 95 means it is 5 percent below the average. These figures indicate where each individual ranked in comparison to the average, but they do not tell why they ranked there. It is only suggested that they may rank where they do because of heredity.

Cows should produce a calf each year once they reach sexual maturity and produce their first calf. For this reason each cow has the opportunity to have several calves in a herd during her productive lifetime. Heifers and steers fattened for slaughter, on the other

hand, have only one record because they do not have the opportunity to repeat this record.

Since cows may have more than one record in the herd, the
comparison of individual cows with their herdmates is more complex
than comparing an individual with the average of its contemporaries
for type and performance. The ability of cows to have more than one
record does make it possible to calculate the repeatability of weaning
weights of calves by the same cow. Weaning weight is about 40 to 45
percent repeatable. This means that heifers can be culled on the basis
of the weaning weight of their first calf with a reasonable degree of
accuracy. A heifer that weans a lighter (or heavier) than average calf
the first year will tend to repeat this record in later calves produced.
Thus, culling heifers that wean lighter than average calves will tend to
raise the weaning weights of calves in the herd in future years.

Cows in most herds are of different ages and have produced
different numbers of calves at a given time. The *most probable producing ability* (MPPA)[1] of each cow in the herd may be calculated
from the following formula:

$$\text{MPPA} = \bar{H} + \frac{nr}{1 + (n-1)r}(\bar{C} - \bar{H})$$

where

$\bar{H}$ = herd average

n = number of calves for each cow

r = 0.40, the repeatability of weaning weight

$\bar{C}$ = average weaning weight for all calves the cow has produced

Weaning weights can be recorded as ratios or actual weights of calves.
Calculating the MPPA for each cow in the herd adjusts for age and
numbers of records, so cows can be culled on a more comparable
basis. The cows with the lowest MPPAs should be culled.

9.2.2   Performance Testing Bulls

The rate and efficiency of gains after weaning have been emphasized in beef cattle breeding in recent years, because postweaning
gains are 40 to 45 percent heritable. These medium to high heritability estimates suggest that finding the best and mating the best to the

---

[1] An example of how the MPPA is calculated is given in Chapter 12, Section 12.2.

best for this trait should cause its improvement. The improvement of rate of gain through selection will also tend to improve the efficiency of gains, although efficiency of gains probably could be improved more rapidly by measuring it in each individual on test. This would require individual feeding, however, which is often not practical. Efficiency of gains appears to be 35 to 40 percent heritable and should be improved by finding the best and mating the best to the best.

The practice of placing a group of individuals on a full feed and measuring their rate and efficiency of gain is known as *performance testing.*

Bulls are performance-tested, whereas heifers are not. Heifers are not performance tested because so many of them would have to be tested to find top female replacements in the herd, and those selected for replacements would have to be gradually let down in flesh, which would be a waste and added expense. Furthermore, the full feeding of heifers would increase fat depositions that could be detrimental to reproductive efficiency and milk production later in life. Selecting heifers that gain the most weight on pasture and roughage as replacements would tend to find those who would gain faster on a full feed of concentrates. Pasture gains appear to be 30 to 35 percent heritable.

Proper procedures should be followed in the performance testing of bull calves. The purpose of testing, of course, is to compare the performance of each individual bull with that of others in the same group (fed at the same time and place and under similar conditions) to find those that perform the best. Contemporary bulls should be the same age (plus or minus 1 or more months). They should be fed the same ration, and the ration should be complete in that it contains both a certain proportion of concentrates and roughages. Bulls on test should be started on a limited ration, and the daily amount should be gradually increased until each bull is on a full feed, usually within 2 or 3 weeks. Some breeders prefer to full-feed almost from the beginning using a high roughage ration, with the proportion of roughage (fiber) to concentrates being gradually reduced to the desired amount after 2 or 3 weeks.

The full-feeding period should be approximately 140 days because this helps average some of the environmental effects such as pen differences, fill, etc. Bulls may be fed individually if facilities for this purpose are available. This allows the measurement of both rate and efficiency of gain in individual bulls. If facilities for individual feeding are not available, bulls may be fed in small groups and the efficiency of gain measured, but this does not allow the measurement of individual feed efficiency. If rate of gain can be measured for a bull, however, the faster-gaining bulls will tend to be more efficient. This correlation is not perfect, however. A conformation score which indicates degree of muscling and possibly frame scores should be measured in each bull at the end of the test. Estimates of carcass quality and quantity should also be made.

Some breeders performance-test their own bull calves on their farms. Weight gains and other records are verified by extension or other state personnel.

Central bull testing stations are in operation in many states. Shortly after weaning breeders bring bull calves to these stations where an entire group is full-fed under similar test conditions for a period of 140 to 150 days. Each station has its own set of rules; in general, the rules agree with those of similar stations, although some differences among stations exist.

Yearling weights (365 days) for both bulls and heifers are important. Bulls are usually full-fed, whereas heifers are not. Yearling weight for bulls on a full feed is about 50 to 55 percent heritable and is associated with more efficient gains and a higher quantity of red meat in the carcass. Yearling weight for heifers not full-fed is about 30 to 35 percent heritable.

The following formula is recommended for calculating the yearling weights of bulls and heifers:

$$\text{Adjusted 365-day wt.} = \frac{\text{actual final wt.} - \text{actual weaning wt.}}{\text{number of days between wts.}} \times 160$$

$$+ \text{ 205-day wt. adjusted for age of dam}$$

Those animals with the heaviest weights should be the ones kept for replacements.

The conformation of each individual should also be recorded at the end of the performance test for bulls or at approximately 550 days for heifers. Conformation scores should include items related to structural soundness, size, thickness of natural fleshing, and possibly outside fat covering. These items are thought to be related in a general way with longevity, performance, and carcass quantity and quality. Since conformation score is an estimated value and is dependent on the opinion of the one making the score, it is not as accurate as actual measurements taken by means of the tape and scales. The accuracy of scores can be improved by averaging scores of a committee of three or more for each individual animal. Several different scoring systems may be used. Some breeders use a score ranging between 1 and 15. Others use a numerical score ranging from 60 to

100; the scores of most animals will range between 70 and 90. In each case the higher the score, the more desirable the conformation. The system used would depend on the wishes and desires of the breeder.

Frame scores are used by some stations and some breeders for breeding animals. Frame scores are of value in predicting the mature size and weight of the individual. They are mainly height measurements taken at the fifth rib or elbow and include a measurement from the ground to the top of the shoulder. Measurements are made near 1 year of age. Data from many sources indicate that an extra inch in height at 1 year of age represents about 30 pounds in additional weight. Information about the frame score used at the Missouri Agricultural Experimental Station is given in Table 9-2.

9.2.3   Selection for Carcass Quality and Quantity

The main reason for producing beef is meat for the table. Therefore, the goals of the breeder should be to produce the largest quantity of good-quality meat with the greatest efficiency.

Quality of beef includes such items as texture and color of the lean, marbling, firmness, tenderness, and palatability. Such traits can be measured with accuracy only in the carcass. *Quantity* of beef refers to the amount of salable meat the carcass will yield at slaughter. Fortunately, the use of the 40K counter and to a certain extent visual inspection by a trained operator are helpful in determining carcass quantity in the live animal. As mentioned previously, a beef animal is not likely to have full sibs that can be slaughtered to gain information on the possible carcass quality and quantity of a live individual. More half-sibs could be used for this purpose, but information on carcass measures in numerous half-sibs would be required to equal the accuracy of information based on a single record of the individual.

9.3   Selection on the Basis of Ancestors

A *pedigree* is a record (for two to four generations) of an individual's ancestors. The usual pedigree gives only the name and registration number of the individual's ancestors. In recent years more emphasis has been placed on *performance pedigrees*. A performance pedigree includes information on the records of performance of the individual and at least its sire and dam. Basic information recommended in-

TABLE 9-2   Missouri Agricultural Experimental Station frame score for beef cattle [a,b]

Missouri frame score, height (in.)

Height measurement is taken over the shoulder at the fifth rib or elbow

Potential slaughter weight (lb)	Frame	Age (months)																			
		5	6	7	8	9	10	11	12	13	14	15	16	17	18	19	20	21	22	23	24
750–850	1	32	33	34	35	36	37	38	39	39.75	40.5	41	41.5	41.75	42	42.25	42.5	42.75	43	43.25	43.5
851–950	2	34	35	36	37	38	39	40	41	41.75	42.5	43	43.5	43.75	44	44.25	44.5	44.75	45	45.25	45.5
951–1,050	3[c]	36	37	38	39	40	41	42	43	43.75	44.5	45	45.5	45.75	46	46.25	46.5	46.75	47	47.25	47.5
1,051–1,150	4	38	39	40	41	42	43	44	45	45.75	46.5	47	47.5	47.75	48	48.25	48.5	48.75	49	49.25	49.5
1,151–1,250	5[d]	40	41	42	43	44	45	46	47	47.75	48.5	49	49.5	49.75	50	50.25	50.5	50.75	51	51.25	51.5
1,251–1,350	6	42	43	44	45	46	47	48	49	49.75	50.5	51	51.5	51.75	52	52.25	52.5	52.75	53	53.25	53.5
1,351 & above	7	44	45	46	47	48	49	50	51	51.75	52.5	53	53.5	53.75	54	54.25	54.5	54.75	55	55.25	55.5

[a] The Missouri frame score is a valuable measurement of frame and is used in predicting mature size. The boldface columns show the three commonly used adjustment ages: weaning, yearling, and 18 months.
[b] From Manual 104, Extension Division, University of Missouri, Columbia, Extension Division, *Beef Cow/Calf Manual*, 19, p. 12.
[c] Frame 3 is average for British breeds.
[d] Frame 5 is average for continental breeds.

TABLE 9-3 Predicting the breeding values of individuals from information on the phenotype of the individual combined with that of certain ancestors

---

A. Prediction based on individual's own record.

B. Prediction based on the individual's own record plus that of one parent:

Partial regression coefficients		Heritability of the trait			
		0.10	0.30	0.50	0.70
Individual's record	b_1	0.098	0.284	0.467	0.658
Parent's record	b_2	0.045	0.107	0.133	0.120

$$\text{PBV} = \bar{P}_{ic} + b_1(P_i - \bar{P}_{ic}) + b_2(P_p - \bar{P}_{pc})$$

C. Predictions based on the individual's own record plus that of one parent, one grandparent, and one great-grandparent all in one line of descent:

Individual's record	b_1	0.097	0.282	0.464	0.657
Parent's record	b_2	0.044	0.101	0.124	0.113
Grandparent's record	b_3	0.020	0.037	0.034	0.019
Great-grandparent's record	b_4	0.009	0.014	0.010	0.004

$$\text{PBV} = \bar{P}_{ic} + b_1(P_i - \bar{P}_{ic}) + b_2(P_p - \bar{P}_{pc}) + b_3(P_{gp} - \bar{P}_{gpc}) + b_4(P_{ggp} - \bar{P}_{ggpc})$$

D. Prediction based on the individual's own record plus that of both parents:

Individual's record	b_1	0.096	0.267	0.429	0.603
Sire's record	b_2	0.045	0.110	0.143	0.139
Dam's record	b_2	0.045	0.110	0.143	0.139

$$\text{PBV} = \bar{P}_{ic} + b_1(P_i - \bar{P}_{ic}) + b_2(P_s - \bar{P}_{sc}) + b_2(P_d - \bar{P}_{dc})$$

E. Prediction based on the performance records of the parents and grandparents:

Sire's record	b_2	0.048	0.134	0.214	0.301
Dam's record	b_2	0.048	0.134	0.214	0.301
Maternal record					
Grandsire's	b_3	0.023	0.055	0.071	0.070
Granddam's	b_3	0.023	0.055	0.071	0.070
Paternal record					
Grandsire's	b_3	0.023	0.055	0.071	0.070
Granddam's	b_3	0.023	0.055	0.071	0.070

$$\text{PBV} = \bar{P}_{ic} + b_2(P_p - \bar{P}_{pc}) + b_2(P_d - \bar{P}_{dc}) + b_3(P_{mgs} - \bar{P}_{mgsc}) + b_3(P_{mgd} - \bar{P}_{mgdc}) + b_3(P_{pgs} - \bar{P}_{pgsc}) + b_3(P_{pgd} - \bar{P}_{pgdc})$$

Note: The PBV is predicted by multiplying the partial regression coefficient b_x by the selection differential for that particular individual or ancestor and then adding this to the average of their contemporaries. Where the letter c is included in the subscript, it indicates the individual's contemporaries. For example, $\bar{P}_{sc}$ means the average of the contemporaries of the sire.

cludes the individual's birth weight, 205-day weight, adjusted weaning weight, weaning weight ratio, 365-day adjusted yearling weight,
yearling weight ratio, and numbers of contemporaries at weaning and
1 year of age. When available, information on the number and average
weight ratios of individuals in the pedigree should be included.

Accuracy of selection that is based on the individual's own
record is greater than when based on information on the combined
records of the sire, dam, and all four grandparents. This indicates
that information on the individual's ancestors cannot replace records
on the individual. A combination of records on the individual plus
those of the sire and dam does improve the accuracy of selection as
compared to records on the individual alone. Information necessary
to calculate the probable breeding value (PBV) of an individual based
on the individual's own records plus those of certain ancestors is
given in Table 9-3. The regression coefficients given in this table
along with information given in the pedigree which follows may be
used to calculate the PBV for yearling weight (heritability of 0.50)
listed as ratios using information on the individual and both parents:

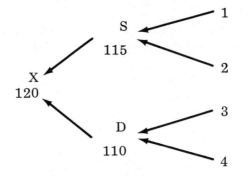

$$PBV = 100 + 0.429(120 - 100) + 0.143(115 - 100)$$
$$+ 0.143(110 - 100)$$

$$= 100 + 8.58 + 2.145 + 1.430$$

$$= 112.16$$

If information on the individual's record alone had been used, the
PBV would have been

$$PBV = \bar{P} + b_1(P_i - \bar{P})$$

$$= 100 + 0.50(120 - 100)$$

$$= 110.00$$

These calculations show that using information on the individual plus

TABLE 9-4   Regression coefficients that may be used to predict the PBV of individuals from information on the phenotype of the individual or that of certain of its ancestors

Regression coefficients		Heritability of the trait					
		0.10	*0.30*	*0.50*	*0.70*	*0.90*	*1.00*
Individual's record	b_1	0.100	0.300	0.500	0.700	0.900	1.000
Parent's record	b_2	0.050	0.150	0.250	0.350	0.450	0.500
Grandparent's record	b_3	0.025	0.075	0.125	0.175	0.225	0.250
Great-grand-parent's record	b_4	0.012	0.038	0.062	0.088	0.112	0.125

Note: The PBV of an individual is equal to $\bar{P_i} + b_i(P_j - \bar{P_i})$, where b_i is the regression coefficient for trait i, P_j the phenotypic record of the individual j, and $\bar{P_i}$ the phenotypic average of that individual's contemporaries.

both parents gives some improvement in the accuracy of selection as compared to selection on the basis of the individual's records alone.

Information given in Table 9-4 may be used to calculate the PBV of an individual based on the individual's records alone or in combination with those of various ancestors. Using records on the individual plus those of some ancestors to calculate the PBV is more accurate when the heritability of the trait is lower and less accurate when it is higher.

9.4   Selection on the Basis of Progeny

As mentioned earlier, progeny tests are quite useful in determining whether or not a bull is a carrier of an undesirable recessive gene. Progeny tests may also be used to estimate the genetic worth of a sire for quantitative traits of economic importance determined by polygenes.

Progeny testing is not practical for determining the probable breeding values of cows for economic traits because five to ten progeny would be required in most instances, and many cows would not produce this many calves in their lifetime. Progeny tests may be helpful in determining the PBV of sires for various traits if properly planned and conducted.

Using progeny test information in addition to that obtained on the individual increases the accuracy of selection as compared to information based on the individual's records alone. This is

shown in Table 9-5. If information on twenty or more progeny
is available, progeny test information is more valuable for lowly
heritable traits (Table 9-6). This number of progeny would usually
be available for bulls used for artificial insemination.

A progeny test for economic traits requires that the record of a
bull's calves be compared with a particular standard. This standard
could be the average of contemporary groups on the same test, which
is the most desirable, or a breed average. Progeny-testing several sires,
of course, increases the time and expense involved and quite often is
not practical.

A progeny test does not change the genetic makeup of a sire.
It merely shows where the progeny of that parent rank as compared

TABLE 9-5   Accuracy of selection attained by using information on a certain number of progeny plus the individual's own record

No. progeny + individ-ual's own record	Accuracy of selection when the heritability of trait is		
	0.10	0.30	0.50
10	1.63	1.36	1.19
20	1.96	1.50	1.26
40	2.31	1.62	1.32
100	2.70	1.73	1.37

Note: Accuracy of selection on the basis of the individual's record alone
would be 1.00.

TABLE 9-6   Number of progeny required for a progeny test to be as accurate
for selection as information on the individual's records alone

Heritability of trait	No. progeny required to equal the accuracy of the individual alone[a]
0.10	5
0.20	5
0.30	6
0.40	6
0.50	7
0.60	8
0.70	9

[a]Number of progeny required
to equal the accuracy of selection
based on the individual's record
alone.

to the progeny of other sires. Information on a sire's progeny is of little or no value unless selection is practiced among sires so tested. Differences among averages for progeny groups are likely to be disappointingly small. If a sire does prove superior on the basis of a progeny test, he should be used for breeding as much and as long as possible or until another sire superior to him is identified.

In recent years some breed associations have developed national sire evaluation programs following guidelines developed by the Beef Improvement Federation. The guidelines used are based on records of progeny of several sires used in the same herds by means of artificial insemination. These sires are designated as *reference sires.* When reference sires produce progeny in several herds through the use of artificial insemination, records of the progeny of other sires used in a herd can be compared with the records of the progeny of reference sires used in the same herd.

National sire evaluation summaries are published in a national sire summary publication. Results published are reported in terms of expected progeny differences for certain economic traits. These expected progeny differences estimate the transmitting ability of sires evaluated in this program. The expected progeny difference estimates only one half of the sire's breeding value because a sire contributes only one half of the inheritance to each of his progeny. The expected progeny difference estimates how the future progeny of a sire should perform as compared to the future progeny of the reference sires used in a breed.

9.5   Selection on the Basis of Collateral Relatives

Collateral relatives of an individual are those relatives other than ancestors or descendants. In beef cattle, information is available as a general rule only on the paternal half-sibs. The half-sibs of an individual are related to it by only 25 percent if no inbreeding has been practiced. Selection based on information on 100 or more half-sibs is necessary to equal the accuracy of selection based on the records of the individual. Information on half-sibs, however, can be used along with that on the individual to increase the accuracy of selection. For example, information on the individual plus that on 20 paternal half-sibs is 1.07 to 1.15 times as accurate as information on the individual alone when the traits measured are 30 to 50 percent heritable. Records on half-sibs have the advantage that they are available at the same time as records on the individual and therefore not as much time for completion is required as in progeny tests.

9.6   Factors Determining Progress Made in Mass Selection

Mass selection is the term used to describe the selection of superior individuals in a group to be used for breeding purposes merely by selecting the top-ranking individuals. The amount of progress expected in mass selection is expressed in the following formula:

Expected genetic progress = heritability estimate $\times$ selection differential

Heritability estimates for various traits are given in Table 9-1.

The selection differential may be calculated for cows and bulls selected for breeding. The selection differential, which is sometimes called the *reach*, may be defined as the amount the individuals selected for breeding exceed the average of the group from which they are selected. For example, let us assume that a bull weighing 1,300 pounds at 1 year of age is selected from a group whose average yearling weight is 900 pounds. The selection differential for this bull would be 400 pounds (1,300 – 900). In other words, this bull weighed 400 pounds more than the average of his contemporaries. Since yearling weight appears to be about 50 percent heritable, only about 50 percent of this superiority (or 200 pounds) will be transmitted to his offspring. In addition, each offspring receives only about one half of its inheritance from the sire, so the genetic improvement in his offspring for yearling weight would be 50 percent of 200 pounds or 100 pounds. This assumes that the cows to which this bull was bred were average in yearling weight.

These calculations will not be absolutely correct for a single sire and his progeny, but they should be approximately correct for the average of several sire-progeny pairs. Results of selection experiments for the improvement of economic traits are usually close to, or a little lower than, the calculated average genetic improvement.

The actual heritability of a trait in selection experiments, called *realized heritability*, may be calculated by dividing the genetic improvement in the offspring by the selection differential of the parents which produced these offspring.

9.7   The Selection Index

Beef cattle breeders seldom select for only one trait at a time; they usually select for several traits. It is best, as a general rule, to select for a limited number of important traits included in an index. If numerous traits are selected for at one time, less progress will be made in a particular trait than if selection for this trait was made with several other traits in an index.

The weaning weight ratio is a good measure of cow productivity, and selection of replacement stock with the heaviest weaning weights should improve this trait. More complex indexes may be used for postweaning traits.

Various indexes have been used or suggested for beef cattle. A suggested index is as follows (T. J. Marlow et al., *Va. J. Sci.* 15: 168, 1964):

$$I = (40 \times \text{Average Daily Gain (ADG)} - 18) + (\text{type score} \times 5)$$

Several others (H. B. Lindholm and H. W. Stonaker, *J. Anim. Sci.* 16:998, 1957) have been suggested:

1. I = weaning weight.
2. I = weaning weight + 72 (ADG).
3. I = 0.58 (weaning weight) + 18.64 (ADG) = 0.73 (days to choice grade) – 5.87 (feed per pound of gain).

If a breeder desires to use an index, it would be best to contact the extension specialists within his or her state or the experimental station geneticist. An index including the traits the breeder wants to select for and the proportional weight each trait in the index should be given could then be used for computations.

9.8   Breeding Plan for Purebred Cattle

Purebred breeders are the seed stock producers for the cattle business. They should have a definite breeding plan and follow it regardless of the breed produced. Several steps should be followed in producing seed stock.

All animals in the herd should be identified by a tattoo and ear tag number or by some other means. This should be a permanent means of identification.

A definite record-keeping system should be developed, and these records should be used to locate superior breeding stock for a herd. The records should include the date of birth, birth weight, 205-day adjusted weaning weight, weaning score, and sex of each calf produced. A lifetime record for each cow including the items previously mentioned as well as the date of disposal and the reason for disposal of each animal in the herd should also be kept.

Superior bull calves at weaning should be performance-tested by full-feeding them for at least 140 days on the farm or in a central testing station. Rate of gain, efficiency of gain, 365-day weight, and yearling score should be recorded if possible. The bulls should then be ranked in order of their superiority and priced according to their rank. The measurement of loin eye area and backfat thickness by means of the sonoray and the percentage of lean estimated by means of the 40K counter should be recorded if possible The use of the 40K counter is limited and less practical, however.

Heifers kept for replacements should be from the best-producing cows. They should have the heaviest 365-day weights based on their performance on a growing ration, on pasture, and during the winter months. Only those with the best conformation should be kept for replacements.

Records on cows and heifers should be used to locate and cull below-average performers. When considerable culling is done in older cows, the most probable producing ability (MPPA) should be calculated for each female, and culling should be done on this basis.

Outbreeding would be the mating system of choice unless an outstanding sire is located by the performance of his progeny. In this case it should be possible to use linebreeding to concentrate the inheritance of this sire in the herd.

Superior performance-tested bulls should be purchased unless an outstanding individual is produced and identified by records kept within the herd. Such a bull may produce calves superior to any bull purchased if close inbreeding (linebreeding) is avoided.

The purebreeder should advertise and promote his best breeding stock. Culls should never be offered for sale to other cattle producers because their sale may cost the breeder more in the long run than might be gained if they prove to be inferior breeders. On the other hand, the breeder should always keep the top individuals, especially females, for his or her own herd.

STUDY QUESTIONS

1. What is the raw material that beef cattle breeders have to work with to improve animals through breeding?

2. What are the tools beef cattle breeders have at their disposal to mold this raw material?

3. What is meant by a heritability estimate? Of what use are they to the beef cattle breeder?

4. Define *realized heritability*.

5. What traits appear to be the lowest in heritability? The highest?

6. What are two major problems in finding superior breeding stock?

7. Why does the average of the offspring of extreme parents tend to regress toward the mean of the population?

8. What can beef cattle breeders do to minimize the problems mentioned in your answer to question 7?

9. What is meant by a genetic limit? Give an illustration of such a limit. Why does such a limit exist?

10. Define *selection*. What are the two major kinds of selection? Which one is the most effective?

11. What kinds of relatives may be used to help in selection of superior individuals?

12. What is meant by a trait ratio? What does a trait ratio of 90 percent mean? What does one of 115 percent mean?

13. Of what use are trait ratios in selection?

14. What is meant by the most probable producing ability of a cow? Why would one use such a figure?

15. You have two cows in your herd with different numbers of records. The 205-day weaning weight of all calves in the herd is 400 pounds. Cow A has three calves with an average weaning weight of 450 pounds, whereas cow B has had six calves with an average weaning weight at 205 days of 435 pounds. Which cow would be the best from the standpoint of her most probable producing ability? Use the repeatability of one record as 0.40.

16. What is performance testing? Outline a procedure that should be followed in running a performance test on forty bull calves.

17. Which are performance tested as a general rule, bulls, heifers, or both? Why?

18. What traits should be measured when a performance test is conducted?

19. What is meant by a central performance testing station and on-the-farm testing?

20. A bull calf weighed 1,250 pounds at 390 days of age and 450 pounds at 205 days (adjusted for age of dam). What is his adjusted 365-day weight?

21. What is meant by a frame score? Of what value is it in selection?

22. Why is it seldom practical to use information on full sibs in selecting a beef animal? Can information on half-sibs be used for this purpose? Explain.

23. What is a performance pedigree? What is meant by accuracy of selection? What is the most accurate in selection, information on the individual alone or information on the ancestors of an individual?

24. What is meant by an individual's PBV?

25. What is a progeny test? Why is it more practical to progeny-test a bull than a cow?

26. Explain the meaning of the following statement: "A progeny or a performance test does not change the genetic makeup of the individual tested." Why?

27. Under what circumstances would you recommend a progeny test for a beef animal?

28. What is meant by a reference sire? Explain the meaning of national sire evaluation summaries.

29. What are the collateral relatives of an individual? What are some advantages of using them in selection?

30. Why is the selection differential sometimes referred to as the *reach*?

31. Assume that yearling weight in beef cattle is 50 percent heritable. A bull weighing 1,400 pounds at 365 days was selected for breeding from a herd where the average weight of all calves was 1,000 pounds. Ten heifers weighing 1,000 pounds were also selected for breeding. What would be the selection differential for the bulls? The heifers? For both bulls and heifers? What would be the expected genetic progress in the offspring of this bull and the ten heifers?

32. What is meant by a selection index? How many traits should be included in an index?

33. Outline a breeding plan for the production of your favorite breed of beef cattle.

mating systems for beef cattle production

Mating systems that may be used to improve beef cattle through breeding include inbreeding, linebreeding, outbreeding, and crossbreeding. All have been used in the past for producing purebreds or commercial cattle. The principles involved in using each of these mating systems have been the subject of much research in the past and are still being studied at the present. The use of each system depends on its genetic effects and practical application.

10.1 Inbreeding

Inbreeding may be defined as the production of offspring by parents that are related. The more closely related they are, the higher the degree of inbreeding. This system of breeding was used in the initial development of many cattle breeds, probably because each breed started from a few individuals most of whom were related. Inbreeding has been used to a considerable degree when exotic breeds such as the Charolais were first introduced into the Americas. The amount of inbreeding was rather intense in Charolais when they were first introduced into Mexico and the southern portion of the United States. Inbreeding is used less and less as the numbers in a breed increase from new importations or from the rapid increase in numbers by grading up. Grading up means the repeated use of purebred sires on percentage cows. Inbreeding is sometimes used in a herd where breed numbers are large but the inbreeding practiced is not intense.

10.1.1   Genetic Effects of Inbreeding

The main genetic effect of inbreeding is to increase the number of pairs of homozygous genes in the individual. Inbreeding increases the homozygosity of genes because relatives tend to carry the same gene, or genes, and the inbred offspring is more likely to receive the same gene from each parent as compared to the outbred individual. The more closely related the parents, the more likely they are to be carrying the same genes and the greater the degree of homozygosity in their inbred offspring.

10.1.2   Some Effects of Increased Homozygosity

Increased inbreeding (and thus increased homozygosity) usually causes a decline in traits related to physical fitness. Such traits include the survival ability of the young and the performance of the inbred individuals. The more intense the inbreeding, the more adverse the effect on physical fitness.

Increased inbreeding tends to pair recessive genes. This causes them to be expressed phenotypically in the inbred individual. Inbreeding does not create new recessive genes, nor does it change the frequency of recessive genes in a population. It does increase the frequency of the appearance of homozygous recessive individuals if recessive genes are present in the original noninbred population. Dominant genes are not uncovered by inbreeding because they are expressed phenotypically even in the heterozygous individual. Inbreeding tends to make all pairs of genes homozygous regardless of their phenotypic expression.

Increased homozygosity tends to increase breeding purity. Breeding purity means that the individual is more likely to transmit the same genes to each of its offspring. This is illustrated in the following genotypes:

Individual 1	Genotype *AABBCCDD*
Individual 2	Genotype *aabbccdd*
Individual 3	Genotype *AaBbCcDd*

Individual 1 can transmit only genes *ABCD* through its sex cells. Individual 2 can transmit only genes *abcd*. Individual 3 can transmit any one of sixteen different combinations of genes to its offspring, which shows that it does not breed true.

Although individuals 1 and 2 transmit only one combination of genes to each of their offspring, individual 1 would be preferred for two reasons. It is homozygous dominant for four pairs of genes, and dominant genes are usually desirable in their effects, whereas recessive genes are usually undesirable. In addition, all of the offspring of individual 1 will be of the same dominant phenotype as that parent for the four pairs of genes involved. Therefore, individual 1 is prepotent because it is homozygous dominant. A prepotent individual stamps its characteristics on its offspring so that they resemble that parent or each other more than usual.

10.1.3   Uses of Inbreeding

The main reason for using inbreeding is to produce inbred lines for crossing. The same principles involved in the production of hybrid seed corn also apply to the breeding of beef cattle, but they are much more difficult and expensive to use on a practical basis. Several inbred lines are formed by inbreeding for the production of hybrid corn because they become more homozygous when inbred. The crossing of two or more inbred lines that are homozygous will tend to produce the same genotype of offspring at one time or in different generations.

Inbred lines of corn are developed by several generations of self-fertilization so that the line is almost 100 percent inbred or homozygous. As inbreeding progresses, some lines are lost due to sterility and the occurrence of lethal defects, and even those that survive are less vigorous and yield less seed than noninbred cornstalks. Once the inbred lines are formed, they are crossed with each other to find those that produce the best-performing line-cross offspring. The inbred lines of corn that cross the best are no doubt homozygous for different alleles. The inbred lines are reproduced as such, and the line-cross progeny are sold to plant for the commercial production of corn. Because they are homozygous, the inbred lines that cross well at one time would be expected to cross well at later times. Single-cross corn hybrids come from seed resulting from the first cross of two inbred lines. Triple-cross hybrid seed corn results from crossing three inbred lines, whereas double-cross hybrid seed corn results from crossing two different single-cross hybrids. Single-cross hybrids are more expensive, as a general rule, because the line-cross seed is produced by an inbred parent that produces low yields per acre and is more expensive to produce.

A few inbred lines of beef cattle have been developed in experimental herds. They are used to only a limited extent for the production of commercial cattle. The main reason that few inbred lines of beef cattle have been developed is because of the time and expense involved in producing them. Beef cattle, of course, usually produce only one young per year as compared to an ear of corn that may produce dozens of grains of corn. The closest matings with beef cattle give only about one half of the degree of inbreeding as self-fertilization. Several inbred lines of cattle may be started, but only a few will be productive by the time inbreeding reaches 40 to 50 percent. This limits the number of inbred lines that will be tested in crosses. When two inbred lines are found that produce superior line-cross progeny, they have to be kept as pure inbred lines, and continued inbreeding would be necessary if they were to be used in the future for the production of line-cross progeny.

Inbreeding can be used to identify genetically superior or inferior breeding stock. This works best for assessing the genetic status of sires to be used for artificial insemination. As mentioned previously, mating a sire to thirty-five of his own unselected daughters tests him for any detrimental recessive gene he might be carrying and would transmit at the 99 percent level of probability. Very few sires prove to be superior and free of detrimental recessive genes when tested in this manner. Those that do pass such a progeny test and prove to be superior may be used by artificial insemination to sire thousands of calves per year. They could also be used to establish a new inbred line.

10.2   Results of Inbreeding

In general, the breeding efficiency of inbred cows and bulls is lower than in noninbred stock, although this varies among inbred lines.

Weaning weight of beef calves decreases by 5 to 15 pounds for each 10 percent increase in inbreeding, but this, too, varies among different inbred lines. Inbreeding in the cows generally lowers the weaning weight of calves, and inbred cows wean fewer calves than do those that are not inbred. Postweaning performance does not appear to be adversely affected by inbreeding in most cases.

Results from a study of the effects of inbreeding on performance traits of beef cattle in the western region (J. S. Brinks et al., *Colo.*

Agric. Exp. Stn. Tech. Bull. 123, 1975) showed that the response to inbreeding varied with different inbred lines. Some lines showed little or no detrimental effects of inbreeding for certain traits, whereas in others inbreeding was very detrimental. These results were probably due to the presence, or absence, of detrimental recessive genes in the original breeding stock.

Most results suggest that relatively small inbred lines can be developed and maintained without great effects of inbreeding on fitness traits. Using four or more bulls per line combined with selection for important economic traits could help to establish acceptable inbred lines. Lower performance of the inbred lines would tend to increase production costs, but these costs probably would not be too great if there were ever a demand for inbred seed stock.

10.3   Linebreeding

Linebreeding is a form of inbreeding in which an attempt is made to concentrate the inheritance of some one ancestor, or line of ancestors, in individuals within a herd. It differs from ordinary inbreeding in that several ancestors may be responsible for the increase in inbreeding when linebreeding is not practiced. This is illustrated in the following pedigree:

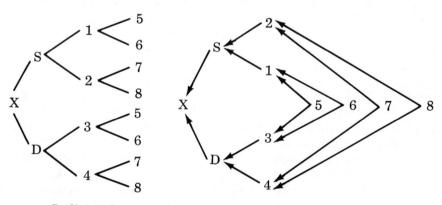

Pedigree 1   Arrow diagram of pedigree 1

In this pedigree four ancestors are responsible for the inbreeding of X, which would be about 12.5 percent. Since all four ancestors of X (5, 6, 7, and 8) trace to individual X through the sire and dam, each one of them could be responsible for a recessive genetic defect showing up in the inbred descendants. It would be difficult to progeny-test this many ancestors and use only those tested free (at a certain level of probability) of detrimental recessive genes.

A form of linebreeding is illustrated in pedigree 2:

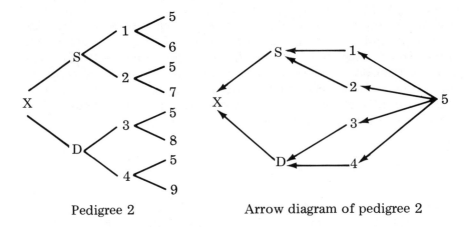

Pedigree 2   Arrow diagram of pedigree 2

In this form of linebreeding both the sire and dam are half-sib matings, but the half-sibs producing the sire were different half-sibs from those producing the dam. In this pedigree, ancestor 5, which is the only great-grandsire individual X possesses, contributes approximately 50 percent of the inheritance of individual X. Usually the sire contributes 50 percent of the inheritance of an offspring and a grand-sire 12.5 percent if no inbreeding is practiced. Thus, it is possible to use this form of linebreeding to concentrate the inheritance of some outstanding ancestor that is dead or whose services are not available. Theoretically it would be possible to produce a linebred individual that had a better combination of genes than ancestor 5. The inbreeding in this pedigree is 12.5 percent, which is not intense enough to cause too much lowered vigor in most instances. If ancestor 5 carried no detrimental recessive genes, the linebred individuals should be superior for many of their economic traits. The disadvantage of this form of linebreeding is that at least two superior half-sib sons and two superior half-sib daughters of ancestor 5 must be available for breeding.

Linebreeding practiced over a period of many years will require that it be continued through a descendant of the ancestor toward which linebreeding is directed. This will usually be through a son so the same line of breeding can be maintained (pedigree 3).

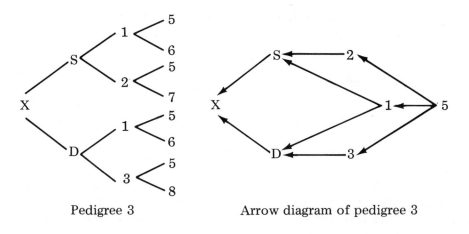

Pedigree 3   Arrow diagram of pedigree 3

In this pedigree linebreeding is continued through a son of ancestor 5, which is number 1. In this form of linebreeding either 1 or 5 contributes about 50 percent of the inheritance of individual X. Since ancestor 1 carries one half of the inheritance of his sire, number 5, and his dam, number 6, he could possess and transmit undesirable recessive genes received from his dam. This form of linebreeding increases inbreeding about twice as much as the matings shown in pedigree 2. One advantage of this form of linebreeding, however, is that only one superior son of ancestor 5 would be required.

Linebreeding should be used only when an outstanding ancestor is found, and even then it should be used only if the breeder is familiar with the possible consequences of linebreeding and inbreeding.

10.4   Outbreeding

Outbreeding is the mating system most often used by producers of purebred seed stock. Outbreeding may be defined as the production of offspring by mating individuals that are not related, at least within the last three or four generations.

The genotypic and phenotypic effects of outbreeding are opposite to those of inbreeding. Outbreeding increases the number of pairs of heterozygous genes in the individual. By increasing heterozygosity there is a tendency to cover up detrimental recessive genes, to decrease breeding purity, and to improve traits related to physical fitness. These are probably the main reasons that purebred breeders usually follow an outbreeding program to avoid inbreeding.

10.5   Crossbreeding

Crossbreeding is the production of offspring by mating parents belonging to different breeds. It is more extreme than outbreeding in its genotypic and phenotypic effects. Crossbreeding also tends to increase heterozygosity of genes, and in doing this it also covers up detrimental recessive genes, decreases breeding purity, and improves traits related to physical fitness. This improvement in crossbred individuals as compared to the average of the parent breeds that produced the cross is known as *hybrid vigor*.

The farther apart the breeds are in relationship, the more hybrid vigor one can expect from crossbreeding in traits normally showing hybrid vigor. The reason for this is that one breed tends to carry a high frequency of one allele, whereas another breed tends to carry a low frequency of this same allele. In other words, the two breeds tend to be homozygous in opposite ways. Thus, when two or more breeds are crossed, more pairs of genes will be heterozygous in their crossbred offspring. For some genes one breed may be homozygous for one allele and another breed homozygous for another, but it seems more likely that the differences between the breeds is due to differences in frequencies of certain genes rather than to differences in homozygosity.

All breeds of cattle carry the same number of chromosomes (thirty pairs). Chromosome abnormalities do exist in some breeds, however.

10.5.1   Traits in Beef Cattle which Express Hybrid Vigor

Hybrid vigor may be defined as the amount the average of crossbreds for a trait exceeds the average of purebreds which produce that cross. Hybrid vigor may be expressed in traits in cows, bulls, and calves. The percentage of hybrid vigor for a particular trait (or traits) may be calculated as follows:

$$\% \text{ hybrid vigor} = \frac{\text{avg. of crossbreds} - \text{avg. of purebreds}}{\text{avg. of purebreds}} \times 100$$

The approximate percentage of hybrid vigor expressed for

various traits in beef cattle is shown in Table 10-1. These are average figures from many experiments.

The effects of crossbreeding are twofold. First, traits which do not express hybrid vigor will equal the average of the parent breeds for that trait in the crossbred offspring. If the two breeds crossed vary widely for a trait and little or no hybrid vigor is expressed in the crossbred offspring, the average for the offspring will be midway between the two breeds. The average for a trait will be higher in one parent breed than the other. The second effect of crossbreeding will be the expression of hybrid vigor. Traits not expressing hybrid vigor will be affected by additive gene action and should be improved by finding the best and mating the best to the best. This kind of selecting and mating system is adapted to improving traits within a breed. To take advantage of hybrid vigor that is due to nonadditive gene action, the superior individuals from two or more lines within the same breed or from two or more breeds should be crossed.

TABLE 10-1   Percentage of hybrid vigor expressed by various traits in beef cattle

Trait	*Percentage of hybrid vigor*
Conception rate	0–1[a]
Percentage of calf crop born	5–6
Fetal death loss	25–30
Length of gestation	0–1
Percentage of calf crop weaned	5–6
Survival rate from birth to weaning	5–6
Weaning weight	4–5
Weaning score	4–5
Pounds of calf weaned per cow exposed to bull	10–12
Postweaning rate of gain	5–6
Postweaning efficiency of gains	5–6
Slaughter grade	0–5[b]
Carcass grade	0–5[b]
Yearling weight	5–6
Percentage of retail cuts	2–3
Ribeye area	4–5
Percentage of fat in carcass	5–8
Tenderness	5–6
Marbling score	2–3

[a]Conception rate may show negative heterosis when some breeds are crossed.
[b]This may be considered a negative heterosis because the crossbred animals tend to carry more fat than the purebreds.

10.5.2 Expression of Hybrid Vigor in Calves

Hybrid vigor in calves is largely expressed in many traits of economic importance. Results from the Missouri Agricultural Experimental Station heterosis experiment indicated that slightly more inseminations (1.43 versus 1.38) were required per conception when purebred cows produced purebred calves than when they produced crossbred calves. Other experiments show little or no hybrid vigor for this trait. The Missouri data suggest the possibility that in occasional crossbred matings there may be some incompatibility between the mother and fetus, or egg and sperm, early in pregnancy. This is not a widespread occurrence but occurs often enough to affect the overall conception rate when large numbers of matings are made.

Fetal death losses from 60 days to birth were 32 percent less in crossbred than in purebred calves. This suggests that some fetal death losses in cattle after implantation are genetically determined.

Purebred cows produce about 5 to 6 percent more calves at birth and weaning when they produce crossbred than when they produce purebred calves. Crossbred calves will also be 4 to 5 percent heavier at weaning. The pounds of calf weaned per cow exposed to the bull is usually 10 to 12 percent greater when purebred cows produce crossbred calves, and these crossbred calves gained 5 to 6 percent faster and were 5 to 6 percent more efficient in their gains to market weight. Crossbred calves are usually slightly fatter at slaughter than purebred calves.

Crossbred cows appear to be superior to purebred cows in mothering ability. This advantage averages between 10 and 15 percent in the pounds of calf weaned per cow exposed to the bull. Most of this advantage is due to lower fetal death losses and better survival rate in calves from crossbred cows. Crossbred heifers appear to reach the age of puberty sooner than purebred heifers.

10.6 Crossbreeding Systems

Several different crossbreeding systems may be used for commercial beef cattle production. Certain systems are better adapted to some conditions than others.

10.6.1 The Two-Breed Cross

The two-breed system of crossbreeding has been widely used in the past. It involves breeding cows of one breed to bulls of another.

This produces crossbred calves, but the parents are still purebreds. Replacement heifers would need to be purchased when needed, or some purebred matings would have to be made from time to time within the herd to supply needed replacements.

The two-breed cross may be used to advantage by the commercial producer that has a small herd. Superior crossbred calves have a good sale value as feeder calves. If the program calls for staying with the two-breed cross, obtaining replacement heifers becomes a problem sooner or later.

The two-breed cross produces hybrid vigor only in the crossbred calves. This is approximately one half of that possible when both crossbred cows and calves are produced.

10.6.2   The Two-Breed Backcross

The two-breed backcross crossbreeding system is also referred to as the crisscross system. In this system of crossbreeding only two breeds are used. Crossbred cows from the first two-breed cross are then mated to an unrelated bull of one of the original breeds. The mating of crossbred cows back to a bull of one of the original breeds gives the system the name of backcross. The second-generation heifers are then mated to a bull from the second breed. Thus, the breed of bull

Figure 10.1.   *Two-Breed Crisscross or Back Cross System*: Heifers from cross $F_1 AB$ back crossed on sire B and then those BAB heifers mated to sire A, continuously crisscrossing. Heifer BA mated to sire A as a back cross and then heifers from this mating mated back to sire B as crisscross mating. This system can be managed with two pastures and natural mating for 50 to 80 cows or multiples. The two-breed cross gives about 67 percent heterosis if dominance is basis of heterosis. This assumes that recombination effects in offspring from crossbreed dams are negligible.

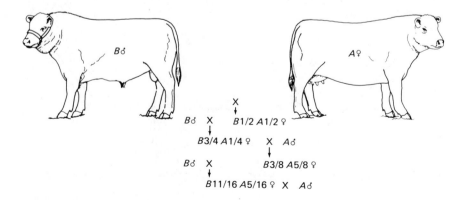

crisscrosses from one generation to another. This is illustrated in Figure 10.1.

The advantage of this system is that both the cow and calf are crossbred and express hybrid vigor. After two or more generations of backcrossing the amount of hybrid vigor in the calves stabilizes at about 65 to 70 percent of the original amount.

10.6.3   The Three-Breed Sire Rotation Cross

The three-breed cross involves the mating of crossbred females of two breeds with a purebred male of a third breed. This gives hybrid vigor in both the cow and calf since both are crossbred. This results in optimum hybrid vigor. Most experimental data are concerned with the first generation of the three-breed cross. This cross can be continued generation after generation by using purebred sires of three breeds on crossbred cows from previous generations. This

Figure 10.2.   *Three-Breed Rotational System*: This system will be effective with 75 to 120 cows, or multiples, or three producers using natural mating. This system gives about 20 percent more heterosis than the two-breed crisscross or back cross system. The maximum potential heterosis expected in this system is approximately 87 percent. The key to this system is finding three complementary breeds.

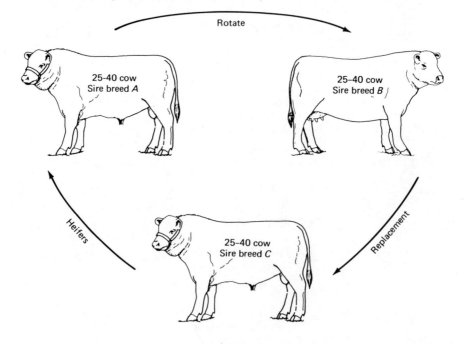

procedure is often referred to as the three-breed sire rotation cross. If care is taken to use sires of the three breeds in exact rotation in consecutive generations, about 87 percent of the heterosis is maintained from generation to generation.

A four-breed sire rotation on selected crossbred cows may be used. It gives little added heterosis as compared to the three-breed sire rotation cross but maintains a slightly higher percentage of heterosis in successive generations. The three-breed sire rotation cross is illustrated in Figure 10.2.

10.6.4   The Three-Breed Terminal Sire Cross

The three-breed terminal sire crossing system is one in which a purebred sire is mated to crossbred cows and none of the calves produced is kept for breeding purposes. Heifers as well as bulls are purchased or produced by making particular matings within the breeding herd. This crossbreeding system is illustrated in Figure 10.3.

The terminal sire cross gives hybrid vigor in both the cow and calf since both are crossbred. The advantage of this system is that crossbred cows used can be from a cross that is superior in milk producing and mothering ability and possesses acceptable carcass quantity and quality. The sire used may be selected from a breed that introduces fast and efficient gains and good carcass traits into the crossbred calves. Maximum heterosis is maintained continuously in this crossbreeding system.

The main disadvantage of this crossing system is that replacements have to be purchased periodically. If replacements are produced within the herd, this means that sires and dams of different breeds would have to be maintained at least periodically.

10.6.5   The Crossbred Sire

The possible use of crossbred bulls for commercial beef calf production has been seriously considered in recent years. This idea meets with considerable opposition from purebred breeders because it might reduce the number of purebred bulls sold.

The use of a crossbred bull could offer some advantages and some disadvantages. Each commercial producer should seriously consider these points before deciding to use a crossbred sire. (Interest in the use of crossbred sires has probably arisen because of the production of crossbred offspring and grading up to increase numbers of animals of the exotic breeds recently introduced into the Americas.)

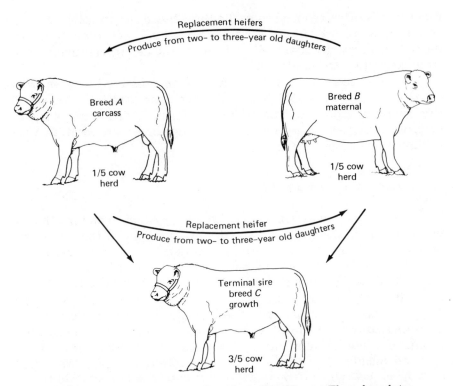

Figure 10.3.   *Three-Breed Terminal Sire Cross*: Three-breed terminal cross breeding system uses a two-breed criss- or back cross to
produce females for a terminal sire line. This system requires at least
three pastures and adaptation to 100 or more cow herd. The progeny in two-fifths of herd crisscross line will give 67 percent heterosis
with limited calving difficulty. There will be three-fifths of cows going in the terminal line as mature cows with limited calving difficulty,
making maximum use of genetic germ plasma reach and giving 87
percent of heterosis potential. Use sires from performance-tested
herd since the cow lines will be closed.

The main advantage of the crossbred bull is his improved fertility
and libido (on the average) as compared to purebred bulls. The main
disadvantage could be the increased variation in the offspring that
could result from mating crossbred bulls with crossbred cows.

   If a crossbred bull is used, he should rank very high in the
highly heritable traits such as rate and efficiency of gains and carcass
quality and quantity. Such traits are affected mostly by additive
genes and show little heterosis. A crossbred bull superior for these
traits would be expected to transmit about the same amount of
superiority for such traits to his offspring as a purebred bull. A cross-

bred bull would be more likely to look better than pure breeds for preweaning traits that are lowly heritable and which express medium to large amounts of hybrid vigor.

Several points should be considered in deciding whether or not to use a crossbred bull for breeding purposes. Only the top-performing crossbred bulls of the best type and conformation should be used for breeding. In addition a crossbred bull would be expected to give best results if he is used on purebred cows or crossbred cows not carrying a breed composition similar to what he is carrying. This kind of mating system should give maximum hybrid vigor.

Crossbred bulls may be used in rotation on crossbred cows if the crossing system is planned so that maximum breed differences are maintained generation after generation. This is illustrated in Table 10-2. If this is done, a high percentage of the possible heterosis will be maintained from generation to generation even when crossbred bulls are mated with crossbred cows.

10.7   Breeding Plan for Commercial Beef Production

Commercial beef cattle producers should, in general, keep the same kinds of records as purebred breeders, although fewer in some in-

TABLE 10-2   Amount of hybrid vigor expected in successive generations when crossbred cows are mated with crossbred bulls[a]

Generation number	Percentage of breeds in		Percentage of hybrid vigor[b] in calves
	Cows	*Bulls*	
1	50 A 25 B 25 C	50 B 50 C	75
2	25 A 37.5 B 37.5 C	50 A 50 B	68.8
3	37.5 A 43.75 B 18.75 C	50 C 50 A	71.88
4	43.75 A 21.87 B 34.38 C	50 D 50 E	100.00

[a]Close relationship of cows and bulls is purposely avoided. Breeds A, B, C, D, and E are used.

[b]Percentage of inheritance from different breeds.

stances. This will allow them to select heifer replacements and to cull unprofitable females from a herd.

Commercial producers usually purchase bulls to use in a herd. These bulls should be selected because they ranked high on a performance test and are of good conformation and show a high degree of muscling.

A crossbreeding system should be followed for commercial beef cattle production. The crossbreeding system used would be the one best suited to a particular herd. It should be kept in mind that a crossbreeding system which takes advantage of hybrid vigor in both the cow and calf gives optimum results. This would involve a three-breed sire rotation cross or a three-breed terminal sire cross. For commercial cow producers with a small herd, the two-breed backcross or crisscross might be the most practical. The breed of sires used should be incorporated into a systematic scheme in order to make maximum use of superior breeding stock in addition to hybrid vigor.

10.8   Some Management Difficulties Associated with Crossbreeding

Several management difficulties may be encountered in trying to follow a crossbreeding program. This is especially true for small commercial producers. The number of breeding pastures available may often limit the number of sires that can be used. The number of replacement heifers of a certain breed combination may be too small to justify ownership of another sire. In addition, the annual purchase of new bulls creates extra management problems and higher expense and increases the risk of disease being introduced into the herd.

STUDY QUESTIONS

1. Name and define the mating systems that may be used for improving livestock through breeding.

2. When is inbreeding generally used the most in the development of a breed and why?

3. What are the genetic effects of inbreeding?

4. Why does inbreeding tend to uncover recessive genes? Why doesn't it uncover dominant genes?

5. What traits usually decline when inbreeding is practiced? Why?

6. Why does inbreeding tend to increase breeding purity? Explain.

7. What genetic effects are responsible for prepotency? What is prepotency?

8. What is the main reason for inbreeding? What is another reason for using inbreeding?

9. Why haven't more inbred lines of beef cattle been developed? Why is it more difficult to develop inbred lines of cattle than inbred lines of corn?

10. What usually happens to important economic traits when inbreeding is practiced?

11. Is it possible to develop inbred lines of cattle on a practical basis?

12. What is linebreeding? How does it differ from ordinary inbreeding?

13. Which seems to be the best method of producing inbred lines, inbreeding or linebreeding? Why?

14. Under what conditions should linebreeding be practiced?

15. Define *outbreeding*. What are the genetic effects of this form of breeding?

16. What is crossbreeding, and what are its genetic effects?

17. Why is it usually said that the farther apart the parents are in relationship, the more hybrid vigor can be expected in traits that show hybrid vigor?

18. What traits in beef cattle appear to show hybrid vigor?

19. How many chromosome pairs do domestic cattle normally possess?

20. Cows of breed *A* wean 440-pound calves at 205 days of age, and cows of breed *B* wean calves that weigh 480 pounds. If the crossing of these two breeds gives calves that weigh 520 pounds, what is the percentage of hybrid vigor?

21. What are the two major effects of crossbreeding?

22. How would one produce the most hybrid vigor by crossbreeding in beef cattle?

23. Name four different systems of crossbreeding, and give the advantages and disadvantages of two of them.

24. What are some of the advantages and disadvantages of using a crossbred bull in a crossbreeding program?

25. Outline in detail a breeding plan that should be followed in the commercial production of feeder calves.

26. What are some management difficulties that may be encountered when a crossbreeding system is used for the commercial production of cattle?

PART FOUR

SYSTEMS OF PRODUCTION AND MANAGEMENT

establishing the breeding herd

The kind of herd to be established will depend on the conditions on the farm or ranch and the objectives of the cattle producer. On many small farms cattle are a supplementary source of income, whereas on large farms or ranches they are the main source. Some cattle producers largely produce feeder cattle, but others purchase cattle and market their hay and grain through them. Under some conditions cattle are purchased as feeders and fattened for slaughter (Figure 11.1). Regardless of the kind of cattle operation involved, efficiency of production as well as market price when the cattle are sold are extremely important.

11.1 Objectives in Establishing a Herd

Many cattle operations have been in a certain family for generations. In such cases there is no need to establish a breeding herd. The efforts should be directed mainly toward improving the quality and efficiency of the herd.

The kind of cattle operation is often dictated by the conditions on the farm or ranch. In range country the herd is usually involved in

Figure 11.1.　Good feeder cattle ready for auction. Their value would have been greater had they been dehorned at an early age.

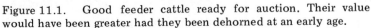

feeder calf production because the main feed is grass or some other forage and little or no grain is available for fattening (Figure 11.2). In such an operation calves may be marketed at weaning or shortly afterward, or they may be grazed during the summer and sold in the fall as yearlings. Keeping calves on the farm or ranch for sale as yearlings reduces the amount of grazing units available for mother cows. For this reason most cattle producers prefer to sell their calves at weaning.

In the Corn Belt, farmers may keep a few cows to utilize grass from land that is not suitable for cultivation. Under such conditions calves produced are usually marketed as feeders, although in some instances they are fattened on the farm.

The production of purebred cattle is a specialized production and management system. The main product of the purebred breeder is seed stock, most of which will be yearling or older bulls of breeding age. Some surplus females may also be offered for sale, however.

Some purebred herds have been in existence for many years,

Figure 11.2.   Feeder cattle coming off the range.

and the breeder is familiar with breeding, production, and management methods. The sale of purebreds requires more advertising and promotion than commercial beef cattle production. For this reason headquarters of the breeding herd must be kept attractive and must appear well kept. Many successful purebred breeders enter their superior breeding animals in local and national livestock shows. The purebred business is very competitive and requires much effort to keep abreast of the activities in it.

11.2   Foundation Stock for the Commercial Herd

Females for establishing a commercial herd may be obtained from several sources. The source decided upon will depend on availability and the program the producer wishes to follow. Some decision must be made as to the breed to be used or whether purebreds or crossbreds will be used for foundation stock.

A large supply of heifer calves is available in the fall of the year on many farms and ranches. In many instances no records are available on such heifers. If this is the case, each heifer selected and purchased should be weighed and identified as soon as possible after purchase. If records are available, they should be used for selection purposes. A high roughage growing ration may be fed the first winter.

Before the beginning of the breeding season the heifers should all be weighed again, and only the larger ones that have made the fastest gains should be selected for breeding. The top 70 to 80 percent should be kept for breeding, and the remainder should be sold. The main disadvantage of purchasing heifer calves is that it will be at least 2 years from the date of purchase before they will wean their first calf. An advantage is that usually they can be purchased at a reasonable price in large enough numbers so that culling can be done before the breeding season and only the top heifers retained. In addition, the producer will have the choice of what bull to breed them to, and the heifers will have plenty of time to become adapted to conditions on the farm.

Open (nonpregnant) yearling heifers may also be purchased. Careful attention should be given to large size and good conformation when they are purchased. Heifers of this kind will not be as plentiful as heifer calves at weaning, so selections may be more limited. The purchase of open yearling heifers does give cattle producers the opportunity to select bulls to which they will be bred.

Bred yearling heifers may be purchased for the establishment of a herd. The return on the investment will come more quickly when females of this kind are purchased. All should be pregnancy-tested to make certain they are pregnant, however. A disadvantage of purchasing bred heifers is that one may not know the kind or breed of bull to which they were bred, and more difficulty may be experienced at calving, especially if they have been bred to large-type bulls or bulls of some of the exotic breeds.

Mature cows may also be purchased to establish a commercial herd. Occasionally these may be obtained from a complete herd dispersion sale, at private treaty from a cattle producer who wishes to sell surplus stock, or they may be purchased at a sale barn. This is a way to get into the cattle business in a hurry, especially if the cows purchased have large calves at their sides and are rebred. A cow with a calf at her side may be considered to be progeny-tested to a certain extent. Bred cows without a calf may also be purchased. A pregnancy test should be made to determine that such cows are pregnant.

The main problem with purchasing mature cows is that they often are culls from someone's herd and may have been culled because of their poor record, age, or other causes. Obviously cattle producers remaining in the cattle business will not sell their best-producing cows. This will not be true of cattle in a dispersion sale and those produced by a cattle producer with a good reputation or if the source of the stock is well known. All stock should be closely observed for health, obvious defects, and internal and external parasites. They should be tested by a veterinarian for diseases such as leptospirosis, tuberculosis, and brucellosis. It is well to isolate newly acquired female stock for 30 to 60 days after purchase to make certain they are not diseased.

Other factors should be considered in selecting breeding stock. They should have desirable feet and legs and a good frame, indicated by length of body and height at the shoulders. Heavily muscled females should be avoided because this condition is sometimes associated with poor reproductive performance. Overfat cows and heifers should also be avoided because this condition may be associated with past poor reproductive performance in cows and poor future reproductive performance in heifers.

One other point should be considered when mature cows are purchased. Young cows (3 to 4 years of age) have a long reproductive life ahead of them and should reach their maximum efficiency of production at 7 to 9 years of age. Older cows, of course, may have passed their peak in performance and will have fewer remaining years of production ahead of them.

Purchasing herd sires for the commercial herd requires a lot of thought and attention because each sire contributes one half of the inheritance of each of its offspring, and the sire must be able to settle a large percentage of the cows exposed to him in a short period of time. The breed of bull is also important because in many commercial herds a crossbreeding program is followed in order to take advantage of hybrid vigor. Performance records on each potential herd sire are important as well as information on both the sire and dam of the prospective herd sire.

The herd sire should have desirable length of side and height at the shoulders. He should be smooth in the shoulders and hips with smooth and not excessively heavy muscling. The scrotum and testicles should be well developed because larger testicles appear to be associated with greater sperm production and fertility. A herd sire should have good feet and legs and should be active and alert but not excessively so (Figure 11.3). It is also well to emphasize selecting bulls that are especially desirable in traits that are somewhat lacking, in general, in the cows in the herd.

Bulls of breeding age (1.5 to 2.0 years of age) should be considered. Bulls of this age are old enough to have completed a performance test, or at least they are old enough to have expressed their growing ability, potential size, and other required characteristics. Bulls of desirable type and conformation with heavy yearling weights should be considered. Yearling weight is a highly heritable trait (50 to 60 percent) and is associated with efficiency of gains and leanness.

Figure 11.3.   An excellent herd sire. (Courtesy American International Charolais Association)

A few commercial cattle producers purchase several bull calves shortly after they are weaned and performance-test these calves on their own farm or ranch. The best-performing bulls of the most desirable type and conformation are retained for breeding, and the remainder are sold. Occasionally, a mature bull may be purchased if he is of superior type and performance and records on his progeny are available. Otherwise, it would be more desirable to purchase young bulls.

11.3   Foundation Stock for the Purebred Herd

Purebred herds are the seed stock producers for the beef cattle business. This is mainly through bulls that are produced and sold to other purebred herds or to the commercial producer. Because of this fact, it is important to choose the kind of breeding stock that is highly productive when a purebred herd is established.

Heifers may be purchased from a reputable breeder to be used in establishing a purebred herd. Heifers purchased can be calves or yearlings, depending on what is available. It is just as important to select good-performing purebreds as it is to select good-performing commercial females. Ten to 15 percent more heifers should be purchased than are needed in the herd. Some may not grow to breeding age as well as desired, others may not prove to be of good type and conformation by breeding age, a few may not conceive when bred, and still others may prove to be poor milkers and poor mothers when they calve.

Type and conformation are very important and should receive considerable attention in establishing a purebred herd, because the animals produced must meet certain standards of perfection if they are to sell at a profit. Heifers should be long and smooth-muscled without excessive muscling. They should have a clean-cut feminine head, sound legs and feet, and desirable size for their age. The exact date of birth should be known for all registered females. As much information as possible should be obtained on the parents of each calf to ascertain whether or not they have produced offspring free of recessive genetic defects. Heifers, insofar as possible, should be selected from dams that are good milkers and good mothers.

Cows may be purchased from established breeding herds. A great deal of care should be exercised in making such purchases. Cows for sale may be those that have performance records of average or below, or if they are above average to superior, they may command an excessive price. Purchasing purebred cows from a dispersion sale gives a buyer the opportunity to purchase cows of good type and conformation and with good performance records.

Points to consider in purchasing a herd sire for the purebred herd include all those previously discussed for the commercial herd. In addition, a prospective herd sire for a purebred herd should be particularly outstanding for desirable traits and should be from bloodlines that are popular at that particular time. Because of these characteristics, the purebred herd sire prospect will usually cost more than top commercial bulls because top bulls command top prices.

Even though an outstanding herd sire prospect commands a high price, he may be worth it. Experimental results show that 85 to 90 percent of the genetic improvement in a herd over a period of years comes from herd bull selection. Improvement through the cow side is less, on the average, because more female replacements must be kept and selection for females is less intense than for bulls.

Purebred breeders are more likely to sell their best bull calves but more than likely to retain their best heifers for herd replacements. This is another reason why emphasis should be placed on obtaining the best purebred herd sire possible.

The price commanded by a purebred herd bull prospect depends on its own type and performance as well as that of its sire. The record of a bull's sire is important because a good record from a well-promoted sire helps sell the purebred calves produced. A popular bull may not be genetically superior to one that is less popular, but popularity of bloodlines helps sell more breeding stock at a higher price.

11.4   Maintenance of the Breeding Herd

Once the herd is established, other factors must be considered for maintaining efficiency of production. One of these is the replacement rate for breeding stock in the herd. The number of females kept for breeding will depend on whether or not the herd is increasing or decreasing in numbers or remaining the same.

The replacement rate depends on the degree to which females in the herd are culled and on the number of losses due to death and injuries. In one study (H. B. Sewell et al., *Mo. Agric. Exp. Stn. Res. Bull.* 823, 1963) an average of 16.8 percent of the Hereford cows wintered left the herd annually over an 11-year period when cow numbers in the herd were remaining constant. About 47.5 percent of the cows removed from the herd left because of death, accidents, diseases, cancer eye, age, and failure to conceive. The remaining 52.5 percent left the herd because of lower than average weaning weights of their calves. These results suggest a 15 to 20 percent replacement rate of cows when a herd is remaining constant in size. In the well-managed herd where a 90 percent calf crop was weaned each year over an 11-year period, this would require that about 40 to 50 percent of the heifers produced each year be kept for replacement. If a herd were expanding in numbers, a larger percentage of the heifers produced each year would have to be kept as replacements.

11.4.1   Age to Cull

The age at which cows and bulls should be culled from the herd will depend on several circumstances, and those culled may be of many different ages. For example, some heifer calves kept as replacements may be culled because of poor health, accidents, poor conformation, or poor performance. This would represent a small percentage, however, if they were well fed and cared for and if careful selections were made as weaning calves. This would suggest that 5 to 6 percent more heifer calves should be saved in the beginning than will be needed. This will allow for some attrition before calving.

About 3 to 5 percent of the heifers exposed to the bull will fail to conceive in some herds. They should be culled. Some will have calving difficulties when they calve for the first time, and some may prove to be poor milkers. It would be necessary to cull some of these as young cows also.

Cows generally reach their peak of productivity between 7 to 9 years of age (Table 11-1). Their productivity declines after that time. Productivity will decline much more rapidly in some cows than in others, but when it is obvious that old cows are rapidly failing in performance as indicated by greater than average weight losses while nursing a calf and during the winter months, they should be sold and replaced by younger cows or heifers. Some old cows, however, remain productive many years beyond the usual peak of productivity.

Injuries and disease may affect cows of any age. If their performance is reduced because of these causes, they should be culled regardless of age.

The age at which bulls should be replaced varies with different individuals and in different herds. In a small herd where only one herd sire is used each year, a bull may have to be replaced when his daughters reach breeding age to avoid inbreeding. In many herds this would be after he has sired two calf crops, and he would be between 3 and 4 years of age. This is a relatively young age, and a bull

TABLE 11-1   Weaning weight of calves from cows of different ages[a]

Age of dam (yr)	Average weaning weights of calves (lb)
2	353.1
3	376.5
4	395.8
5	411.4
6	425.1
7	430.9
8	434.9
9	435.0
10	431.2
11	423.5
12	412.0
13	396.6

[a]Average 180-day weaning weights adjusted to a steer basis and fitted to a second-degree parabola. From H. B. Sewell et al., *No. Agric. Exp. Stn. Res. Bull.* 823, 1963.

could be highly fertile and very productive for several more years. A bull could be kept separate from his daughters during the breeding season, and they could be bred to another bull by artificial insemination. This would prolong the productive life span of a bull in a small herd. In herds where two or more sires are used for breeding each year and at least two breeding pastures are available, a bull may be used for several years longer than when used in a one-sire herd. The daughters of one bull in a two-sire herd could be bred to the other bull and vice versa. This could extend the productive life of each bull for at least another 2 years. Only bulls that sire poor-quality calves of average or below-average performance should be culled and replaced regardless of age, as would be true for bulls which are injured, become ill, or prove to be of lowered fertility.

In large herds where many sires are used during the breeding season, the possibility of inbreeding becomes less of a problem. In such herds a productive bull may be used until he reaches 8 to 10 years of age if he remains an active breeder and produces desirable offspring. In one range herd where pastures were large and many bulls ran with the cows, each bull was branded to indicate his age. All bulls were gathered and sold as they reached 7 to 8 years of age. Young bulls were turned on the range each year in numbers exceeding those who were sold.

11.4.2   The Generation Interval

The generation interval in cattle may be defined as the average age of the parents when their offspring are born. This would include the age of both the sire and dam during their period of use in the herd.

Research data from beef breeds (D. B. Crenshaw, "Measures of Reproductive Performance in Beef Cattle," M.S. thesis, University of Missouri, Columbia, 1969) indicated that the average generation interval was 4.0 to 4.5 years. This included the cows and bulls used in the breeding herd over a period of several years regardless of the reason for their leaving the herd. Some cows, however, remained in the herd until they were 10 to 12 years of age, and others left as two-year-olds. In this study the generation interval for cows averaged 0.5 to 1.0 year longer than for bulls.

The shortest possible generation interval in beef cattle would be between 2 and 3 years of age. A short generation interval such as this would require that both the sires and dams be sold after producing just one calf crop. This, of course, is not desirable under practical conditions. The longest possible generation interval would be 8 to 9 years, but this would seldom be attained under practical conditions.

11.4.3   Possibility of Inbreeding

The question of whether or not to practice inbreeding often arises in a one-sire herd. Inbreeding or linebreeding is not recommended for the production of commercial cattle, because both are usually associated with a decline in physical fitness of the inbred individual. In commercial production, it would be the goal to produce individuals that are as vigorous as possible.

In one-sire herds, mating a sire to his own daughters would result in at least 25 percent inbreeding in the offspring produced (Figure 11.4). In many cases this could be associated with the appearance of one or more recessive defects and a decline in physical fitness. Therefore, such matings are not recommended. In purebred

Figure 11.4.   Pedigree of a Charolais calf resulting from a sire X daughter mating. Such matings were made early in the history of the breed; because breeding animals were limited in number, this mating probably occurred in North America.

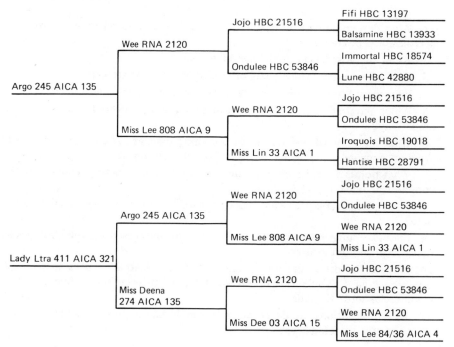

herds, sire-daughter matings could be used to estimate the genetic worth of a particular sire. If such matings are made and the inbred offspring remain vigorous and show no recessive defects, this indicates the sire is superior genetically. About thirty-five such matings with no defective offspring produced are necessary, however, to prove a sire free of recessive defects at the 99 percent level of probability. Few sires would pass such a test in actual practice. If such a sire is found and he is superior, a linebreeding program could be followed by mating his half-sib progeny.

Inbreeding in a two-sire herd is of little concern and can be avoided. If it becomes necessary to inbreed, a sire could be mated to his own granddaughters, which will produce calves that are about 12.5 percent inbred. Such a mating system might be useful in a purebred herd, but would seldom be practiced in a commercial herd.

Inbreeding in multiple-sire herds would probably not increase more than 1 to 2 percent per year over a period of time and would be less the more sires used each breeding season.

STUDY QUESTIONS

1. What determines the profitability of a beef cattle herd regardless of whether it is purebred or commercial?

2. When are most feeder cattle marketed? Why?

3. Why are so many feeder cattle produced in the range states?

4. Why should the purebred breeder consider showing animals in his herd?

5. Which requires more management efforts, a purebred herd or a commercial herd? Why?

6. What are some of the sources of obtaining females for a commercial herd? What are some advantages and disadvantages of each source?

7. What points should receive particular attention in selecting females for the commercial herd?

8. What points should be considered in purchasing herd sires for a commercial herd?

9. Should young bulls or old mature bulls be purchased for the commercial herd? Explain your answer.

10. Outline the procedure you would follow in selecting females to establish a purebred herd. Would this be different from the selection of females for a commercial herd? Explain.

11. What precautions should be taken in purchasing mature cows for the establishment of a purebred herd?

12. What traits would you consider in selecting cows and heifers for the purebred herd?

13. Why is attention to type and conformation of more importance in purebreds than in commercial cattle?

14. What are possible sources of breeding females when a purebred herd is being established?

15. Why is it important to check the record of a bull's sire when purchasing a purebred bull?

16. What proportion of the females produced in a herd each year should be kept as replacements when the herd is remaining at approximately the same size?

17. How old should a cow be when she is culled from the herd?

18. At what age do cows usually reach their peak in productivity?

19. Discuss the length of time a bull can be retained in the herd if the herd is a (a) one-sire herd, (b) two-sire herd, or (c) multiple-sire herd.

20. Define *generation interval*. What is the average generation interval in beef cattle? Is the generation interval usually longer in bulls than cows? Why?

21. Why is it not usually advisable to practice inbreeding or linebreeding in a commercial herd?

22. Under what conditions could linebreeding be practiced in a purebred herd?

care
and management
of the breeding
herd

Environmental conditions are responsible for 50 to 90 percent of the variations in most economic traits in beef cattle. Even the most productive individuals from the genetic standpoint will be disappointing in their performance unless they are properly cared for and managed. The discussion here will be directed toward the identification of individuals in the breeding herd and their care and management.

12.1 Identification Systems for Beef Cattle

All animals in the herd should be permanently identified by some method. Possible methods of identification are illustrated in Figure 12.1. Animals may be identified by an ear tattoo, an ear tag, a horn brand, a hide brand, a neck chain, an electronic implant that emits certain radio waves, or a combination of these methods. A clear, well-placed tattoo in both ears of the individual will be permanent unless the ears are lost. Different-colored tattoo ink may be used for animals with different-colored skin in the ears. Usually a black tattoo ink is used. Plastic ear tags are an excellent means of identification,

Figure 12.1.   Identification systems: cows and calves.

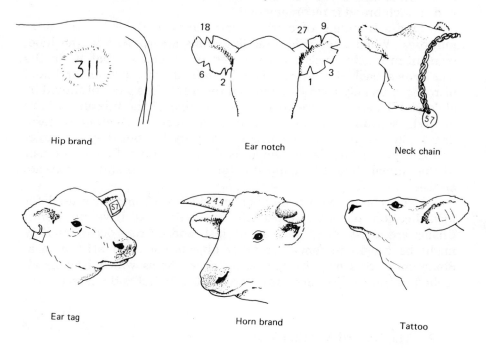

although they may be lost and must be replaced from time to time. Some tags remain in the ears longer than others, and the experience of the cattle producer usually results in personal likes and dislikes of various tags.

Horn brands may be used to identify individuals that possess horns. The brands usually consist of a series of numbers which correspond to tattoo numbers in the ear. Some horn brands may be readable for many years, but with increased exposure to the weather and damage from rubbing feed bunks, fence posts, etc., in older animals they may become worn and blotched and difficult to read.

Numbers branded on the shoulder, side, or hips may be used for identification purposes. The usefulness of such brands varies with how clearly they are branded on the animal. Keeping a hot branding iron on the side too long during the branding operation or using irons that are too hot may cause numbers to become blotched and blurred and not readable. Individual branding irons heated by means of a separate heating source or electric branding irons with a heating element in the surface may be used. The size of numbers that appears to be most useful for calves is about 3 inches. For mature cows a 4- or 5-inch brand is recommended.

Freeze branding of cattle is sometimes used for identification purposes. When supercold branding irons are applied to the hide, pigment-producing cells are destroyed or altered so that when the hair grows back it is white. Dry ice mixed with alcohol or liquid nitrogen is usually used to supercool the irons. The alcohol should be at least 95 percent pure to prevent freezing of the mixture. Either methyl, isopropyl, or ethyl alcohol is satisfactory. Acetone or gasoline could also be used in a mixture with dry ice, but they are explosive and pose some danger when used near an open flame. The hair of the animal should be clipped before the cooled branding irons are applied.

Neck chains with a number attached are sometimes used for identification purposes, especially in cattle that have no horns. Chains are satisfactory but can be lost, and on rare occasions they might be caught in fences or other objects and torn off the neck. Sometimes cows may be caught by neck chains which become attached to some object and may starve unless found and released.

12.2   The Record-Keeping System

Keeping records takes time, but it pays. Even in a small herd records are important because few persons possess the ability to keep accurate records in their heads. The breeding and production of cattle involve repeated comparisons of individuals within the herd. Records make these comparisons valid. If records are kept, they should be used to select future breeding animals and to cull poor producers. Records are of little value unless they are used for these purposes. The kind of records to keep on calves is illustrated in Figure 12.2. The kind of records illustrated in Figure 12.2 may be used to evaluate the performance of each individual animal in the herd as well as sire and dam progeny data.

A record-keeping form for postweaning data is shown in Figure 12.3. This system is designed primarily for on-the-farm performance testing.

Records should be kept on several important economic traits. The kinds of records kept will vary to a certain extent with different cattle producers. The following are suggestions on records to keep and how adjustments may be made for some of these traits when adjustments are necessary.

AHE Form No. 2 P.R. (Rev.) 75

County & Area: OIO-942

Name: UNIVERSITY OF MISSOURI   Address: COLUMBIA, MO 65201

Specialist: SNODFORD

Date of Weaning Weight: MARCH 31, 1976   Year: 1975-76

Sex¹ M   Indicate Management Code² CREEP   Breed: ANGUS

Calf I.D. (1)	Dam (2)	Sire (3)	Age of Dam (4)	1975 Birth Date (5)	Birth Wt. (6)	Weaning Age in Days (7)	Weaning Wt. 160-250 Days of Age (8)	205-Day Wt. (9)	205-day Wt. Adj. for Age of Dam³ (10)	Adj. A.D.G. (11)	Adj. W.W. Ratio (12)	Actual Height / Adj. Height (13)	Feeder Grade / Feeder Grade Ratio (14)	Frame 1-5 / Muscle Score 1-5 (15)	Trimness 1-5 / Soundness 1-5 (16)
5605	3651	D	2	8-3	57	241	450	391	450	1.92	1.00	39 / 388	14 / 1.05	3 / 2	3 / 2
5607	2604	E	3	8-5	57	239	550	480	528	2.30	1.17	405 / 401	14 / 1.05	4 / 2	2 / 2
5612	0604	D	5	8-11	46	233	465	415	415	1.80	.92	38 / 371	13 / 97	2 / 2	2 / 3
5615	0626	E	5	8-13	75	231	570	514	514	2.14	1.14	405 / 39.44	14 / 1.05	3 / 3	2 / 2
5635	5115	E	10	8-22	58	222	365	341	341	1.38	.75	36.5 / 35.44	12 / .90	1 / 2	2 / 3
5636	2609	E	3	8-24	60	220	520	489	537	2.33	1.19	39.5 / 37.76	14 / 1.05	3 / 2	3 / 2
5637	3634	E	2	8-24	62	220	480	452	519	2.23	1.15	39.5 / 40.01	14 / 1.05	4 / 2	2 / 3
5645	0601	E	5	8-29	68	215	540	518	518	2.20	1.15	42 / 41.67	15 / 1.12	4 / 2	2 / 2
5657	1625	D	4	9-23	68	190	430	459	482	2.02	1.07	380 / 39.25	14 / 1.05	3 / 2	2 / 2

¹Use separate form for each sex. ²Management Information: (1) Creep; (2) Noncreep; (3) Specify groups—Sire, Management or Herd (minimum of 10 per group). ³Use col. 9 to figure col. 10 from table on p. 14 of UMC guide MP474.

Figure 12.2.   Record form used by the Missouri Extension Program for cow-calf records.

Name _____

Complete
Address _____

Date _____

Year _____ Sex _____ Breed _____

Period Fed _____ To _____

No. days on feed _____

Calf No. (1)	Birth Date (2)	Sire Dam (3)	Adj. 205-Day Wt. Weight Ratio (4)	Actual Weaning Wt. Date Weaned (5)	Feed Period Final Wt. Initial Wt. (6)	Date Final Wt. Total Gain (7)	Feed Gain Post Wn. A.D.G. (8)	Post Weaning Gain Ratio (9)	Life Daily Gain (10)	Adj. 365-Day Wt. (11)	Adj. 365-Day Wt. Ratio (12)	Conf. Score (13)	Conf. Score Ratio (14)	Remarks (15)

Figure12.3.   Record form for the postweaning period.

A calving ease score should be recorded for each calf at birth. One scoring system used is to assign a score of from 1 to 6, with 1 being assigned when there is no difficulty at birth and 6 being assigned to those very difficult births where the calf is lost. Scores from 2 to 5 represent increasing difficulties at birth. Such records may be used to identify bulls that cause a higher than average percentage of difficult births so they may not be used on first-calf heifers.

Birth weights of calves may be recorded, but they are of limited value in selection. If recorded, they should be taken within 24 hours of birth.

The weaning weight of each calf is very important in selection and should be recorded. The weaning weight reflects the milking and mothering ability of the cow as well as the growth rate and vigor of the calf. Since calves may be born on different days over a 90-day (or longer) calving period, it is not often practical or convenient to weigh each calf as it reaches a predetermined age. For this reason all calves within a few days of the desired weaning age may be weighed at weekly or 2-week intervals. The records may then be adjusted to a certain age such as 180 days, 205 days, etc. The following formula may be used to adjust records to a 205-day basis:

Adjusted 205-day wt. =

$$\frac{\text{actual weaning wt.} - \text{birth wt.}}{\text{actual weaning age}} \times 205 + \text{birth wt.}$$

The weaning weight may be adjusted to any desired age by replacing the 205-day figure by the desired weaning age. If birth weights are not recorded, an average weight figure of 70 pounds may be used. It would be more desirable to use a birth weight figure close to the average of the breed for which weaning weights are being recorded. Calves from some breeds weigh more at birth than those from others.

The following example illustrates how to adjust weaning weight records to 205 days by using this formula. We shall assume that a calf had a birth weight of 75 pounds and an actual weaning weight at 220 days of age of 550 pounds. The adjusted 205-day weight would be equal to

$$\frac{550 - 75}{220} \times 205 + 75 = 517.1 \text{ pounds}$$

When the record of each calf is adjusted in this manner, it is possible to calculate the weaning weight ratio for each calf by dividing each calf's adjusted 205-day weight by the average of all calves of the same sex in the herd. For example, if a calf had an adjusted 205-day weight of 518 pounds and all calves in that calf crop averaged 400 pounds, the weaning weight ratio for that calf would be (518/400)100 or 129.50. All calves above average would have a weaning weight ratio of 100 or above, whereas those below average would have a weaning weight ratio below 100. Weaning weight ratios for bull calves should be determined from the average weight of all bull calves and for heifer calves from the average of all heifer calves. In making selections, weaning weights of bull calves should be compared with other bull calves and those of heifers with other heifers without adjustments for sex differences.

An adjustment for the effect of age of dam on the weaning weight of each calf is often desirable. This is especially true when cows of different ages are within the herd, which is true in most herds. The following adjustments for age of dam and sex are recommended by the Beef Improvement Federation for use on a nation-wide basis:

Age of dam	Additive factors	
	Male calves (lb)	Female calves (lb)
2-yr-old-cows	60	54
3-yr-old-cows	40	36
4-yr-old-cows	20	18
5- to 10-yr-old cows	0	0
11-yr-old-cows	20	18

Other conditions may affect the age of dam adjustment factors. A cow nursing twin calves should be figured as a two-year-old dam for that lactation period regardless of her age. If a cow has twin calves and nurses only one, the nursing calf should receive the regular adjustment for the age of its dam. Calves which nurse dairy cows should receive no age of dam adjustment. Calves weaned before 120 days of age and then placed on a self-feeder should receive no adjustment for age of dam.

Other adjustments for age of dam are recommended in different beef cattle state programs. Conditions vary so much in the United States that a specific adjustment factor may be more appropriate for the area in which it is developed.

In the comparison of cows for selecting and culling purposes, the weaning weight of each calf should be adjusted for age of dam as well as for the sex of the calf. These adjustments are necessary because the age of the cow as well as the sex of the calf produced varies during each cow's productive lifetime. If a cow produces mostly bull calves, the average weaning weight of all her calves should be heavier because bull calves weigh more at weaning than heifer calves. These adjustments make within-herd comparisons of cows more accurate. Some breeders adjust the sex of the calf to a steer basis by adjusting records of bull calves downward by 5 percent and those of heifers upward by 5 percent. Adjustment factors within a particular herd may be obtained by dividing average bull weights by average steer (or heifer) weights. For example, if the average weight of all bull calves was 450 pounds and that of all heifer calves was 420 pounds, 450/420 would give a factor of 1.07. The adjusted 205-day weight of a heifer calf multiplied by 1.07 would give the heifer's weaning weight adjusted to a bull basis.

Production records for all cows in the herd may be compared more accurately by calculating the most probable producing ability (MPPA) for each cow in the herd. The formula is

MPPA =

$$\text{herd avg. ratio} + \frac{nr}{1 + (n-1)r} \times (\text{cow's avg. ratio} - \text{herd avg. ratio})$$

In this formula, n is the number of records for each cow and r is the repeatability of one record for a certain trait or several traits included in an index. The repeatability (r) of one record for weaning weight is about 0.40 and for conformation about 0.30. The part of the formula $nr/[1 + (n-1)r]$ is the repeatability of two or more records and is referred to as R. The repeatability for various other numbers of records for weaning weight and weaning conformation score is given in Table 12-1.

The following illustrates the calculation of the MPPA for a cow that has produced five calves (repeatability of 0.77 from Table 12-1) with a weaning weight ratio adjusted for sex and age of dam of 110:

$$\text{MPPA} = 100 + 0.77(110 - 100)$$

$$= 100 + 7.70$$

$$= 107.70$$

The MPPA should be calculated for each cow in the herd, and cows with the lowest scores should be culled.

The recording of a conformation score or grade at weaning is optional. Such a score should be based strictly on skeletal sound-

TABLE 12-1   **Repeatability (R) of two or more records for weaning weight and conformation score**

	Repeatability of	
No. of records	Weaning weight	Conformation score
1	0.40	0.30
2	0.57	0.46
3	0.67	0.56
4	0.73	0.63
5	0.77	0.68
6	0.80	0.72
7	0.82	0.75
8	0.84	0.77
9	0.86	0.79
10	0.87	0.81

ness and carcass desirability. It is best to make visual evaluations after other measurements are taken. An example of a scoring system is as follows:

Prime	High prime = 17
	Prime = 16
	Low prime = 15
Choice	High choice = 14
	Choice = 13
	Low choice = 12
Good	High good = 11
	Good = 10
	Low good = 9
Medium	High medium = 8
	Medium = 7
	Low medium = 6
Common	High common = 5
	Common = 4
	Low common = 3
Inferior	2 to 0

Postweaning records are of value in selection for replacements and progeny tests. Postweaning records should include yearling weights (365-day weights). This trait is highly heritable and is genetically associated with efficiency of gains and the pounds of retail boneless beef produced.

Yearling weights should be computed separately for each sex and comparisons made within each sex. The formula for adjusting to a yearling weight basis is as follows:

Adjusted 365-day wt. =

$$\frac{\text{actual final wt. - actual weaning wt.}}{\text{no. of days between wts.}} \times 160$$

+ 205-day weaning wt. adjusted for age of dam

Yearling weight ratios may be calculated for each animal in the herd in the same way as described for weaning weight.

12.3   Gentling Cattle

Cattle within the same herd vary widely in disposition. Some of this variation is due to heredity, but much of it is due to environment. Regardless of the cause, any wild, stubborn, quick-tempered animal should be culled from the herd. The actions of a single animal may be contagious, under certain conditions, and may cause difficulty in the entire herd. Gentleness of cattle in any herd is an important goal to strive for, especially on the range where pastures may include thousands of acres, part of which is covered by thick brush or timber. Some range cattle have little contact with humans, and what contact they do have is when the calves are branded, vaccinated, or castrated. These experiences, of course, are harrowing to the entire herd and promote wildness.

Cattle on smaller farms of limited acreages have more contact with humans than cattle on the range. Usually farm cattle are fed roughage in the winter months, which brings them into close contact with the persons doing the feeding. This is an appropriate time to gentle cattle and improve their trust of people and their close relationship with them. Quiet moves together with a low voice can do much to win the confidence and trust of cattle in the herd. Quick, threatening actions combined with wild yells and the cracking of whips should be carefully avoided unless they become absolutely necessary. On the range with moderately gentle cattle, it is best to leave the lariat in camp.

12.4 The Breeding Season

The breeding season is one of the main segments of management of beef cattle production. Poor management here could result in losses in the operation, whereas good management often results in profits.

12.4.1 Time of the Breeding Season

Many species of wild animals have a definite breeding season. Almost always the breeding season occurs at a time which ensures that the young will be born when conditions for survival are optimum. In species which are seasonal breeders the onset of reproductive activity is correlated with length of daylight and to a lesser extent with temperature. For cattle the breeding season should be at such a time that the calves are born when green grass is plentiful and temperatures moderate. Under domestication, people have controlled the breeding season of cattle so that it is possible for calves to be born almost any time of the year. This is done by regulating the time that bulls run with the cows on the pasture or range.

The majority of calves in the United States are born in the early spring or late winter. Spring-born calves have several advantages. They are born during the period of the year when most farming operations are at a standstill, which allows the cattle producer to spend more time with the cow herd during the calving season. This allows the assistance of cows and heifers at calving, if needed, and it helps get calves off to a good start in the most critical period of their life. Many cows are being fed hay or other roughages during this calving period, and because of this, they are already subjected to daily observation. Cows calving during this period are usually not stimulated to produce a maximum amount of milk. Very young calves do not need large amounts of milk at first but require just enough to keep them healthy and growing rapidly. Spring calves will usually be large enough to use the greater milk production of their dams when placed on lush, green grass early in the spring. Spring calves will also consume increasingly large amounts of grass as they grow older and should have heavy weaning weights when marketed. This allows the cows to go through the fall and winter months while not lactating and when the feed supply is not optimum and the weather is wet and cold. They are better able to withstand such stress conditions while not lactating. Cows on green grass, even though they are nursing a calf, will usually rebreed on schedule so a calf will be produced each year. One disadvantage of spring calves is that they may be born in extremely wet and cold weather conditions, especially in the northern states. This requires more care and protection to prevent higher than average death losses.

Fall calves are born during the period from August through October. Weather conditions for fall calving are usually favorable. Also, the cows are usually in good condition and will rebreed on schedule. When fall calves are weaned in the spring, they are old enough to use large amounts of grass during the normal grazing season. Cows calving in the fall may require more feed than dry cows because they are producing milk in addition to maintaining their body weight and functions. Fall calves are usually lighter at weaning than spring calves.

In some parts of the United States cows are bred to calve in the winter months of December, January, and February. These calves are born during the coldest winter months and require more attention at birth than calves born during warmer seasons. In extremely cold weather an occasional cow may suffer from frozen teats, and the teats may become cracked and sore. Calves born at this time will usually be heavier at weaning than calves born later in the year. If cows calving in the winter are not fed adequate amounts of feed during January and February, some difficulty may be encountered in getting some of them to rebreed on schedule.

The production of fall and winter calves may be more practical in the southern states where grazing may be almost year-round. In the northern states where winters are long and cold, it may be undesirable to calve at this time. When it is possible to have more than one calving season during the year, more use can be made of herd sires because they can be run with the cows over a longer period of time with a rest between breeding seasons.

12.4.2   Management of the Breeding Herd

Beef cows convert grass and other roughages that cannot be utilized directly for food by people into a highly nutritious and desirable food — meat. Beef is very high in protein, which is one of the very important nutrients required by people and all other animals. Good management is necessary for the cow herd to be the most productive, whether it is a purebred or commercial herd.

The percentage of calf crop weaned is closely related to net returns per cow per year over production costs. This is shown in Table 12-2. Net returns per cow, of course, will vary with the cost of production, especially feed costs, and the price per pound received for

TABLE 12-2   Relationship of percentage of calf crop weaned to the net returns per cow per year[a]

Percentage of calf crop weaned	Returns over variable costs ($)	Net return per cow ($)
90–100	52.72	24.96
80–89	40.98	12.76
70–79	15.65	–8.11
Below 70	13.66	–13.43

[a]From *Missouri Agricultural Extension Division Guide Sheet 2006.*

market animals. Cattle producers may exercise some control over production costs, but they have little or no control over the price received for the market animals.

Most commercial herds in the United States will not be profitable unless an 80 to 85 percent calf crop is weaned each year for cows and heifers of breeding age in the herd. The goal should be at least a 90 to 95 percent calf crop weaned. Some herds attain this goal year after year through good feeding and management practices.

Several factors may be responsible for a low percentage of calf crop weaned. These include the use of a bull of low fertility, disease in the herd such as brucellosis and leptospirosis, extremely thin condition of some cows, vitamin and phosphorus deficiencies, and structural or physiological defects which prevent cows and heifers from becoming pregnant.

Adequate nutrients to ensure that cows will remain in thrifty condition throughout the year may be supplied by using a well-planned feeding and pasture program. These programs include the supplying, in season, of a good-quality pasture that includes a good grass-legume mixture. Fertilization of pastures will increase the per acre carrying capacity in regions of adequate rainfall. The rotation of pastures to prevent overgrazing keeps grazing cows in good condition, in addition to keeping pastures more productive. Overgrazing results in thin cattle and poor pasture growth.

Mineral requirements may be met by supplying mineral mixtures, in block or granular form, which include calcium, phosphorus, and salt. Vitamin A requirements are usually adequate when cows are fed good-quality green pasture, hay, or silage. Stabilized vitamin A concentrates may be purchased and added to protein supplements or rations if needed. The addition of 500,000 International Units of stabilized vitamin A per pound of salt should meet the cow's requirements for this vitamin. The mixture must be kept dry to prevent the destruction of vitamin A, and a fresh mixture should be supplied at 10- to 14-day intervals.

Enough bulls should be run with the cow herd to ensure that each cow or heifer will be bred when she comes into heat. The breeding ability of bulls varies widely, but certain recommendations are made for the number of cows per bull when pasture mating is practiced. A yearling bull should be exposed to ten to fifteen cows, a two-year-old bull to twenty-five to thirty cows, and a mature bull to thirty to forty cows. Hand mating will increase these numbers, but it involves the added problem of finding cows when they are in heat. Cows in the breeding pasture should be closely observed during the breeding season to make certain there is not an excessive number of repeat breeding cows. If several cows are repeat breeders, this suggests the bull may be of low fertility or even sterile. A physical examination of the bull and a semen test before the breeding season will give a good indication of his breeding soundness. It is not always a good indication that he will remain fertile throughout the breeding season, however. Close observation of cows to detect repeat breeders is a good method of determining the bull's fertility even while he is in the breeding pastures.

Regular blood tests to detect possible diseases combined with a vaccination program to prevent diseases is a good management practice. The recommended testing and vaccination program to follow may be determined by contacting the extension beef cattle specialist and/or local veterinarian.

A definite breeding season should be decided upon so that calves will be dropped at an appropriate time. This will help ensure adequate care at calving time and will establish the best time that calves may be weaned and marketed.

Sixty to 90 days after the end of the breeding season all cows exposed to the bull should be examined for pregnancy. Cows that are open should be culled to avoid the expense of feeding them during the winter months and until they calve again more than a year later. An average of 5 to 10 percent of the cows pregnancy-tested fail to be with calf. The cause of the failure to conceive may not be determined, but culling open cows will help prevent this problem in the future if it is chronic with certain cows. A pregnancy test, of course, does not account for other losses which occur during fetal life after the pregnancy test is run or from deaths from birth to weaning.

Cull cows bring considerable income to the beef cattle producer.

All things considered, it is probably best to sell cull cows in the fall of the year when they have weaned their calf and have been found to be open after a pregnancy test.

Several facts may be used to determine which cows to cull. As mentioned previously, all open cows and heifers should be culled after the end of the breeding season. Cows with defects due to injury or disease, such as cancer eye, which may affect their future performance should be culled even if they are pregnant. Cows which wean a lighter than average calf should also be culled. Such cows and heifers may be located by a close examination of each cow's record as compared to the herd average. Aged cows declining in performance may also be culled. Their performance as compared to the herd average may also be determined by an examination of their lifetime and most recent production records. Cows that remain above average in production even to an old age could remain in the herd. When aged cows go through the winter and the lactation period in very poor condition, they should be culled.

Bulls with good progeny, good conformation, and a good growth rate and which are highly fertile should be used for breeding as long as possible. Those with poorer than average progeny records and defects due to injuries which lower their reproductive capacity should be culled and replaced by a young herd bull prospect. In pure-bred herds where an outstanding sire is located through progeny tests, some linebreeding may be practiced, but it is probably best not to mate a sire to his own daughters. Inbreeding should be avoided in commercial herds.

The breeding herd is more productive when external and internal parasites are controlled. They may be controlled during the grazing season by means of sprays, dust bags, or back rubbers containing an appropriate insecticide. Lice should be controlled, especially during the winter months, as should internal parasites such as grubs and worms. These will be discussed in more detail in a later chapter.

Care at parturition. Details of parturition were discussed in Chapter 5. It was pointed out that parturition is a very critical period because difficult parturitions can result in the death of the calf and sometimes in the death of both the calf and its mother. The time of parturition cannot be predicted exactly several days in advance, but it can be estimated within a few hours when certain events occur. Parturition may occur at any hour of the day or night. This is the reason pregnant cows in the herd should be observed at least two or three times during each 24-hour period. Pregnant heifers expecting their first calf need to be observed every 4 to 5 hours throughout the day and night if parturition appears to be close at hand.

When calving time approaches, the cow becomes nervous, her udder is greatly distended, and she seeks seclusion from the rest of the herd. When she leaves the herd, she should be observed every hour until calving is complete. Older cows seldom have trouble at calving, but first-calf heifers do, and heifers must be observed much more closely than older cows as calving time approaches.

In the first stages of normal calving, the water bag appears. This is soon followed by the appearance of the two forefeet, with the nose and head on top of the legs. If the delivery makes no further progress for an hour or more after the forelegs can be seen, assistance may be needed. Placing obstetrical chains on each foreleg and pulling one foreleg ahead of the other makes delivery easier. Sometimes placing a halter or cord around the head and nose of the calf and pulling on the head and forelegs at the same time aids delivery. The head and shoulders may pass through the pelvic opening, but the hips may become "locked" in the pelvic cavity. Hip locks can often be eased if the calf is rotated so that one stifle of the leg is at the top and one at the bottom of the pelvic opening. The calf should be pulled downward toward the cow's hocks when pressure is applied. Difficult delivery is often due to abnormal presentations such as the head or one or both legs turned back. These must be straightened to a normal position before delivery is possible. It is a good rule to keep one's hands out of the uterus of the cow, but if it is necessary to insert the hands or arms into the uterus, one should be certain they are washed with an antiseptic solution. If the delivery is very difficult and help results in no progress, a veterinarian should be contacted at once. A Caesarean operation may be necessary to deliver the calf.

Occasionally calves are born hind feet first. This is called a *posterior presentation.* Delivery must be rapid in such cases, because when the navel cord is ruptured, the calf tries to breathe and is smothered. As long as the navel cord is intact, the calf receives oxygen from the mother through the umbilical cord and will remain alive.

As soon as the calf is delivered, mucus should be removed from the calf's mouth and nostrils so it can breathe more easily. Lifting the calf by the hind legs and swinging it back and forth a few times aids in starting the breathing process and clears mucus from the nose and throat.

Calves born in extremely cold weather often become chilled. They may die unless they are warmed. The calf may be warmed by placing it in warm water, covering all of the body but the head. As soon as the calf revives, it should be dried off and returned to its mother.

Colostrum should be given to the newborn calf as soon as possible. This is necessary because the calf obtains antibodies from the colostrum which aid it in resisting infections. Most calves will die if they do not get colostrum. Vigorous calves soon stand and nurse and obtain the required colostrum in this way. Weak calves that cannot stand and nurse should be given colostrum through the use of a bottle or a stomach tube. It is helpful to have a store of frozen colostrum on hand during the calving season. This may be obtained from dairy or older beef cows in the herd that have previously calved. If colostrum is not available, the calf may be injected with blood serum obtained from the blood of its mother, or it may receive a blood transfusion from her. Antibiotics and vitamins A, D, and E should be injected into calves who have received little or no colostrum.

Calves dropped on pasture usually have less disease problems than those born in confinement. All newborn calves should be frequently observed to make certain they are nursing normally and do not develop scours (diarrhea). Early detection and treatment of scours are important if the calf is to recover and make normal gains thereafter. Calf scours and its treatment will be discussed in a later chapter.

Cows should shed the placental membranes within 24 hours after calving. If they do not, treatment may be necessary. Uterine capsules (or boluses) may be inserted into the uterus to cause the placenta to separate from the uterus and to combat possible associated infections.

12.4.3   Management of Suckling Calves

Weanling calves are the main source of income for the commercial producer. Several points of management are of importance.

Castration.   The commercial cattle producer should castrate all male calves to be sold as feeders before they are 3 months of age. Although bull calves gain faster than steer calves, bull calves that are castrated late will lose enough weight during the operation and recovery period that they will weigh about the same as steer calves of the same age.

Several methods of castration may be used. If the knife is used, there is no doubt that castration will be successful, but the possibility of excessive bleeding always exists when the knife is used on older calves. All calves should be watched closely for a few days to make certain bleeding is not excessive. Instruments that cut and crush the spermatic cords at the same time are available and reduce the amount of bleeding when properly used. The scrotum, instruments, and the hands of the operator should be washed with an antiseptic solution before the operation to avoid infections. When larger bull calves and yearlings are castrated with the knife, it is sometimes advisable to give an intramuscular injection of antibiotics in addition to using an antiseptic solution locally. In fly season, a good repellent should be applied to the wound.

Bloodless methods of castration may be used by persons familiar with their use. Burdizzo forceps (clamps) may be used to sever both spermatic cords without any cutting (Figure 12.4). In the hands of unskilled operators the cord is sometimes not completely severed, and the calf may become "staggy" or bullish in appearance unless reclamping is again done as soon as it is discovered the first effort was not successful. The elastrator, which consists of placing a rubber band around the upper part of the scrotum, eventually causes the scrotum containing the testicles to dry up and slough off. The advantage of these methods, in addition to being bloodless, is that

Figure 12.4.   Burdizzo forceps used for bloodless castration.

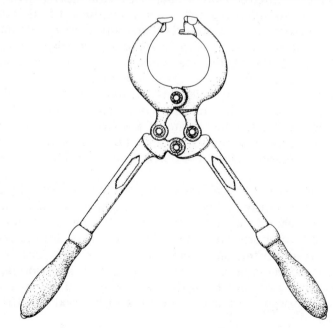

they can be used for castration during the fly season. A longer period for recovery is required when such methods are used as compared to using the knife.

Dehorning.   Dehorned cattle have several advantages over horned cattle. They sell for more as feeders and are easier to handle. They are also less aggressive in the feedlot. Dehorned cattle fed together have fewer bruises at slaughter than horned cattle.

The presence or absence of horns is a genetic trait. In the British breeds such as the Angus, Hereford, and Shorthorn, polledness is dominant to horns, with a single pair of genes involved in most cases. In Charolais and some of the other breeds such as the Santa Gertrudis, Brangus, and Simmental, more pairs of genes may be involved. More research needs to be done on the inheritance of horns and polledness in these breeds. Some breeds may possess the African horn gene (A_f), which is dominant to polledness in males and recessive in females. It is present at a low frequency in some of these breeds.

Scurs are rudimentary horns not attached to the skull bones. They vary in size and length. Any horny projection continuous with the head bones should probably be classified as a horn. Scurs are due to a separate pair of genes (not the horned gene), at least in the British breeds. Scurs are dominant to polledness in males and recessive in females. Horned individuals may carry the scurred gene, which cannot be detected unless a progeny test is conducted to determine if this gene is present. In some breeds scurs appear to be dominant to polledness in both sexes, at least in some individuals. Scurs do not have all the disadvantages of long, sharp horns, but sometimes they are unsightly and are undesirable for this reason.

Horned cattle may have their horns removed (dehorned) at almost any age by mechanical means. Young calves may be dehorned by using certain commercial chemical preparations. If used strictly according to the manufacturer's recommendations, some newer compounds on the market cause less burning to the calf and its mother when applied.

Older cattle may be dehorned by mechanical means by placing them in a head chute. A Barnes-type dehorner is often used to remove the horns; one is shown in Figure 12.5. The dehorner should be placed over the horn, and a portion of the skin around the horn should be cut and lifted out along with the horn. The cut should be deep enough to remove all of the horn. If this is not done, horny projections may grow out from the head as the animal grows older. The Barnes dehorner comes in three sizes to be used on cattle up to 1 year of age.

Figure 12.5.   Dehorning a calf with a Barnes dehorner.

Dehorning tubes can also be used to dehorn calves up to 4 to 5 months of age. Older cattle may be dehorned with the Leavitt-type dehorner or a saw. Often when older cattle are dehorned, bleeding may be excessive unless the arteries are tied or pulled out with a pair of forceps. If the cut is made very close to the base of the horn, the arteries are exposed and easily removed or tied. Instruments used for dehorning should be dipped in an antiseptic solution before use, and a fly repellent should be placed on the wound during the fly season. If older calves and cattle are not cared for properly when dehorned, a few may actually bleed to death.

Vaccinations.   Vaccinations may be administered as a preventative for many diseases. The vaccination program to follow should be discussed with the local veterinarian and/or extension specialist and planned in advance. The kinds and number of vaccinations given may depend on conditions under which the cattle are kept and the part of the country where the farm or ranch is located. Diseases and vaccination programs will be discussed in greater detail in a later chapter. All calves in any herd should be vaccinated for blackleg and malignant edema regardless of the conditions within the herd. Such vaccinations are cheap and easy to administer.

Creep feeding.   Calves nursing their mothers are sometimes given access to a grain mixture. This mixture is placed in an enclosure to which calves have access but cows do not. The calves can creep through a small opening to the grain, and this is where the term creep feeding originated.

The calf receives less than one half of the nutrients it needs for maximum growth from its mother's milk after it is about 3 months of age. If abundant green grass is available, the calf obtains added nutrients from the grass it consumes. To weaning, creep-fed calves, however, will usually gain 30 to 60 pounds more than calves not creep-fed. It is estimated that each pound of gain on a creep ration will require 7 to 12 pounds of feed.

Whether or not it will pay to creep-feed calves depends on whether or not the added gains from the creep feed will pay for the cost of gains made. Creep feeding may be more profitable for calves from first-calf heifers or aged cows or for all calves who are dropped in the fall and early winter. Creep feeding may also pay when drought conditions are extreme. In years when grass is plentiful throughout the grazing season, creep feeding may not be profitable in the commercial herd.

Several creep rations may be fed. One that is recommended consists of six parts of cracked shelled corn, three parts of whole oats, and one part soybean meal; replacing 10 percent of the corn with molasses often increases consumption. Another recommended ration consists of two parts of whole shelled corn and one part oats or oats alone.

When calves are placed in the feedlot immediately after weaning, creep-fed calves will go on a full feed with a minimum of delay. They will usually stand the stress of weaning better than non-creep-fed calves. Large framed, thin non-creep-fed calves will make compensatory gains, however, during the first part of the full-feeding period. These compensatory gains are usually cheap gains.

Weaning calves.   The right time to wean calves varies with different farms. The milk production of the beef cow usually reaches its peak 2 or 3 months after calving and gradually declines thereafter. During the sixth month of lactation milk production of the beef cow has declined to about one third of her peak milk production. Weaning any time after 4 or 5 months of lactation is possible, but some producers prefer to leave calves on the cows until sale time. Even though the calf may not be obtaining much milk from its mother at that time, it remains quiet and retains its "bloom" until taken off the cow.

Under certain conditions early weaning of calves may be desirable. Early weaning of calves from first-calf heifers may cause them to rebreed earlier than if they continue to lactate. This is especially true if they are heavy milkers and if grazing conditions are poor. Early weaning of calves may also be considered in aged cows under certain conditions. It may be considered for all cows in the herd under extreme drought conditions.

Calves are weaned by simply separating them from their mothers. Weaned calves should be placed in a pasture with very good fences because they will walk the fences for 2 or 3 days, seeking a weak spot in the fence through which they can pass to return to their mothers. Both cows and calves will bawl periodically for 2 or 3 days after the calves are weaned, and it is quite disconcerting to people at night when they are trying to sleep if weaned calves are kept near the farm or ranch home. The bawling gradually subsides and usually ceases by the third or fourth day after the calves are weaned.

Calves on creep feed during the lactation period suffer little weight loss at weaning. Even calves who have not been creep-fed will suffer only slight weight losses at this time if offered good-quality feeds or grass. Commercial feeds containing antibiotics are available for calves at weaning to reduce sickness. Dehorning, castration, and vaccination should be done during the early to middle part of lactation and not at weaning time. Combining all these operations with weaning adds to the stress the calves are subjected to and may increase weight losses and cases of sickness.

12.5   Care and Management of Lactating Cows

Lactating beef cows need little attention while grazing green, lush pastures. They should be fed plenty of good-quality roughage in the winter months or during extreme drought. Under these conditions range cubes containing protein may be fed. Cows should have access at all times to water and a mineral mixture containing salt, calcium, and phosphorus.

Immediately following parturition the udder of the cow is swollen and distended with milk. The swelling usually disappears after a few days, and the udder regresses to a more normal size. Part of this distension is due to an accumulation of water (edema) in the mammary tissues. In heavy milking cows the young calf cannot utilize all of its mother's milk, and one or more teats may not be nursed and will remain full and distended. Some cattle producers prefer not to milk out the distended quarters. Eventually the calf will nurse all teats as it grows older and larger and has a larger stomach capacity. One danger of not milking out the teats is that they may become so large on repeated lactation periods with the increasing age of the cow that very young calves cannot or will not nurse them. It is not known if failing to milk distended quarters early in lactation causes permanent damage to the milk production of that quarter.

STUDY QUESTIONS

1. How much does environment affect most economic traits in beef cattle?

2. What are some of the ways to identify individual beef animals?

3. What is freeze branding? What is the principle involved in the use of this method of identification?

4. Why is it important to keep records on individual beef animals?

5. What records should be kept on the herd?

6. What is the formula recommended for adjusting weaning weights to a 205-day basis?

7. Why should weaning weight adjustment factors be used? Why doesn't the beef cattle producer weigh each calf as it reaches 205 days instead of using adjustment factors?

8. Assume you have a calf that had a birth weight of 80 pounds and weighed 600 pounds at 230 days of age. What would be its adjusted 205-day weight?

9. Under what conditions would it be necessary to correct a calf's weaning weight for age of dam and sex?

10. Assume you have a bull calf out of a three-year-old cow that weighed 500 pounds at 210 days of age. What would be the 205-day weaning weight of this calf adjusted for age of dam and to a steer basis?

11. Assume that cow *A* has produced three calves whose average 205-day weight adjusted for age of dam and to a steer basis is 480 pounds. Cow *B* has produced eight calves whose average 205-day weight adjusted for age of dam and to a steer basis was 450 pounds. Which cow has the better record based on her MPPA?

12. What is the repeatability (*r*) of conformation score for a cow that has produced five calves if the repeatability of one record is 0.30?

13. Why does the repeatability of a trait increase as the number of records increases?

14. The actual weight of a bull calf at 392 days of age was 1,250 pounds, and his adjusted 205-day weight was 465 pounds. What is his adjusted 365-day weight?

15. What is the quickest way to make cattle wild?

16. Describe the advantages and disadvantages of (a) spring calves, (b) fall calves, and (c) winter calves. Is the time of calving the same for all parts of the United States? Explain.

17. What factors may be responsible for a low calf crop?

18. Outline management practices to follow in getting a high-percentage calf crop year after year.

19. Why is a high-percentage calf crop so important in profitable beef production?

20. What factors should be taken into account when cows are to be culled?

21. On the average, which will remain longer in the herd, a cow or a bull? Why?

22. What is the normal presentation of a calf at birth?

23. What is meant by a *posterior presentation*? What are the dangers involved in such a presentation?

24. Why is it necessary for the calf to receive colostrum shortly after birth? What is likely to happen if it doesn't?

25. What is frozen colostrum? Suppose no frozen colostrum is available and the cow giving birth to a calf has no milk. What should be done to save her calf?

26. Discuss the advantages and disadvantages of various methods of castration.

27. What is meant by bloodless castration? When is such a method used and why?

28. Discuss the inheritance of polledness and scurs in beef cattle. What is a scur?

29. List some of the methods available for dehorning cattle.

30. How does one prevent excessive bleeding when older cattle are dehorned?

31. What is meant by creep feeding? Under what conditions might creep feeding pay?

32. Under what conditions should early weaning of calves be considered?

33. Can a cow give too much milk for her calf? Explain.

CHAPTER 13

fitting and showing beef animals

Show animals are just that — they are animals which are fed, fattened, and groomed to appear their best and to be shown in competition with other similarly prepared animals. They compete with other animals of approximately the same age and breed to determine which one better fits the ideal of a particular breed or class. In recent years there has been some criticism of the show ring, but nothing has been developed to completely replace it. The criticisms have dealt mostly with the failure of the appearance of a beef animal to be closely correlated with individual performance. Some criticism has also been leveled at a few individuals who use unfair practices to fit and show their animals.

Efforts to avoid criticisms of showing beef cattle have resulted in some changes in livestock shows. Size and scale of animals of a certain age give some indication of the individual's growth rate to that age. This has been emphasized. Research in which the live animals are judged and then their placings compared again for carcass quality and quantity when slaughtered has helped to identify meatier animals on the hoof and to train breeders to evaluate better the live animal. Although some progress has been made in locating meaty animals from their visual appearance, the best way to get actual carcass quantity and quality for an individual is to measure these after slaughter. This can be done, of course, only with finished animals and not with breeding animals. Estimates of the amount of lean or fat in the live animal by means of low-level radiation (40K) counters and the sonoray have been quite helpful in many instances. These measurements together with eye appraisal have greatly improved the ability of judges and breeders to identify the meaty individuals while they are still alive.

The show ring has made other improvements in recent years. These improvements include more get-of-sire classes and evaluations of the carcass quantity and quality of live animals as compared to these traits after slaughter. Much of the accuracy of evaluating the merits of the live animal by eye appraisal depends on the competency of the person doing the judging. Eye appraisal, of course, is of little value in predicting the future performance of an individual in many instances.

Livestock shows have some good points. They identify the kind and type of individual that should be selected at a particular time. They also serve as a good means of advertising and promoting breeding stock. In addition, livestock shows allow breeders to compare their animals with those of other breeders. Finally, fitting animals for show and actually showing them is a game in which one breeder (or exhibitor) tests his or her skill with that of others in competition. This is valuable experience, especially for young boys and girls who are interested in the livestock business.

Methods of showing animals are subject to continual change and vary on a regional basis. It is advisable to attend major shows and/or contact extension agents for current styles or fads in a particular region.

13.1   Body and Carcass Parts of the Beef Animal

To the beginner showing livestock in shows, the first requirement is to become familiar with the various parts of the live animal and the carcass. The body parts of a live beef animal are shown in Figure 13.1. By knowing the various body parts of an animal, a comparison can be made among different individuals, especially in body parts that have the highest economic value. It also helps exhibitors to select for strong points in the animals they plan to fit and show and to talk about the various parts on a knowledgeable basis.

The various parts of the beef carcass are shown in Figure 13.2.

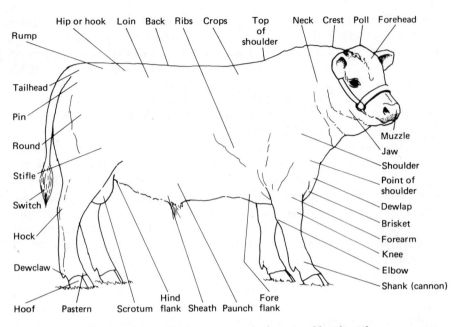

Figure 13.1.   Various anatomical parts of beef cattle.

The loin, rib, round, and rump are the parts of greatest value. Emphasis should be placed on superiority in these parts when selecting and showing superior animals.

The U.S. Department of Agriculture (U.S.D.A.) has developed quality grades for beef carcasses. The showperson should also become familiar with these grades. The reason for establishing such grades is to help identify the desired eating qualities of beef. These include juiciness, tenderness, and flavor.

The conformation of the carcass and the amount of marbling largely determine quality grades. Marbling refers to the deposits of fat between the lean fibers which are scattered throughout the muscles. The quality grades, beginning with those that are most desirable, include (1) prime, (2) choice, (3) good, and (4) standard (if under 48 months of age) or commercial (if over 48 months of age).

U.S.D.A. carcass yield grades have been developed to estimate the percentage of salable beef in the carcass. Yield grades are designated by numbers 1 through 5, with 1 being the highest grade. The chief factors determining the yield of the carcass include the amount of fat that must be trimmed from the carcass in preparing cuts of beef and the thickness and fullness of the muscles. Many pounds

	Saleable Beef–lbs	Other lbs
CHUCK *164.8 lbs (26.8% of total carcass)*		
Blade pot roast	59.3	
Stew or ground beef	32.1	
Arm pot roast	22.3	
Cross rib pot roast	10.7	
Boston cut	9.9	
Fat and bone		30.5
TOTAL	134.3 lbs	30.5 lbs

	Saleable Beef–lbs	Other lbs
RIB *59.0 lbs (9.6% of total carcass)*		
Standing rib roast	24.2	
Rib steak	12.4	
Short ribs	4.7	
Braising beef	2.7	
Ground beef	3.5	
Fat and bone		11.5
TOTAL	47.5 lbs	11.5 lbs

	Saleable Beef–lbs	Other lbs
BRISKET *23.4 lbs (3.8% of total carcass)*		
Boneless	9.4	
Fat and bone		14.0
TOTAL	9.4 lbs	14.0 lbs

SHANK *19.1 lbs (3.1% of total carcass)*

	Saleable Beef–lbs	Other lbs
LOIN *105.8 lbs (17.2% of total carcass)*		
Porterhouse steak	18.7	
T-bone steak	9.5	
Club steak	5.2	
Sirloin steak	41.4	
Ground beef	2.9	
Fat and bone		28.1
TOTAL	77.7 lbs	28.1 lbs

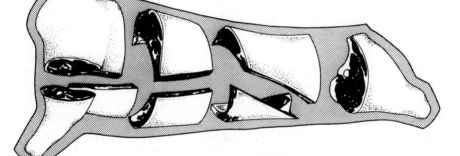

	Saleable Beef–lbs	Other lbs
SHORT PLATE *51.0 lbs (8.3% of total carcass)*		
Plate, stew, short ribs	40.8	
Fat and bone		10.2
TOTAL	40.8 lbs	10.2 lbs

	Saleable Beef–lbs	Other lbs
FLANK *32.0 lbs (5.2% of total carcass)*		
Flank	3.2	
Ground beef	12.6	
Fat		16.2
TOTAL	15.8 lbs	16.2 lbs

	Saleable Beef–lbs	Other lbs
MISC. *22.1 lbs (3.6% of total carcass)*		
Kidney, hanging tender	3.6	
Fat, suet, cutting losses		18.5
TOTAL	3.6 lbs	18.5 lbs

	Saleable Beef–lbs	Other lbs
ROUND *137.8 lbs (22.4% of total carcass)*		
Top round (inside)	21.0	
Bottom round (outside)	20.3	
Tip	13.1	
Stew	8.3	
Rump	4.8	
Kabobs or cubes	2.1	
Ground beef	14.2	
Fat and bone		54.0
TOTAL	83.8 lbs	54.0 lbs

SUMMARY
(1000 lb choice steer)

Dresses out 61.5%	615 lbs
Less fat, bone, and loss	183 lbs
Saleable beef	432 lbs

Figure 13.2. Parts of the beef carcass. (Courtesy National Livestock and Meat Board)

of fat must be trimmed from the retail cuts of extremely fat animals. This is undesirable because the fat (tallow) sells for much less per pound than the lean meat and is discriminated against by the consumer.

A detailed description of carcass and yield grades for beef animals may be found in *U.S.D.A. Bulletin S.R.A. 112*, official U.S. standards for grades of slaughter cattle, and the *Meat Evaluation Handbook*, published by the National Livestock and Meat Board, 444 North Michigan Avenue, Chicago, Ill. 60611.

13.2   Selection of Prospective Show Animals

Two general classes of beef animals are shown in competition: animals which have been finished for slaughter, which includes mostly steers, and breeding animals, which includes both heifers and bulls.

13.2.1   Selecting the Steer Calf

One of the first decisions that must be made in selecting a steer calf to fit for show is the breed from which it should be selected. The decision of what breed to select from will depend on the likes and dislikes of the person who plans to do the showing or on the excellence of available animals. Usually a calf should be selected from the showperson's favorite breed, but if only mediocre animals are available in this breed, a calf with the greatest show potential should be selected regardless of breed. A calf from one of the old, established breeds such as the Angus or Hereford may result in stiffer competition when animals are shown because so many will be shown. In recent years even crossbred steers have been shown in competition, and one of these might be considered if an excellent one is found.

Calves to be shown will often be selected at weaning time or in the fall of the year. The weight of the calf when selected is important because it should weigh 950 to 1,200 pounds when shown or sold. If a calf weighing 400 to 500 pounds is selected in the fall, he will usually gain 2.0 to 2.5 pounds per day until show time. The time one expects the calf to be shown should be considered ahead of time, if possible, so that it will be properly finished and will produce a choice-grade carcass when sold. Some very growthy calves may be too heavy when shown unless they are fed and cared for properly.

A steer calf should be selected from a herd for which performance records have been kept and used as a basis of selection for a number of years. He should be selected from parents who also possessed good performance records and have produced good-performing progeny. The performance records of an individual's half-sibs may also be considered. A superior-performing individual with superior-performing relatives is more likely to perform better than one who has mediocre- or inferior-performing relatives.

A calf that will produce a steer of modern type when shown should be selected. Emphasis in recent years has been toward growthy, long-bodied, large-framed animals with heavy muscling. The legs should be straight, with heavy but not coarse bones. A tall, narrow, shallow-bodied calf should be avoided, as should one who is thick, compact, short, low-set, and deep-bodied, with a big paunch or middle. Attendance at one or more of the major shows, when possible, will help fix in mind the desired show type in a particular breed.

13.2.2   Selection of Potential Breeding Animals for Show

The selection of potential breeding animals whether for the show ring or for production purposes adds one more objective to those considered in selecting steers for show. The added consideration is potential reproductive efficiency, which is probably the most important single trait in beef cattle production. Animals to be shown in breeding classes should have strong indications of high fertility.

Heifer calves should be selected from a herd that has high reproductive efficiency. This is indicated by a high calving percentage of 90 to 95 percent or more year after year. The herd should also have a short calving season of 60 days or less and a calving interval of 12 months or less.

Inspection of the heifer or her close female relatives gives some idea of her potential reproductive efficiency. Lean, smooth-muscled heifers showing a high degree of femininity should be selected. The same characteristics should be present in her close female relatives. Coarse, heavy-muscled, masculine-appearing females should be avoided because they tend to be less fertile.

Modern-type heifers should be growthy, tall, trim, long-bodied individuals with a high degree of trim muscling. They should stand on strong legs and feet. They should also have good performance records, weighing 500 pounds or more at 205 days of age and 700 to 750 pounds or more at 365 days of age. Short, dumpy, heavily muscled heifers should be avoided. Heifers of ideal type should be trim in the flank, twist, and brisket. They should also possess good spring and depth of rib, with a large capacity and roominess of the middle. They should not be paunchy, however.

Selection of a bull calf should take into consideration all of the characteristics mentioned in the selection of heifers that apply to both sexes. The bull should be especially growthy, large-framed, long-bodied, well-muscled, and trim in his muscling. He should be masculine in appearance and structurally sound in the legs, feet, and external genitalia. He should weigh 550 to 600 pounds or more at 205 days of age and should weigh 1,000 pounds or more at 365 days of age. He should also possess a well-developed scrotum which contains two normal-sized and well-developed testicles.

13.3   Feeding Show Animals

A few years ago bull and heifer beef calves were taken from their mothers and placed on a nurse cow which would give large amounts of milk. Nurse cows were usually dairy cows, many of which were Holstein breeding. The purpose was to grow and fatten the calves as quickly as possible. Today the use of nurse cows for show calves is looked upon with disfavor, because this does not reflect the milk producing and mothering ability of the beef cow that was the mother of the calf. No doubt this is the proper viewpoint. Furthermore, with emphasis on growth rate and meatiness today, a calf should make it to show condition and a respectable yearling weight without a nurse cow. If it does not, it is not a desirable show prospect.

Young beef cattle require a ration that includes 65 to 75 percent energy or total digestible nutrients and 10 to 14 percent crude protein on a dry matter basis. In addition, the ration should contain 0.2 to 0.6 percent calcium, 0.2 to 0.4 percent phosphorus, and 0.25 percent salt. Vitamin A should be added in the stabilized form to supply 1,000 International Units per pound of ration. Rations composed of high-energy grains and dry roughages together with protein, mineral, and vitamin supplements will meet these requirements. Rations for show animals may be more expensive than the usual rations used for fattening cattle. Linseed meal is sometimes added to the ration to give the hair coat a glossier appearance. This protein source may be somewhat higher in price than other high-protein supplements. Extra bulk may sometimes be introduced into a ration to keep a steer from gaining too fast and becoming too fat by show time. The addition of bulk feeds may increase the costs of a pound of gain because more feed is required per pound of gain.

High-energy grains such as corn, barley, oats, and milo should be included in a ration for developing show cattle. The one used often depends on the cost, the availability, and the desires of the showperson. These high-energy grains are usually high in phosphorus and low in calcium. Grains are usually fed rolled, crimped, or coarsely ground because finely ground grains may cause digestive disturbances. Excessive amounts of high-energy grains in the ration may lead to founder and digestive upsets of the show animal.

Several different protein feeds may be added to the ration to meet protein requirements. Vegetable proteins such as soybean meal, cottonseed meal, and linseed meal are most commonly used. Generally, they contain from 40 to 55 percent crude protein. Wheat bran contains 15 to 18 percent crude protein, but it is usually added to the ration to increase its bulk and not its protein content. Wheat bran also has laxative qualities, which are often desirable in a ration. Urea may be fed to supply up to one third of the protein requirements for the sake of economy, but excessive amounts should be avoided. Excessive amounts of urea in the ration may cause ammonia toxicity and may slightly decrease the rate and efficiency of gains even if no toxicity results.

Mineral supplements may be supplied by means of a commercial mixture or by one mixed at home. Ground limestone contains 34 to 38 percent calcium and is a good source of this mineral. Dicalcium phosphate and steamed bone meal which contain 20 to 30 percent calcium also contain 14 to 18 percent phosphorus. Some trace mineral salts supply salt and traces of iodine, cobalt, copper, iron, manganese, and zinc. Magnesium and sulfur may be supplied in the form of magnesium oxide and calcium sulfate.

Dry roughages may be supplied from several sources. Good-quality legume hays contain 55 to 60 percent TDN (total digestible nutrients) and 12 to 20 percent crude protein. Grass hays are usually lower in TDN and crude protein than legumes. Both tend to be high in calcium and low in phosphorus. Green, sun-cured grass and legume hays should contain large amounts of vitamins A and D. Bulky feeds such as cottonseed hulls, ground corn cobs, and wheat or oat straw are low in TDN, crude protein, and minerals. They may be included in a ration to increase its bulk rather than its nutritive content.

Corn silage contains about 70 percent TDN on a dry basis. It is low in crude protein and minerals but usually high in vitamin A content. It is often used as a roughage in rations of cattle of all classes and ages. Silage from grasses or legumes is high in water content but has about the same nutrient content as hay made from the same materials.

Beet pulp may be fed dry or soaked in water to increase the bulk of the ration. It also serves as a good appetizer. Molasses, fed in either the liquid or dry form, is also a good appetizer and is high in energy.

Adequate quantities of water should be kept before the animals at all times. An 800- to 900-pound animal will drink 8 to 10 gallons of water per day in warm weather.

Many different rations may be fed to show cattle. The feeder may prefer one over another, or certain feeds may be available in larger quantities and cheaper in some areas. The inexperienced feeder may gain valuable information by consulting an experienced feeder as to the ration he or she has had the most success with over the years.

A ration containing high-energy grains, a protein supplement, minerals, and vitamins may be fed along with hay or roughage that is fed free-choice. Some feeders prefer to grind the hay and mix it with the grain to form a complete ration. For such rations the amount of grain should not exceed 80 percent of the ration by weight, with the remainder of the ration some form of roughage. Feeders with considerable experience may successfully feed an all-concentrate ration, but when some roughage is fed, it will help avoid some management problems. Grains such as ground whole ear corn are often included in the ration to add bulk. Sample rations for feeding show cattle are given in Table 13-1.

A simple mineral mixture to feed free-choice is one part trace mineralized salt and two parts of either dicalcium phosphate or bone meal. This mixture is fed with legume hay. If grass hay is fed, one part of trace mineralized salt to one part of ground limestone or either dicalcium phosphate or bone meal should be fed.

Show cattle are usually fed twice each day, with stale feed removed from the bunks before each feeding. Stale feed may cause the animals to go off feed.

Certain procedures are recommended in starting animals on feed. They should be given a full feed of hay or silage the first week. The animal should then be started on 2 to 3 pounds of grain per day. The grain should be increased by about 0.5 pound per day and the hay decreased until the animal is on a full feed in 2 or 3 weeks. A young animal (calf) on feed should eat 2 to 2.5 pounds of grain per day per 100 pounds of body weight along with 3 to 5 pounds of hay. Hay may be restricted to about 20 percent of the total ration during the latter part of the feeding period. If an animal goes off feed after being on a full feed, the grain fed per day may be decreased by about

TABLE 13-1   Some rations for growing and finishing beef cattle[a,b]

1. Silage and protein supplement:
   Silage — full fed 30–50
   Protein (44%) .... $1\frac{1}{2}$
   Mineral mix[c] .... Free choice
   Salt .... 1 part
   Dicalcium phosphate .... 1 part

2. Hay and grain:
   Hay (at least $\frac{1}{2}$ legume) 10
   Grain (1 lb/100 lb of body wt.) 4–6
   Protein (44%) .... $\frac{1}{2}$
   Mineral mix[c] .... Free choice
     Salt .... 1 part
     Dicalcium phosphate .... 1 part
     Monosodium phosphate .... 1 part

Finishing Rations

1. Shelled corn-corn silage:
   Ground shelled corn
     (1–$1\frac{1}{2}$ lb/100 of body wt.) ... 8–15
   Protein (44%) .... $1\frac{1}{2}$
   Corn silage .... Full-fed
   Mineral mix[c] .... Free choice
     Limestone .... 1 part
     Salt .... 1 part
     Dicalcium phosphate .... 1 part

2. Shelled corn-corn silage:
   Ground shelled corn .... Full-fed
   Protein (44%) .... $1\frac{1}{2}$
   Corn silage 5–10
   Mineral mix[c] .... Free choice
     Limestone .... 2 parts
     Salt .... 1 part
     Dicalcium phosphate .... 1 part

3. Shelled corn-grass hay:
   Shelled corn .... Full-fed
   Hay (grass) 4–6
   Protein (44%) .... $1\frac{1}{2}$
   Mineral mix[c] .... Free choice
     Limestone .... 2 parts
     Salt .... 1 part
     Dicalcium phosphate .... 1 part

4. Shelled corn-legume hay:
   Ground shelled corn .... Full-fed
   Hay (legume, good-quality) 4–6
   Protein (44%) 1
   Mineral mix[c] .... Free choice

TABLE 13-1　(continued)

Salt	1 part
Dicalcium phosphate	1 part
5. Rations for self-feeding:	
Ground ear corn	1,700 to 1,800 lb
Protein supplement	300 lb of 32% or 200 lb of 44%
Salt	10
Limestone	5 (unless commercial supplement contains these minerals)
Dicalcium phosphate[c]	5 (unless commercial supplement contains these minerals)

[a]Courtesy of Dr. Homer B. Sewell, Department of Animal Husbandry, University of Missouri.
[b]Rations given on pounds of feed per head daily unless otherwise stated.
[c]Use trace mineral salt. You may substitute bonemeal for dicalcium phosphate and tripolyphosphate for monosodium phosphate.

one half and then gradually increased again until the animal is back on a full feed.

13.4　Fitting and Showing

The animal must be of good conformation and quality to win in the show ring, but it must also be properly groomed and trained to make the best impression.

13.4.1　Equipment Needed

Certain equipment is needed to properly train, fit, and show an animal. Included are the following:

1. An adjustable rope halter
2. A leather show halter
3. A scotch comb
4. A round curry comb
5. A grooming comb
6. A rice root brush
7. A scrub brush for washing
8. A water bucket
9. Feed pans

10. A neck rope
11. A pitchfork
12. A broom
13. An electric blower
14. A show stick
15. Coat dressing (a mixture of equal parts of denatured alcohol and mineral oil or one that is commercially prepared)
16. Rags and a sprayer for oiling
17. Saddle soap
18. A hair set
19. Butch wax
20. Towels
21. Clippers
22. Scissors

13.4.2   Training to Lead

It is best to start training a calf to lead when it is young and easy to handle. The first step is to get it accustomed to a halter. This can be done by tying it to a manger or to a stall for a few days. The tie should be made 1 foot or 1.5 feet above the ground, with enough halter length to allow the calf to lie down. The knot used to tie the calf should be one that is easily untied.

The next step is to break the calf to lead. The first few times it is led, another person may be necessary to help move it along. Leading the calf to water or to a manger where the feed is usually kept and where it wants to go should make leading easier. Be certain the calf does not break loose when training it to lead, or this may become a bad habit later on. Lead the calf at least once a week, and preferably more often, during the winter months. Then as show time approaches, lead it 0.5 mile or more each day. This gives added training as well as desirable exercise and favors gentleness. Early training to lead should include actions as close as possible to those that will be followed in the show ring at a later date.

The showperson should walk on the left side of the calf, with the halter rope or lead strap in the right hand. In leading, a short rein should be used, with the extra part of the lead strap or rope rolled into small loops and held in the right hand. The head of the calf should be kept up so its top line is level as it walks. This is the procedure to follow in the show ring; practicing it will get the calf accustomed to following this approach, and it will not be a new procedure for the calf to be exposed to when show time arrives.

When the calf is stopped in the show ring, the lead strap should be changed to the left hand, and the showperson should turn and face the calf. The calf's head should be up, and its feet should be placed squarely under the body, with the back level. It is best to place the front feet on higher ground when possible. The showperson can place the front feet with his or her foot or with a long show stick. It is more appropriate to use the stick to place the hind feet. The calf should be trained to stand in a straight line, with the halter being used as a guide.

It is helpful to show and handle the calf often as if it is in the show ring so it will not be a new experience to it when show time arrives.

13.4.3   Grooming the Hair

A show animal must be clean and the hair as glossy as possible. Repeated washings before show time remove dandruff and keep the skin and hair clean. Washing should begin about the month of May, and the calf should be washed once a month until the last month before show time; then it should be washed and groomed once each week.

Several steps should be followed in grooming the calf's hair. Replace the halter with a chain when the calf is to be washed, because when a rope halter gets wet, it swells and becomes difficult to adjust or remove. The chain to be used should have a snap so it can be adjusted to size and will not be too tight around the neck.

The calf should then be thoroughly soaked in water all over the body, with care taken to keep water out of the ears and eyes. Apply a mild soap or detergent to the hair all over the body of the wet animal. Many showpeople use a special livestock soap for this purpose. Then rub enough to work up a good lather, being certain to lather all parts of the body so all of them can be cleaned. Rubbing with the hand will work up a lather, but some people prefer to use a brush which also cleans the skin. The animal should then be thoroughly rinsed with clean water, making certain all soap is removed from the hair coat and skin. A Scotch comb is then used to comb the hair up and forward in the same motion, combing the hair the same direction all over the body. An electric blower should then be used to completely dry the hair. The hair should be blown in the same direction as it was combed.

Some showpeople prefer to clip the hair on the poll, but others prefer not to clip it and sort of "fluff" it up. If the hair is clipped from the head, it should be clipped when the hair is dry. With steers, some showpeople clip the hair in front of the ears, following an imaginary line from the ear down and from one ear to the other, with everything forward being clipped. Some prefer to clip hair from the ears also; others do not. Only long hair should be clipped so as to give a smooth appearance. Long hair on the dewlap, on the rear flank, and behind the elbows of the front legs is usually clipped. Clip the tail upward, beginning at the twist, and to the point where the tail-head starts to round at the rump.

Blocking the animal consists of clipping the long guard hairs on the coat, blending in the points where hair has been closely clipped. Blocking should be done to emphasize the animal's strong points. Watching how experienced showpeople block their animals may be helpful. The top line should be straight and level as viewed from the side and should be clipped to give a round, muscular shape. A final clipping may be given just before the animal goes into the show ring.

The tail should be fluffed or ratted to form a neat-appearing ball, covering up the space between the twist and the hocks. Some showpeople use a hair set on the tail to help maintain its fluffy appearance. Bulls and more mature animals should have their tails fluffed in a larger ball than younger animals.

Just before entering the show ring, oil should be applied to the hair, using just enough to give it a shiny appearance. Too much oil may cause the hair to become sticky and matted. After oiling, the hair may be brushed up or down, depending on how the animal is to be shown. Oil should not be applied too early before show time because it tends to pick up dust and dirt, which gives the hair a dull appearance. If animals are entered in more than one show, all additives should be washed out of the hair immediately after the show so a good coat of hair is maintained for later shows.

Good sportsmanship should be displayed in the show ring, and the showperson should be polite and display a good temper. If the animal being shown becomes excited, work patiently and quietly with it to bring it under control. When possible, help other show-people. If the animal in front fails to lead properly, give it a nudge with your show stick. When the animal stops, leave about 3 feet between it and the next animal to give the judge room to walk between animals.

The hooves of show cattle must be trimmed and kept in good condition so they can walk correctly in the show ring. Trimming the hooves in the spring and again 3 to 4 weeks before show time is usually sufficient. The hooves should be trimmed before they grow too long, which makes it difficult to trim them back to normal shape. If the trimming of the hooves is delayed until too near show time, the animal may be lame and may not be able to walk normally when in the show ring.

The animal must be tied securely so it cannot move too much when the trimming is being done. This avoids injuries to the animal and makes it possible to trim the hooves more easily. The best method is to secure the animal on a trimming table, which turns it on its side. When such a table is not available, a holding stock or chute can be used. The trimming operation involves several steps.

The dirt and manure should be removed from the bottom of the hoof with a farrier's knife. The next step is to remove the outer wall of the hoof with a pair of hoof nippers until it is level with the bottom of the hoof. The blunt edge of the nippers should be on the outside of the hoof and the cutting edge inside so that the cutting edge cuts through the hoof to the blunt edge; this helps to avoid cutting too deep.

The excess hoof growth on the bottom of the hoof may be removed with a hammer and wood chisel. Another way is to use a sander attached to an electric drill. If a chisel is used, cut slowly, being careful not to cut too deep. The beveled edge of the chisel should be out when the cutting is done. The bottom of the hoof should be trimmed so that its entire surface touches the ground at the same time. The outer wall of the hoof should then be smoothed with a rasp or sander, with the latter being attached to an electric drill. The toes of the hooves usually curve inward, and they should be smoothed with a rasp. A hammer and chisel should never be used to chop off the ends of the toes on grown-out hooves. The hooves should not be trimmed so close that the feet become tender and sore.

13.5   Letting down Show Animals

Show animals with good records sooner or later are usually used for breeding purposes. Because show cattle are highly finished, the over-fatness may interfere with future milk production, breeding performance, and the reproductive life span of some females. As a general rule, fertility in bulls does not appear to be affected as adversely by fat as in females.

Letting down refers to the removal of show animals from a full feed after the show is over. The idea is to remove most of the fat and return the animals to average condition by reducing the amount of ration fed. This process is best accomplished by gradually reducing the grain ration over a period of 2 to 4 weeks so that it is eventually eliminated or reduced to the point where small amounts are fed daily. Animals which are let down should be maintained on good-quality roughage or grass with a minimum of grain. If bulls used for breeding adapt poorly to the *let-down* process, they may be given a 10 to 12 percent protein ration at the rate of 0.5 to 1 percent of their body weight.

STUDY QUESTIONS

1. What have been some criticisms of the show ring in recent years?

2. What has been done to avoid some of these criticisms?

3. What are some advantages of the show ring?

4. Name the body parts of the live animal.

5. Name the various parts of the beef carcass. Which parts are of the greatest value?

6. What determines the quality grade of a beef carcass? What are the grades?

7. What are U.S.D.A. carcass yield grades, and why were they developed?

8. What two classes of beef animals are usually shown in competition?

9. Describe points to consider in selecting a steer calf to be prepared for the show.

10. Why should a calf's performance to that point be considered in selecting a show calf? What indications of the calf's future performance are to be considered?

11. What is a *modern-type* calf?

12. Discuss points to consider in selecting a heifer calf for showing in a breeding class.

13. Describe a modern-type heifer and bull calf.

14. Discuss points to be considered in selecting a bull calf to be shown in a breeding class.

15. Should a nurse cow be used in fitting calves for show? Why?

16. Discuss important components of a ration for young show animals. What is a good ration for this purpose?

17. What equipment is needed to fit a calf for the show ring?
18. Describe how a calf may be broken to lead.
19. Describe points to consider in showing a calf.
20. Outline the procedures to follow in grooming the hair of a show animal and preparing it for the show ring.
21. Describe when and how to trim the hooves of a show animal.
22. What is meant by *letting down* a show animal? How may this be done?

CHAPTER 14

methods
of beef production

About 10 percent of the land area of the world is suitable for crop production. Of the remainder, 20 percent consists of pastures and meadows, and another 30 percent is covered by forests. The remaining 40 percent is made up of deserts, rocks, and mountains, only a small portion of which produces some forage for livestock.

In the United States about 17 percent of the land area is used for crop production (Figure 14.1). An additional 30.6 percent is grassland pasture and range, and 8.7 percent is forest land suitable for grazing. Thus, of the 2,264 million acres of land in the United States, only about 56 percent is suitable for any kind of agricultural purposes. Of this land area, livestock harvest forages on about two thirds of the land not suitable for growing crops.

Beef cattle and other livestock are of great value because they convert forages that cannot be directly utilized by humans into highly nutritious foods such as beef and beef products. These animal products are extremely valuable to humans because of their high protein content. Protein is the nutrient in shortest supply on a worldwide basis of all the major nutrients consumed by humans and livestock. Humans require large amounts of good-quality protein for growth, reproduction, repair of tissues, and the proper functioning of

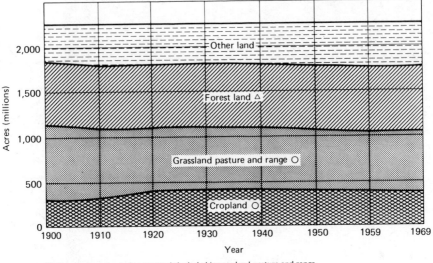

O Cropland used only for pasture is included in grassland pasture and range.
△ Excludes forest land reserved for parks and other special uses of land.

Figure 14.1.   Land use in the United States. (Courtesy U.S.D.A. miscellaneous publication 1290, "Our Land and Water Resources Current and Prospective Supplies and Uses," 1974, p. 2)

the machinery of the body. Proteins in meat and milk are of excellent quality and are efficiently utilized by humans.

The beef cow utilizes forage from millions of acres of land in the world where it is not practical to raise dairy cattle. Dairy cows must be confined to an area where they can be milked twice daily. Daily observation of beef cows is not necessary and often not possible. In the range areas they are often observed only a few times per year. In high-producing dairy cows it is necessary to supply high-energy rations for optimum production. The beef cow survives largely on pasture and forages with little or no grain. In some areas in the southwestern United States the beef cow runs on the range year-round with a minimum of supplementary feeding. Beef cows, for this reason, utilize forages in desert and mountainous country that would be impossible to utilize with cattle in any other way (Figure 14.2). Both beef and dairy cows can utilize by-products from grain crops such as straw, cornstalks, etc., which otherwise would be of little value except to plow back into the soil to supply humus.

Beef cattle production holds a great deal of glamour for many people. This is illustrated by the large numbers of books, movies,

Figure 14.2.   Cattle can harvest grass and browse from semi-arid land not suitable for crop production. This is a view of a western range which has a carrying capacity of eight to twelve head per section of land.

television shows, and songs associated with beef cattle production. Beef is also highly prized as a luxury food. Many people believe that a T-bone steak is the ultimate in luxurious dining. Even the lowly hamburger is a staple food for many persons, not only in the United States but all over the world. All signs point to a great demand for beef and beef products for centuries to come.

In the United States, certain methods of beef production are best adapted to certain parts of the country. (See Table 14-1.) Cow-calf operations are best adapted to regions of large grazing and small grain crop production areas. The fattening of cattle is best suited to areas where grain crops are produced in abundance or where grains for fattening are available nearby in adequate supplies. It is very costly to ship large amounts of grain for long distances.

14.1   Cow-Calf Production

The brood cow is actually a calf factory. How well she is fed and managed has a great deal to do with her efficiency of production. Beef cows may be maintained during their lifetime on a minimum of grain.

TABLE 14-1   Major feeder cattle producing states in the United States[a]

Beef cows and heifers that have calved
(major state feeder cattle producers)

State	Head
Texas	6,236,000
Missouri	2,418,000
Oklahoma	2,186,000
Nebraska	2,032,000
Iowa	1,800,000
Kansas	1,660,000
South Dakota	1,478,000
Montana	1,338,000
Kentucky	1,253,000
Florida	1,212,000
Tennessee	1,155,000
Mississippi	1,100,000
Arkansas	1,080,000
Alabama	1,015,000

Replacement heifers
500 lb and over

State	Head
Texas	900,000
Missouri	360,000
Oklahoma	288,000
Nebraska	260,000
Montana	206,000
South Dakota	204,000
Tennessee	199,000
Arkansas	187,000
Iowa	187,000
Mississippi	184,000
Florida	181,000
Kansas	180,000
Kentucky	180,000
Alabama	138,000

[a]From *Agricultural Statistics*, U.S. Department of Agriculture, Washington, D.C., 1978, p. 302.

14.1.1 Managing Replacement Heifers

Heifer calves for replacement are usually selected when they are weaned in the fall months. In some areas such as the West Coast, however, a different sequence of calving may be followed. Those selected for this purpose should have heavy weaning weights for their age and should be superior in conformation. Heavy weaning weights show that the dam of the calf gave sufficient quantities of milk to stimulate rapid growth in the calf and that the calf has the ability to grow rapidly.

Replacement heifers weaned in the fall must go through their first winter and make sufficient growth to be bred and calve when they are about 2 years of age. This requires that they must come into heat at 12 to 14 months of age. Replacement heifers calved in the fall and weaned in the spring will usually make acceptable gains on pasture and will come into estrus and become pregnant and calve at 2 years of age without additional feed.

Heifers from the British breeds such as the Angus, Hereford, and Shorthorn will usually reach the age of puberty (sexual maturity) when they weigh 600 to 700 pounds and are 350 to 400 days of age. Heifers from some of the larger exotic breeds such as the Charolais and Simmental may be a little slower in reaching the age of puberty and will usually weigh more than heifers of the British breeds at this time. Heifers weaned in the fall should gain 1 to 1.5 pounds per day during the winter months. Twenty-five to 35 pounds of corn silage per day with about 1 pound of a 40 to 45 percent protein supplement and access to a mineral mixture free-choice would be sufficient for such gains. Twelve to 15 pounds of legume hay (or about 50 percent good-quality legume hay) per day plus access to mineral mixture also will be adequate for such gains. If more daily gain is desired, it may be necessary to feed 4 to 6 pounds of grain per day plus about 1 pound of a protein supplement. Under range conditions, range cubes at 1 to 2 pounds per day should be fed.

Research indicates that it is more profitable to breed heifers to produce their first calves at 2 years of age rather than at 3 years. Heifers calving at 2 years of age will produce more pounds of calf at weaning during their lifetime. Heifers calving at 2 years of age are likely to experience more calving difficulties than those calving at 3. If two-year-old heifers are of large size and of moderate fleshing and are bred to a small-type bull (or one for which heifers bred to him in the past have experienced a minimum of difficulty at calving), this can help avoid much of the calving troubles.

Two-year-old heifers nursing a calf are sometimes slow in rebreeding to calve again at 3 years of age. Feeding pregnant heifers in the winter to gain 0.5 to 1 pound per day favors only rebreeding Breeding heifers in the herd to calve 4 weeks earlier than older cows is also helpful in getting them to rebreed to produce an early calf the next calving seasons. If drought conditions exist, calves from two-year-old heifers may be weaned early, which hastens rebreeding, but added feed must be supplied to the calves to keep them making desirable gains. This may not be practical. One other possibility is to allow the calves to nurse just once each day (*Feed Management*, July 1978, p. 17) until the heifers come in heat, after which the calves can again be run with their mothers.

14.1.2   Feeding and Managing Pregnant Cows

Most cows will lose some weight during the winter months when not on pasture. Mature cows in medium condition in the fall of the year can lose 10 to 15 percent of their body weight without adversely affecting their ability to nurse a calf and rebreed the next breeding season while lactating. Pregnant cows which go into the winter in poor condition should be separated from other cows and given extra feed and attention.

Many cows will lose 100 to 140 pounds when they calve. These weight losses include the birth weight of the calf and the weight losses in placental fluids and membranes. Approximately 50 percent of the weight gains of the fetus will be made in the last one third of pregnancy. Some try to reduce the size of the calf at birth to avoid calving difficulties by reducing the feed intake but this is not a recommended practice. When a cow is starved during this period of pregnancy, she draws on nutrient stores in her own bone and muscle tissue to develop the fetus. Depleted body stores of nutrients might slow the rebreeding process of the cow during the next lactation period. Starvation during pregnancy has little effect on birth weights for this reason.

Pregnant beef cows should not be overfed so they become fat. Cows that are overfat at calving time tend to experience more calving difficulty. From the practical standpoint, pregnant cows should be fed to gain 100 to 140 pounds by calving time because in so doing they will neither gain nor lose body weight and should rebreed on schedule.

Pregnant cows will winter in better condition if they are free of internal and external parasites. Methods of control of both will be discussed in more detail in Chapters 22 and 23.

In some areas, especially the southwestern range area of the

United States, cows are wintered on stockpiled grass supplemented by browse. They are also sometimes given range cubes or a high-protein supplement such as cottonseed cake and a mineral mixture. If ranges are not overstocked and grass is accumulated during the growing season in the summer, cows winter very well in these areas when handled in this manner. In portions of the United States farther north it is necessary to supply hay and/or silage for winter feeding. Even in the northern regions, cattle are grazed as long as possible in the fall before being fed hay or silage. In the midwestern states grasses are often stockpiled in summer and fall for winter grazing (Figure 14.3). Tall fescue is one of the grasses utilized in this way. In the southern states cattle are often grazed yearlong, especially if fall-seeded small grains such as oats, wheat, barley, and rye make enough growth.

14.1.3   Feeding and Managing Lactating Cows

Pasture is the major source of nutrients for the brood cow in the spring, summer, and fall months in most of the United States (Figure 14.4). The pasture grazed varies from western ranges, which carry only eight to twelve cows per section, to irrigated land, where

Figure 14.3.   Round bales of fescue on a midwestern farm. (Courtesy Duane Dailey, University of Missouri, Columbia)

Figure 14.4.   Cattle on a southwestern range. These cows have grazed all year on the range with no supplemental feeds except salt.

one cow or more may be grazed per acre. In highly productive areas many pastures include grass and legume mixtures, mostly legumes, or mostly grasses such as fescue. Growing conditions vary so widely in the United States that it is not possible to make general recommendations of grazing methods and pasture mixtures that would be suitable for all areas. It is recommended that the beef cattle producer keep in contact with the extension service in his or her state for up-to-date recommendations. One general recommendation that applies to most farms is to avoid overgrazing because it damages pastures and prevents adequate reproduction and milk production in cows. The weaning weights of calves produced under these conditions are usually light. The rotation of pastures helps prevent overgrazing. In areas of adequate moisture, pastures may be fertilized in either the spring or fall or both. Fertilization greatly increases forage production per acre and produces better performance in beef cattle. The mature beef cow on pasture is producing milk to nourish her calf and is usually nourishing her next calf in her uterus.

Most calves are weaned when they are 200 to 225 days of age. By the time they reach this age the calves are old enough to grow on roughage and other feeds without their mother's milk. This also gives cows who have weaned a calf time to gain in condition and replenish body stores before winter arrives.

Cows grazing on pasture should have access to an ample water supply and a mineral mixture containing salt. If flies are a problem, insecticides for fly control should be used. These may be supplied by means of back rubbers saturated with recommended insecticides, or all cattle may be sprayed periodically. If internal parasites are a problem, treatment once or twice each year is recommended. Keeping livestock healthy will be discussed in detail in Chapters 21, 22, and 23.

Methods of range cattle management vary from one area to another. A definite breeding season should be established for a particular area, and the bulls should be run with the cows and then removed to control the breeding season. The number of cows grazed in any pasture should be controlled to avoid overgrazing. On the range some distribution of cattle may be brought about by placing salt in areas usually not grazed because they are a considerable distance from the water supply. In areas where some winter feeding is necessary, plans should be made in advance for storing hay and other roughages. All other management requirements such as weaning and vaccination of calves should be practiced as for any cow herd.

Nonproductive cows should be culled and sold soon after the calves are weaned. This will include cows who have obvious defects, failed to wean a calf, weaned a light calf, or are aged. Good records on each cow will help locate those who should be culled.

14.1.4   Feeding and Managing Herd Bulls

Herd bulls should be fed and managed similarly to lactating cows. Since bulls are larger than cows as a general rule, they will require a larger allowance of feed. Mature bulls can be maintained largely on pasture and other roughage supplemented with minerals and vitamins recommended for brood cows. They should be kept in medium condition and should not be fat. Twenty-five to 30 pounds of good-quality hay or 60 to 75 pounds of corn silage per day will usually maintain an 1,800- to 2,000-pound bull. Since a mature bull may lose 200 to 300 pounds during the breeding season, a few individuals may need some added grain to bring them to a medium condition after the breeding season closes.

Young bulls are still growing and need extra care and feed, as do growing heifers. Rations fed pregnant heifers should be satisfactory for young bulls, although more grain may be required to obtain satisfactory growth. Young bulls weighing 700 to 950 pounds should gain about 2 pounds per day. A pound of grain per 100 pounds of body weight plus free-choice access to good-quality hay should produce such gains. Young bulls should be fed enough to keep them growing, but they should not be overfat. Bulls that are too thin or too fat will often do a poor job settling cows.

The breeding herd should be closely observed during the breeding season to make certain cows are being successfully bred. This is indicated by very few cows showing estrus during the latter part of the breeding season or some of the cows showing estrus two or more times during the breeding season. Repeat breeding cows indicate the bull may be of low fertility, and he should be replaced by a fertile bull. Using a bull of low fertility will result in late calves the next season and some cows that fail to calve.

14.2 Stocker and Feeder Cattle Production

Stocker cattle are young animals fed and managed to grow rather than fatten. This includes heifers who are intended for replacements in the breeding herd or steers that will be marketed as feeders or will be fattened for market by their owner. The main objective is to produce as much growth and development as possible on pasture or hay and other roughages such as fodder and silage. Feeder cattle are older calves and yearlings carrying some finish and bloom; they are placed on high-energy rations to feed them to slaughter weight in order to take advantage of their extra condition.

Stocker cattle may be grown out on the farm or ranch where they are produced. Usually this procedure involves weanling calves who are weaned in the fall, grown out on a high-roughage ration, run on pasture the next grazing season, and placed in the feedlot for fattening as yearlings. Stocker cattle are seldom kept beyond 1 year before being placed in the feedlot. Some Corn Belt cattle producers purchase calves in the fall and carry them through the winter on surplus roughage, pasture them during the spring and summer, and fatten them in the feedlot. This is an excellent way to utilize fodder and roughage that would not be used in any other way.

In some range states steers are not sold as weanling calves but are left with the herd until they are yearlings. They are then gathered and sold in the fall as yearlings. Some of them are grazed on irrigated pasture or fed other forage so as to make cheap gains before entering the feedlot. Some range calves are purchased and grazed on wheat pasture in Kansas and Oklahoma and to a lesser extent in some other states. In some areas young cattle are purchased in the spring and grazed during the summer. The gains made may be as much as 200 to 300 pounds per animal. They are then sold in the fall to go into the feedlot. Young cattle are usually thin or in medium condition when purchased in the spring and sell for a high price per pound. If prices hold well into the fall, some money is made by the purchase, but if prices are on the decline, little or no profit and even some loss results.

The production of stocker and feeder cattle has certain advantages and disadvantages. The advantages include the following: (1) The production of stocker and feeder cattle is adapted to farms that produce large amounts of surplus roughage and pasture. (2) Stocker cattle can be produced for sale rather quickly, in many instances in a 4- to 6-month period so that two or more groups of cattle may be processed on the same farm per year. (3) Stockers can utilize large quantities of cornstalk and pasture aftermaths which might otherwise be wasted and thus can make cheap gains. (4) Stockers carried through the winter can be marketed before heavy farm work begins in the spring. (5) Stockers do not need specialized care and equipment as do cows and heifers at calving time. (6) Death losses in stockers are usually less than encountered in the cow-calf program. (7) The size of the operation can be readily adjusted to fit the situation on a particular farm. (8) Good-quality stockers are salable at most times during the year.

Some disadvantages are encountered in following a stocker program. These disadvantages include the following: (1) Much capital or credit is often required for such an operation if the cattle must be purchased outright. (2) A great deal of skill in judging, buying, and management is required. (3) The stocker program involves a certain amount of risk because the cattle cannot be held for long periods, as is the case with cows in the cow-calf operation, in order to wait for higher prices.

14.2.1   Stocker and Feeder Steer Grades

Stocker and feeder cattle going to market vary greatly in conformation and finish. For this reason the U.S.D.A. has developed official standards for placing live cattle into various groups. Group 1 divides them according to use such as slaughter or feeder cattle. Group 2 divides them into classes according to sex and condition. Group 3 divides them into grades determined by their apparent relative desirability for a particular use. These standards will be discussed in detail in Chapter 19.

14.2.2 Supplies of Feeder Cattle

Some feeder cattle are available throughout the year, but the greatest supply is in the fall months when most calves are weaned. Some are available in the spring months in areas where fall calves are produced or where spring pastures come early enough to support increased gains a few weeks before sale time. In the Midwest and other summer grazing areas thin cattle are purchased in the spring to go on grass during the grazing months for additional gains. Pregnant cows or cows with small calves are in good demand at this time of year and bring relatively good prices.

14.2.3 Purchasing Feeder Cattle

In recent years there has been a tendency to refer to both stocker and feeder cattle as feeders. For this reason our discussion will mention only feeders, although both may be involved.

Feeder cattle may be purchased by several methods and from several sources. Each method and each source has certain advantages and disadvantages. Some cattle may be purchased through commission firms or dealers. Arrangements for the purchase are usually made several weeks in advance of the expected delivery time by contacting the dealer or commission firm and explaining the kind of cattle the buyer wishes to purchase. The dealer then locates the cattle, purchases them, and arranges for their delivery to the ranch or farm. The actual purchase price is based on the delivery point weight minus a customary shrink of 3 to 4 percent or some other mutually accepted agreement between buyer and seller. The cost of transporting the cattle, insurance costs, feed costs during transit, commission costs, and all other miscellaneous costs are paid by the buyer. Since the buyer may not see the cattle until they are delivered, it is best for him to deal only with reliable firms of good reputation.

Feeder cattle may be purchased at local sale barns or through special feeder sales and auctions. Special feeder sales are held from time to time in local auction barns, and large numbers are available. In these special sales the cattle are classed into uniform lots according to weight and age and sold to the highest bidder. The lots are usually truckload or smaller in size. Feeder calf sales are often sponsored by local groups where different producers pool their calves for sale at one special location. The calves are then placed into groups of approximately equal size, age, and quality. One of the disadvantages of buying at auctions is that bidders often tend to become too enthusiastic in the heat of the bidding and may pay more than the cattle are worth. The stress of transporting, classing, and actually selling the cattle may increase the possibility of sickness in some calves so the buyer must be alert to this possibility and be ready to treat any animals that are ill. One advantage of such sales is that the cattle can be observed personally by the purchaser, and he or she is more likely to find the number and quality of animals desired.

Feeder cattle may be purchased by contract directly from the rancher or farmer. The contract, which is either oral or written, is usually made in the spring or summer months for fall delivery. The contract should include the price to be paid as well as agreements on methods of transportation, method and date of payment, purchase weighing conditions, and which party will pay other costs involved. Sometimes it is possible to contract for the desired number of cattle produced by a large organization with cattle of known past performance. This is usually a situation developed over a long period of time. If large numbers of cattle are desired, it may be necessary to place contracts with several ranchers or farmers. It is best that all contracts be in writing and include agreements on all major points. One advantage of such an arrangement is that cattle do not have to go through the stress of auctions and are not mixed with animals from other farms, ranches, or herds.

Cattle may be purchased in advance directly from farmers or ranchers without a contract. Purchases would be made under these circumstances near the time the cattle are offered for sale. The main disadvantage is that most producers may have already made arrangements to sell their cattle before that time, and the supply available for purchase in this manner may be very small.

14.2.4   Transporting Feeder Cattle

Cattle may be transported in large numbers in two major ways: by rail or by truck. Few cattle are now shipped by rail. Trucks are generally used for short hauls or in cases where cattle cannot be loaded near a railroad. Long hauls such as those involved in moving cattle from the range area to the Midwest may require shipping by rail, although some trucks are used for long hauls.

The law requires that cattle shipped long distances by rail be unloaded so they can be fed, watered, and rested every 28 hours. If a haul requires less than 36 hours, the owner of the cattle can give written permission for a "36-hour release" so the cattle will not have to be unloaded.

Most cattle shrink (lose weight) when being transported from the source of purchase to the final destination. Calves may lose 8 to 12 percent of their weight, whereas older cattle tend to lose only about one half of this amount. Weight losses are due to the excretion of urine and feces as well as actual losses of fleshing and body water. The longer the haul and the hotter the temperature, the greater the weight losses in the cattle. Much of the shrinkage will be regained in a few days when the cattle reach their final destination, rest, and are placed on good grass with plenty of water. This assumes that the cattle remain healthy.

14.2.5   Diseases in Feeder Cattle

Certain diseases may develop in cattle who are associated with the stress of selling and shipping. This illness is often called *shipping fever* and occurs in animals in many cases 10 to 14 days after reaching their destination. Shipping fever appears to be due to a number of infections, and not all sources are known. Recently, certain viral diseases have been identified which are associated with the shipping fever complex. Three of these are parainfluenza-3 (PI-3), infectious bovine rhinotracheitis (IBR) or "red nose," and bovine virus diarrhea (BVD). Affected animals are listless, usually have a high body temperature of 104°F or above, breathe rapidly, cough, and have a nasal discharge. Diarrhea occurs in some of the affected animals. Death losses may be 3 to 5 percent. Most of the surviving animals completely recover, but some may become "poor doers" and make costly gains.

Treatment of these viral diseases is difficult since they do not respond to antibiotics. The use of antibiotics is usually recommended, however, to prevent secondary bacterial infections.

The prevention and control of the spread of the shipping fever complex are important since treatment is not too effective. Vaccinations according to recommended procedures are very important in the prevention and control of these diseases. The purchaser should inquire about the vaccination program followed in the herd from which the cattle are purchased. The vaccination program should be discussed with the local veterinarian and a program decided upon. It is a good practice to isolate newly purchased cattle from other stock on the farm or ranch for at least 3 weeks.

14.2.6 Preconditioning Feeder Cattle

Preconditioning refers to the preparation of feeder cattle by the producer to eliminate as much as possible the shrink, sickness, and loss of time in getting cattle on a full feed. Preconditioning includes vaccinations, worming, treatment for lice and grubs, dehorning, castration, weaning, and starting cattle on feed. Some states have preconditioning certificates which state the practices, treatments, and immunizations used in the preconditioning process. Some feeders will pay 2 to 4 cents more per pound for preconditioned cattle if the preconditioning practices can be verified.

14.2.7 Backgrounding Feeder Cattle

Backgrounding is a term used in reference to the wintering and spring and summer grazing of feeder cattle before they are placed in the feedlot. A cow-calf and backgrounding programs compete with each other because they utilize the same resources on the farm or ranch. Price margins tend to point toward whether the backgrounding of feeder cattle or the cow-calf program is most profitable.

Twenty-five to 40 percent more gain can usually be produced from the same feed by the backgrounding program than by the cow-calf operation, because backgrounding occurs during the period of the calves' lives when they make their most rapid and efficient gains.

Backgrounded calves usually sell for less per pound than weanling calves. Thus, backgrounded calves usually have a negative price per pound margin. If backgrounded calves would sell for as much per pound as weanling calves, farmers and ranchers would tend to shift away from the cow-calf program.

The negative margin of backgrounded calves can still produce some price gains for the cattle producer. For example, if a 400-pound weanling calf sells for 45 cents per pound, the calf would be worth $180 (400 × 0.45). If the calf is backgrounded and as a yearling weighs 700 pounds and sells for 40 cents per pound, the total value would be $280 (700 × 0.40). Thus, the increased value in dollars would be $100 (280 – 180). If the backgrounded calf would sell for as much per pound as it would have brought at weaning, the gain would have been considerably larger ($315 – $180 = $135).

Prices received for weanling calves and backgrounded yearlings plus the cost of producing pasture and forage governs whether a cow-calf or backgrounding operation would be more profitable. Facilities for management as well as the distribution of labor requirements throughout the year may also help determine which program is most profitable. For example, if the cattle producer cannot be available to properly care for cows and calves at calving time, backgrounding calves may be more desirable. More capital may be required in the backgrounding operation, however, if weanling calves have to be purchased and paid for at the time of delivery.

14.3   Finishing or Fattening Cattle

Few cattle were fed grain and fattened for slaughter in the early days of the United States. Usually cattle for slaughter were grass-fat and in some instances were driven many miles to slaughter pens. These slaughter cattle were usually steers 3 and 4 years of age. The rapid development of grain farming operations in the Corn Belt produced an abundance of grain. The surplus grains were eventually used for fattening livestock for slaughter.

The cattle-feeding enterprise has rapidly developed outside the Corn Belt in recent years. Feedlots where thousands of head of cattle are finished each year have been developed. In the Texas and Oklahoma Panhandle region and adjoining states, cattle finishing operations have greatly increased because of the increased production of grain sorghums. In states such as California and Arizona the production of large amounts of barley as well as other crops on irrigated land together with a greatly increased population growth and a good market demand has made these states favorable areas for this purpose. Thousands of cattle are still finished in the Corn Belt, with cattle being fed on many different farms. The extremely large feedlots, however, are found in states such as Texas and California which are outside the Corn Belt. The ten states in the United States where the largest numbers of cattle are fed are shown in Table 14-2. The ten largest stockyards where the largest numbers of cattle are received are given in Table 14-3. These data show that packing plants are no longer concentrated in Corn Belt cities such as Chicago. In recent years there has been a tendency for packing plants to operate in areas close to the large feedlots where thousands of cattle are fattened each year.

14.3.1   Finishing Cattle as a Business

Finishing cattle for slaughter has become a big business in many areas. The major reason for feeding cattle is to make a profit from

TABLE 14-2   Ten leading states in number of cattle in feedlots[a]

State	Number of cattle
Texas	4,227,000
Nebraska	3,785,000
Kansas	3,287,000
Iowa	2,862,000
Colorado	2,301,000
California	1,612,000
Illinois	940,000
Minnesota	758,000
Oklahoma	732,000
Arizona	646,000

[a]From *Livestock and Meat Statistics*, U.S. Department of Agriculture, Washington, D.C., 1977.

TABLE 14-3   Ten largest stockyards in the United States[a]

Stockyard	Cattle	Calves & vealers
Oklahoma City, Okla.	1,087,537	24,329
Omaha, Nebr.	924,845	23,548
So. St. Paul, Minn.	837,418	155,723
Sioux City, Iowa	722,376	[b]
Springfield, Mo.	580,487	16,401
Sioux Falls, S. Dak.	578,809	28,893
Amarillo, Tex.	576,942	[b]
Joliet, Ill.	406,155	[b]
National Stock Yards, Ill.	385,246	12,627
Torrington, Wyo.	381,816	[b]

[a]From *Livestock and Meat Statistics*, U.S. Department of Agriculture, Washington, D.C., 1977.
[b]Calves' figures are in total marketed.

such an operation and to convert pasture, roughage, and grain into a highly desirable and valuable human food. Large feedlot operations make it possible to use efficient, specialized equipment combined with specialized knowledge and technology to feed cattle for a profit.

Cattle are fed on the farm for several reasons. Even when the farm is largely a grain producer, there is some roughage that would go to waste if not utilized by livestock. When conditions are favorable, more cash may be obtained by marketing these crops through fat cattle rather than by selling them directly on the market. Feeding cattle on the farm also makes more efficient use of labor year-round. In addition, the manure produced may be returned to the land to maintain or increase its fertility. Cattle produced on the farm may be fed out to have a better market for surplus animals produced.

Cattle are fed in large feedlots mainly for business purposes and to make a profit. Much of the feed utilized is purchased. Persons who are well trained and skilled in economics, nutrition, health maintenance, and management are used in such operations. Well-designed feedlots and modern equipment are also utilized. But even possible profits made in these large efficient operations are subject to fluctuations in feed and fat cattle prices. When conditions are favorable, considerable profit is made. When they are unfavorable, large losses may result.

The disposal of manure becomes a problem in large feedlots where thousands of cattle are fed. Thousands of tons of manure are produced with a cost of $1 to $2 per ton for removal. In addition, the accumulation of manure and urine poses a pollution problem. Large amounts of manure near large feedlots have stimulated research into its greater use. Equipment has been developed to produce methane (gas) from cow manure. It has been estimated that a feedlot with a 100,000-head capacity would produce enough energy for a city of 30,000 people. Some cities such as Hereford, Texas are surrounded by many feedlots. It has been estimated that 600,000 head of cattle are fattened each year within a radius of 15 miles of this city. It is also possible to make anhydrous ammonia from cow manure, and manure may be recycled with the liquid portion used as fertilizer and the remainder as a part of the maintenance ration for feeding cows and/or feedlot cattle.

14.3.2   Contract Feeding

Contract feeding means that a rancher, farmer, packer, or speculator enters into a contract with a feedlot operator to fatten or finish cattle. The owner retains ownership of the cattle, but the feedlot operator provides the feed, facilities, and care during the feeding period.

Contracts may vary for costs involved, but these should be clearly stated so that both parties involved are in agreement. The owner usually transports the cattle to the feedlot and absorbs a death loss of 1 percent. Losses greater than this are usually borne equally by both parties. When cattle are ready for market, the owner, or the feedlot operator, may decide when to sell and for what price, or they may both be equally involved. Profits realized may be distributed according to the terms of the contract. Since all costs involved are seldom covered by the contract, it should be entered into after much thought and consideration. If large numbers of cattle are involved, it may be advisable to obtain legal advice in drawing up the contract.

14.3.3   Kinds of Cattle Fed

Cattle feeders vary in the kind of cattle they prefer to feed. The kind fed may be best suited to a particular condition. This is particularly true when profits are likely to be small.

The age of cattle fed is an important consideration to the feeder. Calves could be placed in the feedlot shortly after weaning if they are large and fast-growing individuals. Gains made by weanling calves would require a long feed of concentrates, which would be more expensive than gains made on pasture and forages. For example, a 450-pound calf in the feedlot would have to gain 600 pounds on a concentrated ration to reach a market weight of 1,050 pounds. A yearling steer that had been grazed through its first winter and summer after weaning could weigh 700 to 800 pounds and would have to be fed a concentrated ration to gain 250 to 450 pounds to reach a slaughter weight of 1,050 pounds. Very few two-year-old steers are available for feeding.

The sex of feeder cattle is important. Steers and/or bulls are often fed. Bulls tend to gain faster and more efficiently in the feedlot than steers but sell for less per pound. Heifers and cows may also be fed, but heifers fed seldom exceed 2 years of age. Heifers tend to make slower gains than bulls and steers but tend to fatten more rapidly and at a lighter weight.

Cattle fed may be of straight beef breeding, dairy breeding, or a mixture of the two. With the increased emphasis on crossbreeding dairy and beef breeds, such crosses are often fed. In the South and Southwest, Brahman and British crosses are popular in the feedlot, as are crosses of some of the new breeds such as the Brangus, Beefmasters, and Santa Gertudis. Finished cattle from such crosses sell well in the South and Southwest but may sell for less per pound than cattle of straight beef breeding.

The conformation grade and quality are important to some cattle feeders, as is the condition at the time of purchase. Some prefer to feed low-quality animals if they can be purchased at a reasonable price, whereas others prefer to feed animals of high quality.

STUDY QUESTIONS

1. Will beef cattle always be needed in the United States and the world, or will they someday become extinct? Explain.

2. Why are more beef cattle than dairy cattle found in the range areas of the United States?

3. In what areas of the United States are the huge feedlots located? Why?

4. Is it necessary to feed grain to beef cows used for producing calves? Explain.

5. Assume you have purchased 100 weanling heifers to start a herd for calf production. Outline how you would manage them until they produced their first calf.

6. What states are the largest producers of feeder cattle in the United States? Why?

7. Why are two-year-old heifers nursing their first calves usually slow in re-breeding to produce their second calf as three-year-olds? What can be done to cause them to rebreed more quickly?

8. Outline a recommended system for feeding and managing pregnant cows.

9. Outline a recommended system for feeding and managing lactating cows.

10. Why are calves usually weaned at 200 to 225 days of age?

11. Outline a recommended system of feeding and managing herd bulls.

12. What is meant by stocker cattle? Feeder cattle?

13. What are some of the advantages of producing stocker cattle? Some disadvantages?

14. How may one purchase feeder cattle? Which method would you prefer and why?

15. What are two major ways of transporting feeder cattle? When should a certain one be used?

16. What are some important diseases of feeder cattle, and how can they be prevented?

17. What is meant by *preconditioning* cattle? By *backgrounding* feeder cattle?

18. What are the five largest stockyards in the United States and why are they located in a particular area?

19. What are the leading states in the United States in number of cattle in feedlots? Why are they the largest?

20. How may manure be disposed of in feedlots of 10,000 or more cattle?

21. What kinds of cattle are usually finished in feedlots in the United States?

buildings and equipment

Commercial beef production does not require elaborate equipment and shelter as a general rule. The main facilities and equipment needed include good fences, a corral with a working pen and a holding chute, plus hay storage and feeding facilities.

15.1   Shelter

Beef cattle require little shelter from the weather throughout the year. Usually fewer problems are encountered if calves are born out on the pasture or range than if they are born in barns and pens where a high density of cattle has been located for a number of years. The exception to this is during extremely cold weather when it may be necessary to bring cows into maternity pens before, during, and shortly after calving. Pole barns with an open front usually provide adequate protection for cows and their calves, and the calves are healthier. Tightly closed barns are less suitable for calving than open-front pole barns. In very cold weather, however, confining cows and calves to a clean building permits closer observation, and light and supplemental heat are available if a birth is difficult or if the calf is weak at birth and needs treatment. This is only a temporary arrangement, however.

Pastures or ranges which include timber and some hills and valleys supply shelter for cattle in all seasons of the year. Cattle seek such shelter when it is needed, and this is usually sufficient for their needs (Figure 15.1).

A building or a small lot is useful as a hospital area. Sick animals are readily available for treatment and can be closely observed and treated if restraining equipment is at hand.

Buildings are expensive, and because of the low return per beef cow and because beef cattle are rugged, a commercial beef operation usually does not have them available. For purebred beef cattle operations, however, buildings and pens are necessary and important. Purebred cattle must be cared for in a way to show them to best advantage, and they must be available for inspection by potential buyers.

15.2   Fences

Good fences are necessary for any type of beef operation. Several types are available, and different materials may be used for their construction, depending on their availability and cost.

Figure 15.1.   Some timber in pastures provides shade in summer and protection from the wind in winter.

15.2.1 Fence Posts

Fence posts for supporting fencing material are available in either steel or wood and a variety of sizes and shapes. Wooden posts are available in many parts of the United States and are usually less expensive than steel posts. Wooden posts often can be cut on the ranch or farm where timber is available. For permanent fences where wood posts are used, a decay-resistant variety such as Osage orange, cedar, juniper, black locust, as well as others should be used. These untreated varieties often have a lifetime of 15 to 30 years. Less decay-resistant varieties such as pine, hickory, and oak will have a life of 25 to 30 years if pressure-treated with a preservative such as pentachlorophenol or creosote. Such posts should be pressure-treated (6 to 9 pounds of pressure per cubic foot of wood) to force the preservative into the wood. Surface treatment with a preservative is not desirable because not enough of the preservative is absorbed to give an effective protection from decay.

The length of the post depends on the desired height of the fence and the depth at which the posts are set. Most wooden posts, except corner posts, range in length from $5\frac{1}{2}$ to 8 feet with a top diameter of $2\frac{1}{2}$ to 8 or more inches. The larger the top diameter, the stronger the post. Line posts of $3\frac{1}{2}$ inches or more at the top diameter will provide a strong, durable fence. Corner posts and gate posts should not be less than 8 inches in diameter at the top. Posts used as braces for corner posts or to add strength to line fences made with steel posts should be at least 5 inches in diameter at the top.

The depths at which posts are set depends on their location and use. Corner and gate posts should be set at least $3\frac{1}{2}$ feet into the ground. Line posts should be set 2 to $2\frac{1}{2}$ feet into the ground. The depth at which the posts are set plus the height of the top wire above the ground plus an additional 6 inches will give the desired post length.

Steel posts have certain advantages over wooden posts. They are fireproof, are easier to drive, have a longer life, and are lighter to handle. When in contact with moist soil they also serve as a ground against lightning. The length of steel posts varies from 5 to 8 feet, and steel posts usually cost more than wooden posts. They are more easily bent out of line than wooden posts when crowded by livestock. Wooden anchor posts located every 50 to 75 feet will give added strength to a fence where steel posts are used.

Line posts should be set about 15 to 20 feet apart. The shorter spacing should be used for steel posts or small wooden posts. Posts may be driven or may be set in a hole dug with an auger mounted on a tractor. They may also be dug by hand, but this is laborious. If very much fence is to be built, mechanical power equipment is recommended.

15.2.2   Fence Wire

Several different kinds of wire are available for fencing, but all of them are either barbed or woven wire. A satisfactory fence may be built with either of these alone or in combination with the other.

Most wire today has a galvanized coating to protect it from rust and corrosion. The thickness of the coating determines the amount of protection. Three classes of coatings are used, classes I, II, and III, with class I having the thinnest coating. Nine-gage wire with a class I coating will usually start showing rust in 8 to 10 years, wheras 9-gage wire with a class II coating will not usually show rust until 15 to 20 years.

Barbed wire is cheaper than woven wire and is often easier to use. It comes in several sizes and patterns in rolls of 80 rods each. The most common gages used are $12\frac{1}{2}$ to $14\frac{1}{2}$. Some barbed wire has two points per barb, and others have four, with 4 to 5 inches between barbs. Woven wire varies from 9 to 11 in gage, with the smaller gage being the heaviest. Nine-gage smooth wire with a galvanized coating may be used on brace posts. The wire is fastened to steel posts by means of specially designed metal clips provided by the manufacturer. Wire is usually fastened to wooden posts by means of $1\frac{1}{2}$- to 2-inch staples.

15.2.3   Fence Types

Several types of fences may be constructed, depending on the type of animals to be restrained, the amount of money to be spent, and personal preference. Four- or five-stranded barbed wire fences are recommended for boundary and cross fences. Three strands of barbed wire may be suitable for some cross fences. In areas of high stress such as those where cattle are driven, six- or seven-strand barbed wire or woven wire fences are recommended. A combination of woven wire with one to two strands of barbed wire may be used but is usually more expensive than barbed wire alone. Woven wire comes in different heights, and the higher the wire, the fewer strands of barbed wire needed. Woven wire fences may be used to contain small animals such as swine, sheep, and calves. Good-quality woven wire has tension curves between line wires to take up slack and to prevent the wire from breaking or loosening corner posts in the winter when the wire contracts due to extreme cold.

Suspension fences are sometimes used. This type of fence has line posts 50 to 100 feet apart but closer where the elevation of the ground changes the slope of the fence. Four or more strands of 12½-gage barbed wire are normally used. Twisted wire stays every 15 to 20 feet are often used. Each strand of barbed wire should be stretched so that there is no more than a 3-inch sag between posts. Steel posts may be used and the wire attached to the post by means of special wire clips, mentioned previously. Extra long staples are used to hold the wire in place on wooden posts, but the staples are driven so the wire is free to move across the post so that when cattle attempt to go through the fence there is some "give" to the wire. A properly constructed suspension fence has a certain amount of "live action" that adds to its effectiveness and will turn cattle as effectively as a conventional fence. Suspension fences require fewer posts and therefore are lower in cost than conventional fences.

Electric fences are low in cost and ideal for temporary fencing or to add to the effectiveness of a permanent fence. Light wire (16 to 18 gage) is used, but it is smaller than barbed wire and is not so visible to livestock or as durable. One wire is often sufficient, although two wires may be needed where both cows and calves are to be contained. One-wire fences are usually about 24 inches above the ground, but the height of the wire depends on the size of the animal to be contained and the potential grounding sources such as plants, and so on. A good rule of thumb is to place the wire at about two-thirds the height of the animal to be contained. Two kinds of chargers may be used, one powered by a battery and a type that plugs into a 120-volt electric system. The battery type is used in remote pastures where electricity is not available. The 120-volt type is best for pastures near an electrical supply. When purchasing a charger, one should be certain that it has been approved by the Bureau of Standards or other official organizations. The approval should be printed on the controller near the name plate. A plug-in charger should be installed according to instructions and should be placed in a dry spot shielded from livestock. It should also be well grounded. Improperly installed plug-in chargers can be dangerous to livestock and people. A properly constructed electric fence commands the respect of livestock and is quite effective.

15.2.4   Constructing the Fence

A fence is no better than the brace posts and materials used to construct it. The extension divisions of many states have publications available on how to construct fences and the materials to use. A double-span brace post assembly is shown in Figure 15.2. This type of brace assembly is more than twice as strong as a single-span assembly and should be used when the fence exceeds 200 feet in length. When the fence is more than 650 feet between corner posts, a braced line assembly should be used every 650 feet. A single-brace-line assembly is recommended, but it should have two diagonal brace wires to make the fence pull in opposite directions.

Steel corner post brace assemblies are available and can be used in place of wood assemblies. The steel posts of the assembly should be set in concrete, however. Steel posts must be inserted into the ground so that they are oriented for proper wire placement. One side of the steel posts has projections at regular intervals between which the wire is attached. This side of the post should be the side to which the wire is attached.

Line posts may be set in a straight line by stretching a cord or wire between brace post assemblies. The barbed or woven wire is usually placed on the pasture side of the posts so that when cattle crowd against the wire it is braced against the posts. If placed on the outside of the pasture, staples might be pressed out and the wires loosened when crowded into by cattle. Sometimes the wire is placed on the outside of the posts for fences near highways because it makes the fence more attractive.

One end of the barbed wire should be nailed or tied to the anchor post of the brace assembly. The wire should be stretched in sections consisting of the span between two different brace assemblies. The wire on the spool should be stretched to the anchor post

Figure 15.2.   Double span brace post assembly. Post depths shown are considered to be minimum. (Courtesy Guide Sheet 1192, "Constructing Wire Fences," University of Missouri, Columbia)

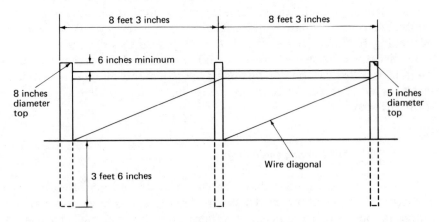

of the brace assembly by wrapping it around the post, wiring it tight, and nailing it to the post. The wire on the spool should then be unrolled and the stretching apparatus fastened to the wire and the anchor post on the next brace assembly. Protective clothing and leather gloves should be worn when working with barbed wire. The wire should be stretched tight but not tight enough to cause it to break. The top wire should be stretched first and the lowest wire last. Woven wire should be stretched in a similar manner using a special stretcher (or jack) for that purpose. The fence should be stretched slowly, with tension being applied equally to all points. Stretching should be continued until the tension curves on the woven wire are straightened about one third. The woven wire should be fastened to the line posts beginning at the posts farthest from the stretcher or jack.

Gates for fences may be constructed or purchased as desired. Gates should be hung to large, well-anchored posts. Some prefer to set gate posts in concrete to make them more secure. Gates are usually 10 to 14 feet in length and should be wide enough for the passage of any equipment to be used.

15.3 Corrals and Cattle-Handling Equipment

An efficient cattle operation requires good corrals and handling equipment. Well-planned facilities make the handling of cattle safer for the cattle and their handlers, and the stress on the cattle will be less when they are worked.

The main working facilities on a farm or ranch are usually located near the headquarters or at a central location where electricity and water are available. Smaller facilities or portable corrals and working chutes may be used in pastures not readily accessible to the main facilities. The working facilities should be located on a well-drained site which is readily accessible to other lots and pastures. They should also be located in an area readily accessible to trucks and pickups. They should be located downwind from the farm or ranch home and from that of close neighbors.

The basic facilities needed include a loading chute, a working chute with a cutting gate, at least two pens to sort cattle into, and a larger holding pen. If cattle are to be sprayed for external parasites, a small crowding pen should be available. Scales for weighing are important, and plans for their location should be included when new facilities are built. Scales vary in size from those that will weigh a single animal to those that will weigh several at one time. The size to use depends on the money available and the preference of the cattle producer.

Many plans are available for building facilities for working cattle. General plans (U.S.D.A. Plan 6230) are given in Figure 15.3. Space requirements for all classes of cattle are given in Table 15-1. Specific and detailed plans for facilities are available to the farmer or rancher through extension centers in his or her state.

Some precautions should be followed in planning facilities to make their operation more efficient. The working chute or loading chute should not be too wide. Proper widths of chutes and other information are presented in Table 15-2. Curved chutes and crowding areas encourage better cattle movement much more than those that are square. Sloping-sided chutes help prevent cattle from turning around, especially when cattle of different sizes are worked.

15.4   Feedlots and Feedlot Equipment

Feedlots vary from the simple to the complex; they are used for small operations and for those where thousands of cattle are fed. The site should be selected in accordance with the suggestions given in Section 15.3. Where thousands of cattle are fed, it is doubly important to study local and state air and water pollution laws.

Figure 15.3.   Plans for circular crowding pens and working chutes. Left figure shows good layout for loading and sorting but is adapted to limited expansion with no ideal scale location. Figure at right also has good loading and sorting layout and could be a hospital area, receiving lot, or both. Consider roofing the working area. (Courtesy U.S.D.A.)

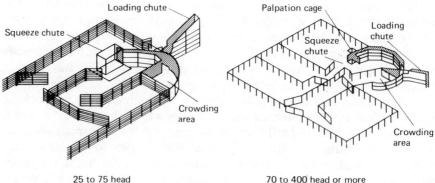

Loading chute

Squeeze chute

Crowding area

25 to 75 head

Palpation cage

Squeeze chute

Loading chute

Crowding area

70 to 400 head or more

TABLE 15-1   Space requirements for all classes of cattle[a]

Feedlot, ft^2/head
20-in. barn and 30-in. lot .... Lot surfaced, cattle have free access to shelter
50 .... Lot surfaced, no shelter
150–800 .... Lot unsurfaced except around waterers and along bunks and open-front buildings, and a connecting strip between them
20–25 .... Sunshade

Buildings with feedlots, ft^2/head
20–25 .... 600 lb to market
15–20 .... Calves to 600 lb
$\frac{1}{2}$ ton/head .... Bedding

Cold confinement buildings, ft^2/head
30 .... Solid floor, bedded
17–18 .... Solid floor, flushing flume
17–18 .... Totally or partly slotted
100 .... Calving pen
1pen/12 cows .... Calving space

Feeders, in./head along feeder
All animals eat at once:
18–22 in. .... Calves to 600 lb
22–26 in. .... 600 lb to market
26–30 in. .... Mature cows
14–18 in. .... Calves

Feed always available:
4–6 in. .... Hay or silage
3–4 in. .... Grain or supplement
6 in. .... Grain or silage
1 space/5 calves .... Creep or supplement

Waterers
40 head/available water space in drylot. Feeders: 15 gal/head/day; 5,500 gal/head/year
Cows: 20 gal/day

Corrals, ft^2/head

600 lb	600–1,200 lb	1,200+ lb
14	17	20 holding
6	10	12 crowding

Isolation & sick pens
40–50 ft^2/head
Pens for 2–5% of herd

Mounds
25 ft^2/head .... Minimum
50 .... If windbreak on top of mound, 25 ft^2/head each side

[a]From *Beef Cow/Calf Manual*, Manual 104, Extension Division, University of Missouri, Columbia, p. 77.

TABLE 15-2   Dimensions for beef cattle corrals and working facilities[a]

1. Holding area	20 ft^2/animal[b]
2. Working chute w/vertical sides	
Width	26 in.[c]
Desirable length (min.)	20 ft
3. Working chute w/sloping sides	
Width @ bottom	16 in.[c]
Width @ top	26 in.
Desirable length (min.)	20 ft.
4. Working chute fence	
Recommended minimum height	50 in.
Depth of posts in ground	30 in.
5. Corral fence	
Recommended height	60 in.
Depth of posts in ground	30 in.
6. Loading chute	
Width	26-30 in.[c]
Length (min.)	12 ft
Rise	$3\frac{1}{2}$ in./ft
Ramp height for	
Gooseneck trailer	15 in.
Pickup truck	28 in.
Van-type truck	40 in.
Tractor-trailer	48 in.
Double deck	100 in.

[a]From *Beef Cow/Calf Manual,* Manual 104, Extension Division, University of Missouri, Columbia, p. 76.
[b]14 ft^2/calf up to 600 lb.
[c]For exotic breeds, the width dimension should be increased 2 in. For exotic bulls, it may be necessary to increase the width 4 in.

In addition, the land should be well drained, and enough land should be available for future expansion. In areas of considerable rainfall, mud can become a problem in lots where the cattle are fed, and this can reduce the rate and efficiency of gains (see Figure 15.4).

The number of cattle per pen may vary widely from 40 to 50 head per pen up to 150 to 200. The size of the pen should be built to accommodate the number of head to be fed. Space requirements per head for various kinds of lots are given in Table 15-1. Heavier cattle, of course, will need more space per head than calves just entering the feedlot.

In areas of the United States where rainfall is light, feedlots are not paved. In areas of moderate to heavy rainfall paving may be a necessity because it is favorable to faster and more efficient gains, it reduces labor costs in feeding, and more manure is saved and is easier

Figure 15.4.   Unpaved feedlots in areas of moderate to heavy rainfall may become very muddy which can interfere with the rate and efficiency of gain of cattle being fed. (Courtesy University of Missouri, Columbia)

to remove. In addition, cattle on paved feedlots do not carry so much dried manure and mud in their coats when shipped to market. If the feedlot is not completely paved, it is advisable to pave 6- to 10-foot strips around water and feed bunks. In some feedlots, mounds are built to favor drainage of water and to make a more mud-free area where the cattle may rest (see Figure 15.5).

15.4.1   Shelter

Shelter should be provided for cattle in the feedlot. Protection from heat and sunshine in the summertime is probably more important than protection from the cold in the winter, providing the lots have a hard, mud-free surface. An open shed is preferred to a closed barn for feeding cattle in either summer or winter. Shades built rather simply and inexpensively are used to protect cattle from the sun and allow for good air circulation, which is especially important in the summer. The heat generated by the digestion and metabolism

Figure 15.5.   Dirt mounds in the center of feedlots help prevent the accumulation of mud and furnish a place where cattle can rest between feedings. (Courtesy University of Missouri, Columbia)

of concentrated rations is enough to keep the cattle warm in the wintertime.

Some cattle in the Midwest are fed in enclosed barns where the environment is controlled year-round. A portion or all of the floor consists of concrete slabs about 6 inches wide and set $1\frac{1}{2}$ inches apart. These are referred to as slotted floors. The slotted floor is usually built over a manure pit 6 to 7 feet deep. A pump is used to empty the pit into special tank wagons which can spread the liquid manure on the fields. Thermostatically controlled fans provide ventilation, with no artificial heating or cooling being supplied. The cattle are fed by an automated feeding system, which reduces the amount of labor involved. Less floor space per animal is needed in such barns than in conventional lots, and the rate and efficiency of gain are usually improved. The high initial cost of such barns, however, limits their use for feeding cattle.

15.4.2   Feeding Systems

Small groups of cattle may be hand-fed in troughs at regular intervals, but such a feeding system requires considerable labor and time. This feeding system may be used to bring cattle to a full feed, although this can also be done by full-feeding and adjusting the amount of roughage in the ration. Hand feeding requires that the feed be put in the feeding troughs once or twice per day and requires extra handling of the feed. The experienced feeder can obtain desirable rate and efficiency of gains by this method, and there is little feed wastage because feeding can be regulated to ensure that cattle are given only the amount of feed they will consume at each feeding. Hand feeding is possible only when small numbers of cattle are fed.

Self-feeders are often used to feed cattle on the farm. They require less labor than hand feeding because enough feed can be placed in them to last 1 to 2 weeks. Feed placed in self-feeders can be mechanically mixed and placed in the feeders by means of trucks specially equipped with augers. Sometimes the ration is ground and delivered directly to the feeders, eliminating much of the handling of grain. Experimental evidence suggests that self-fed cattle gain more rapidly than those that are hand-fed and are less likely to go off feed. One of the reasons for this is that the cattle eat more often because the feed is always available to them, and the cattle often eat at their leisure. The main disadvantages of using self-feeders as compared to hand feeding is that (1) a larger investment in equipment is required, (2) feeders have to be properly adjusted to avoid feed wastage, (3) gains on a self-feeder may cost slightly more, and (4) cattle may not be observed as closely and carefully as those that are hand-fed.

Fence line bunks are used for feeding cattle in many feedlots. An example of such bunks is shown in Figure 15.6. Fence line feed bunks may be made of wood or concrete. One main advantage of such bunks is that they may be placed on the outside of a pen facing an alley. Self-unloading wagons or trucks may be used to fill the troughs with grain, silage, or other feeds. If several pens are side by side and have such troughs, many cattle in several pens may be fed with one trip through the alley. This saves a great deal of labor. Also, manure is more easily removed because pens do not contain obstructions such as feed bunks. In addition, fence line bunks would require less upkeep than bunks in the pen since they are outside the lots and are less likely to be damaged by the cattle. Since cattle fed from fence line bunks feed from only one side, twice as much space should be provided per animal. This space should be 20 to 24 inches per animal if they are fed once each day. Less space would be needed if they are fed two or more times daily. Alleyways for fence line bunks should have a hardened surface (rock or paved) so wagons or trucks have access to the bunks in any kind of weather conditions.

Portable feed bunks made of 2-inch lumber and with 4- by 4-inch legs may be used for feeding cattle. The bunks should be securely bolted and braced. Cattle may feed from both sides of such bunks if they are located away from the fences. Portable bunks should be 30 to 36 inches wide, 20 to 24 inches from the ground, with sides at least 8 to 10 inches deep to reduce feed wastage. The

Figure 15.6.   Fenceline bunks for feeding grain and silages. Such bunks have the advantage that they can be filled when desired by mechanical equipment without entering the pen. (Courtesy Larkin Langford)

legs on some bunks are adjustable to accommodate cattle of different sizes and ages and to raise them as manure accumulates in the pens. The portable feed bunks must be removed from the lots, of course, when the pens are cleaned by using mechanical equipment.

An adequate supply of clean water should be available to the fattening cattle at all times. Commercial automatic water tanks are available with a source of heat, usually electricity or gas, to keep water from freezing in the wintertime. The expense of heating water to prevent it from freezing in the winter has become an expensive item because of the energy shortage. It is suggested that beef cattle feeders keep in close contact with the local and state extension services to obtain advice on the most recent recommendations for feedlot equipment, including automatic waterers and the most efficient use of energy in such operations.

15.5   Confinement Feeding of Beef Cattle

Confinement feeding refers to the feeding of cattle in small, confined quarters during at least some or all of the feeding period. Research has been done on the confinement feeding of beef cows for part or all of the year and finishing cattle in the feedlot.

15.5.1   Beef Cows

The confinement feeding of beef cows all or part of the year requires that the feed be harvested by mechanical means and transported and fed to the cows in mangers or feed bunks. This is done rather than allowing cows to directly harvest the feed in the field. Confinement feeding of beef cows requires well-designed feedlots and equipment for harvesting, transporting, and feeding the feed. It also requires that rations be carefully formulated, especially for cows nursing calves. In addition, the manure they produce must be gathered and spread periodically, whereas on pasture it is usually scattered widely over the field.

The confinement feeding of beef cows helps to gentle them and their calves and allows daily observation. This is an important advantage at calving time and for the detection of cows in heat during the breeding season if artificial insemination is used. Since growing forage is harvested and brought to the cows, they do not trample and waste as much of the feed as they would in the pasture or field.

Experimental results suggest that the confinement feeding of beef cows can be successful. This method of feeding may become more important in the future.

15.5.2   Finishing Cattle

Considerable interest has been shown in recent years for the confinement feeding of finishing cattle. The space per animal in confinement lots may be reduced one third or more as compared to usual feedlots. Confinement lots are usually placed under a roof, and the floors are often slotted. The slotted floors allow the feces and urine produced to collect in a storage area under the slotted floor or nearby where it can be removed periodically by special equipment and spread on the fields.

The advantages of the confinement feeding of steers on slotted floors are several: (1) Labor may be saved in handling feed and other materials, (2) problems with mud and dust are reduced or eliminated, (3) less feed may be wasted, (4) more land is released for farming and other purposes, and (5) properly constructed confinement feeding lots may help in the control of flies and pollution.

The initial cost of constructing buildings and lots for confinement feeding increases the cost per animal fed. Cattle fed in confined quarters during cold weather appear to make more rapid and efficient gains than those fed in larger open lots with dirt floors.

The beef cattle feeder considering the construction of confinement feeding quarters should obtain all of the advice possible from experts in this area and should consider all factors involved before building such facilities.

15.6   Disposal of Manure and Other Wastes

Manure disposal becomes an important problem in large feedlot operations. It represents an important expense, especially in large feedlots with dirt surfaces.

Although manure removal from feedlots is an important problem, it has been of great value as a fertilizer. More recently it has been used for the synthesis of gas (methane) and as a portion of animal feeds.

A 1,000-pound steer on a high-concentrate ration will excrete 55 to 65 pounds of wet manure daily. In feedlots of 10,000 head or more many tons of manure are therefore produced each day, and over a period of time hundreds of tons of manure may be produced and accumulate.

Manure may be collected in feedlots in several ways. It may be done by the use of a wheel-type loader or even a road grader. Other types of equipment may also be used for collection purposes. In large feedlots the manure is loaded into spreader trucks by mechanical loaders. The trucks then transport the manure to the fields where it is spread as a fertilizer. On the farm where smaller numbers of cattle are fed, a loader mounted on a tractor may be used to load manure spreaders, which are pulled by a tractor to the field where it is spread.

STUDY QUESTIONS

1. Why do beef cattle need little shelter throughout the year?
2. At what times is shelter for beef cattle most likely to be required? Explain.
3. What kinds of posts are required for building a good fence, and what are some advantages and disadvantages of each?
4. What is the difference between barbed and woven wire? Explain.
5. What is meant by a suspension fence? Describe how to build one.

6. What is an electric fence? What type should be used in remote areas and why?

7. Why are corner posts so important in building a good fence? Describe how to install a corner post assembly. Would you use two brace posts or one? Why?

8. Why is it so important to have good fences for cattle?

9. What are the basic facilities needed for working cattle?

10. Describe some desirable basic features of equipment for working cattle.

11. Why should feedlots be paved? When is it not necessary to pave feedlots?

12. What are some advantages and disadvantages of self-feeding and hand-feeding beef cattle?

13. What are some advantages of fence line feed bunks?

14. Describe how a good supply of water may be made available to beef cattle throughout all seasons of the year.

15. Define what is meant by *confinement feeding*.

16. What are some of the advantages and disadvantages of the confinement feeding of beef cows?

17. What are some of the advantages and disadvantages of the confinement feeding of finishing cattle?

18. Why is the disposal of manure such an important problem in feedlots of thousands of head of cattle?

19. What are some major uses of manure from feedlots?

PART FIVE

NUTRITION
AND FEEDING

digestion
and utilization
of food

Beef cattle are ruminants. Ruminants possess a complex stomach that is divided into four compartments. For this reason they are called *polygastric* animals. Characteristically, ruminants chew a cud which consists of partially digested food that is regurgitated from the rumen. Because of their complex stomach and digestive system, cattle are able to digest and utilize highly fibrous feeds that cannot be utilized to any extent by single-stomach (monogastric) animals such as swine. Cattle do not directly digest fibrous foods in their own digestive system, but they do this indirectly through the action of various microorganisms found in the digestive tract, especially the rumen.

Beef cattle are sometimes thought of as being very inefficient in converting the feed they eat to meat. It is often not realized that much of the feed converted to meat by beef cattle such as grass and other plants cannot be utilized directly by humans and monogastric animals. Imagine waking up some bright, cool, clear morning with the prospect of having wheat straw, hay, and cornstalks for breakfast, lunch, and dinner. Almost any of us would prefer beef and other products we normally eat. Not only that, but we would soon starve on a diet fed to cattle.

Even though beef cattle can utilize feeds high in fiber, they can also efficiently utilize large amounts of grains which are highly digestible and good sources of energy. Cattle can also utilize feeds that contain large amounts of protein and nonprotein nitrogen. Because cattle are unique in the feed they utilize, it is our purpose here to discuss the anatomy of their digestive tract and how they digest and utilize the feed they eat.

16.1   Anatomy and Physiology of the Digestive Tract

The parts of the digestive tract of cattle include the mouth, esophagus, the four stomach compartments, the small intestine (duodenum), the caecum, the large intestine, and the rectum (see Figure 16.1). Cattle differ from single-stomach animals such as swine in that they possess a four-compartment stomach, which places them in the group of animals called ruminants. Ruminants include cattle, sheep, and goats.

16.1.1   Mouth

Cattle do not have upper incisor or canine teeth. For grazing they utilize the upper dental pad and lower incisors for grasping and conveying food to the mouth (called prehension). The tongue is also

Figure 16.1.   Digestive tract of the beef animal.

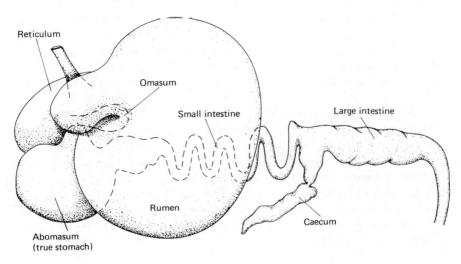

important in the prehension of food. The tongue of cattle is long and covered with rough papillae which are useful in bringing food such as grass and other forage into the mouth when the tongue wraps around it. The grass or forage is then sheared by the closure of the incisors on the dental pad in the upper part of the jaws. After being taken into the mouth, the feed is chewed to a limited extent before it is swallowed. Chewing is done by the back teeth (or molars) located in both the upper and lower jaws. When swallowed, the food passes into the rumen where further steps in the digestion process occur. In cattle a large number of taste buds are present on the tip and the back of the tongue. These buds control, to a certain extent, the intake of food. Grasses or feeds that are unpalatable are not eaten readily probably because of their bitter or unpleasant taste. This is illustrated many times in pastures where grass around dung heaps is tall and green, whereas that between the heaps is grazed close to the ground. Some unpalatable grasses and weeds in a pasture will not be grazed by cattle until more palatable plants are eaten. The tongue also moves the chewed bolus of food to the back of the mouth to initiate the process of swallowing.

16.1.2   The Four-Compartment Stomach

The four compartments of the stomach of cattle include the rumen (paunch), the reticulum (honeycomb), the omasum (many-plies), and the abomasum (or true stomach). Cattle are herbivores, which means that they subsist largely on grasses and herbs. They do not normally consume meat or other animal products. Because of the bulky nature of food consumed by cattle, their digestive system is more complex and much larger than that of animals such as swine. Concentrated rations normally fed to swine are not bulky enough to allow the proper functioning of the complex stomach system of cattle. Cattle, however, can handle large amounts of grains when included in a fattening ration containing some bulky feeds such as hay and silage. Switching from an all-pasture or roughage ration to a fattening ration high in concentrated feeds must be done over a period of time by gradually increasing the concentrated ration until the animals are on a full feed. Otherwise, they may become ill and *founder*, go off feed, recover slowly, and make slow and inefficient gains thereafter.

The rumen is much larger than the other three stomach compartments and contains 50 to 58 percent of the total capacity of the digestive system as compared to 12 to 13 percent for the other three compartments combined. The stomach compartments of mature cattle have a capacity of about 190 quarts as compared to 7 to 8 quarts for the mature pig. The greater stomach capacity in cattle is important in order for them to eat enough roughages to meet their nutrient requirements.

The rumen of cattle harbors billions of microorganisms including bacteria and protozoa. These microorganisms break down fiber (cellulose and hemicellulose) that cannot be digested directly by the individual's own digestive system. The microorganisms break down the fibrous material into mostly acetic, propionic, and butyric acids, which are called the volatile fatty acids (VFAs). The VFAs are largely absorbed through the rumen wall into the blood stream. The action of microorganisms in the rumen is the basic reason why cattle can be maintained on high-roughage diets.

Rumen organisms perform another important function in addition to digesting fibrous material. They synthesize all of the B-complex vitamins and all of the essential amino acids in their growth and reproduction. Microorganisms even synthesize proteins from simple nonprotein nitrogenous compounds such as urea and ammoniated compounds. This is the reason that urea is often used in cattle rations to replace part of the protein. The microorganisms themselves are digested farther along the digestive tract, and in this way their body contents are made available to cattle.

A baby calf does not possess a functional rumen. For this reason, the nutrient requirements of a calf are similar to those of single-stomach animals until the rumen becomes functional at about 6 to 8 weeks of age. As the calf grows older and eats small amounts of hay, certain microorganisms become established in the rumen, which gradually develops to the normal size and functions as in the mature individual.

Cattle do not completely chew their food when it is first eaten. Later when in a quiet and resting state they regurgitate their food in the form of a bolus and rechew it. This is called "chewing the cud" or *rumination* and is peculiar to ruminants. Each bolus of regurgitated food is chewed for about a minute and is then swallowed again. Another bolus is then regurgitated and chewed in the same manner. Cattle may spend up to 8 hours per day in the rumination process, depending on the coarseness and amount of fiber in the ration. The coarser the diet, the more time spent in rumination. Rumination does not appear to improve the digestibility of the food but reduces food particle size so it can pass from the rumen to the other stomach compartments.

The fermentation of hay and roughage in the rumen of cattle by microorganisms results in the production of large amounts of gas. These gases are mostly carbon dioxide and methane. They are normally expelled from the rumen by belching (*eructation*). If they are not expelled in the normal manner, the rumen becomes inflated in what is commonly known as *bloat*, a condition common only to ruminants. Small amounts of the gases produced are absorbed into the bloodstream and eliminated from the body when air is exhaled from the lungs. Bloat can cause the death of the individual.

16.2   Digestion of Food

The digestion of food refers to the process in which carbohydrates, fats, and proteins are broken down into smaller units so that they may be absorbed into the lymph and bloodstream. The digestive process in cattle is largely accomplished by enzymes. Enzymes are organic catalysts that speed up chemical reactions within the body but are not themselves used up in the process. In cattle, the process of digestion, which begins in the abomasum (true stomach), is approximately the same as in other species. The abomasum is the only stomach compartment in cattle that contains digestive glands which secrete digestive enzymes.

16.2.1   Digestion of Carbohydrates

In some species the digestion of carbohydrates begins in the mouth where amylase (ptyalin) is secreted by the salivary glands. *This does not occur in cattle.* Saliva produced by the salivary glands of cattle does perform several functions, however. Saliva lubricates the food eaten so it can be swallowed and regurgitated and aids in the elimination of gases from the rumen in the process of eructation. It also takes several chemicals into solution so they can be detected by the taste buds on the tongue. Finally, saliva keeps the membranes of the mouth moist so they remain functional.

As mentioned previously, complex carbohydrates such as cellulose are digested by microorganisms in the rumen but cannot be digested by the digestive enzymes of cattle. The microorganisms possess their own digestive enzymes for the digestion of these complex carbohydrates.

The pancreas of cattle secretes small amounts of amylopsin (amylase) into the small intestine (duodenum) which acts upon starch and dextrins to form simpler compounds such as maltose and dextrins which can be absorbed into the bloodstream.The walls of the small intestine also secrete small amounts of enzymes which digest carbohydrates. These include sucrase (invertase), which converts sucrose to glucose and fructose, and maltase, which converts maltose to glucose. In young calves, the small intestine produces lactase, which acts upon lactose to form the simpler compounds glucose and galactose. The large intestine (cecum and colon) secretes no digestive enzymes, but microorganisms located there secrete the enzyme cellulase, which digests cellulose, polysaccharides, starches, and sugars to form simpler compounds.

16.2.2   Digestion of Proteins

The true stomach (abomasum) is the site of the beginning of protein digestion in cattle. Cells in the abomasum produce hydrochloric acid, which produces an acid condition that activates the digestive enzymes, pepsin and rennin, and aids in protein digestion. Pepsin is secreted by cells in the walls of the intestine and breaks down proteins into proteases and peptones. Trypsin and chymotrypsin secreted by the pancreas act on proteins, proteases, peptones, and peptides, producing amino acids as end products. Peptidase (eripsin) secreted by the walls of the small intestine produces end products consisting of amino acids and dipeptides. In nursing calves, the walls of the abomasum produce the enzyme renin, which coagulates milk protein (casein) and produces the end product paracasein. Coagulation of casein causes it to remain longer in the digestive tract than if it were liquid. This favors more complete digestion.

16.2.3   Digestion of Fats

Steapsin (lipase), which is secreted by the pancreas, acts upon fats to form end products of higher fatty acids and glycerol. Bile from the liver is secreted into the small intestine. It emulsifies fats and breaks them down into smaller globules. The smaller globules give the fats more surface area, causing steapsin (lipase) to be more effective in the digestion process.

16.2.4   Digestion of Minerals

The acid solutions in the abomasum dissolve minerals from the foods eaten, producing forms that may be absorbed into the bloodstream. Minerals are also released from organic compounds through the action of the digestive enzymes. Vitamins can be utilized in the body without being converted to simpler substances. The same is true of water.

16.3   Absorption of Nutrients

Absorption refers to the transfer of the end products of digestion from the lumen of the digestive tract to the blood and/or lymph. As mentioned previously, a considerable amount of the feed is digested by microorganisms in the rumen, and the nutrients are absorbed there. The primary site of absorption of nutrients digested by the individual's own digestive enzymes is the small intestine, although some absorption takes place in the large intestine. The mucous membrane of the small intestine possesses many very small fingerlike projections called villi where absorption takes place. These villi contain tiny capillary blood vessels and lymph ducts where the nutrients are absorbed and distributed to the rest of the body. Fatty acids and glycerol are absorbed into the lymph ducts, whereas water, inorganic salts, monosaccharides, amino acids, and peptides are mostly absorbed into the bloodstream.

16.3.1   Factors Affecting the Digestibility of Food

Some factors related to the increased, or decreased, digestibility of feed for cattle are important because they may affect the rate and efficiency of gains. At best, cattle do not digest all the nutrients in their feed. Some experiments have been conducted where cattle manure is processed and fed to cows as part of their ration. These experiments were successful. It has been a practice on the farm for many years to run hogs in the same pens as fattening steers so they can recover some of the nutrients in the manure not digested and utilized in the body of the cattle being fed.

The rate of passage of feed through the digestive tract is of some importance in digestion. Some nutritionists feel that the passage must be slow in order for the digestive enzymes to complete their role in digestion. The rate of passage of feed through the digestive tract in monogastric animals may be affected by the amount of feed eaten, with more efficient gains made on limited rations. The capacity of cattle to store large amounts of feed in the rumen and digestive tract may be responsible for the amount of feed eaten having little effect on the rate of food passage through the digestive tract.

Cattle do not chew their grain as well as some other species of animals so grinding may increase the digestibility of grains eaten, because it exposes a larger surface area to the digestive enzymes. Grain definitely should be ground for very young or very old cattle because their teeth may not be developed enough or may not function well enough for them to properly chew it.

Pelleting rations for cattle, especially protein supplements, is favored by many feeders. This process may prevent some feed wastage but may not improve the utilization of the feed enough to pay for the extra costs involved.

Heating or cooking feeds is usually not economical because the cost involved is usually more than it is worth in that the digestibility of the feed is affected very little. Cooking raw soybeans is practiced to destroy a trypsin-inhibiting factor when they are fed to nonruminants. This is not so necessary in ruminants.

STUDY QUESTIONS

1. What is a ruminant? How do they differ from animals which possess a single stomach?

2. Are beef cattle really inefficient in the conversion of the feed they eat to animal products?

3. List the parts of the digestive tract of cattle.

4. Why are some foods unpalatable to cattle? What is prehension?

5. Name the four compartments of the stomach of cattle and give at least two names for each.

6. Why are cattle also called herbivores?

7. What is the total capacity of the stomach compartments of cattle as compared to the total capacity of the digestive system? Which stomach compartment is the largest?

8. What important functions do microorganisms perform in the digestive tract of cattle? Where are they located in largest numbers?

9. Is the rumen functional in the baby calf? Explain.

10. What is meant by rumination? Describe it in cattle. What function does it perform?

11. What is eructation? What is the importance of this process in cattle?

12. What is the purpose of the digestion of food in cattle? Describe the general process involved.

13. Describe the role of saliva in cattle. Does it contain digestive enzymes?

14. Describe the digestion of carbohydrates in mature cattle.

15. Describe the digestion of proteins, fats, and minerals in cattle.

16. What is meant by the absorption of nutrients? Where does it usually take place in cattle?

17. What are some factors that affect the digestibility of food in cattle?

18. Would you recommend the cooking of foods for cattle? Explain.

CHAPTER 17

nutrients and feeds for beef cattle

Plants synthesize the nutrients they require for growth and maturity through a combination of elements obtained from the soil and the action of sunlight on the leaves in the process of *photosynthesis*. Beef cattle, on the other hand, depend mostly on plants for their nutrients. This is the reason they are called *herbivores*. Many plants such as grasses, hay, and grain are used for feeding beef cattle, but the kinds used depend on those available in different parts of the country. The nutrients required by beef include carbohydrates, proteins, fats, minerals, vitamins, and water. The composition of many feeds fed to beef cattle are given in Table 17-1.

17.1 Water

Water is absolutely essential for the proper growth and functioning of the bodies of animals and plants. Without water life could not exist. The body of a newborn calf is slightly more than 90 percent water. The body of a steer ready for market contains 40 percent water.

TABLE 17-1 Composition of feeds commonly used in beef cattle ration (as-fed basis)[a]

	Dry matter (%)	Total protein (%)	Dig. protein (%)	NE_m[b] (mcal/lb)	NE_g[c] (mcal/lb)	TDN (%)	Fat (%)	Crude fiber (%)	Calcium (%)	Phosphorus (%)	Potassium (%)	Sodium (%)	Sulfur (%)
Concentrates													
Barley	89.0	11.6	8.7	0.86	0.57	74	1.9	5.0	0.08	0.42	0.56		
Corn													
Corn & cob meal	87.0	7.5	3.8	0.81	0.54	73	3.0	8.6	0.04	0.22	0.42		
Yellow Dent #2	89.0	8.9	6.7	0.92	0.60	81	3.9	2.0	0.02	0.31	0.30		0.14
Yellow Dent #3	86.0	8.7	6.5	0.89	0.58	78	3.7	2.0	0.02	0.25	0.27		0.13
Molasses													
Beet	77.0	6.7	3.8	0.71	0.48	68	0.2	0.0	0.16	0.03	4.77	1.17	
Cane	75.0	3.2	1.8	0.65	0.41	54	0.1	0.0	0.89	0.08	2.38		
Cane, dried	96.0	4.1	2.3	0.84	0.52	69	0.1	0.0	1.14	0.10	3.04		
Oats	89.0	11.7	8.8	0.70	0.46	68	4.5	11.0	0.10	0.35	0.37		
Sorghum													
Grain	89.0	11.1	6.0	0.79	0.53	74	3.0	2.0	0.04	0.31	0.34	0.04	
Grain (6–9% protein)	88.0	7.0	4.0	0.76	0.50	71	2.6	2.0					
Wheat													
Bran	89.0	16.0	12.5	0.62	0.39	62	4.1	10.0	0.14	1.17	1.24	0.06	
Hard winter	89.1	13.0	10.2	0.87	0.57	78	1.6	2.7	0.05	0.40	0.51		
Soft winter	89.1	10.9	8.2	0.87	0.57	78	1.6	2.5	0.09	0.29	0.39		
Protein supplements													
Cotton seed meal													
Expeller	94.0	41.0	33.2	0.77	0.50	78	4.3	12.0	0.16	1.20	1.40		
Solvent	91.5	41.0	33.2	0.71	0.46	69	2.0	12.0	0.16	1.20	1.40	0.14	
Linseed meal	91.0	35.1	30.9	0.71	0.47	69	5.7	9.0	0.40	0.83	1.38		
Soybean													
Whole seed	90.0	37.9	34.1	0.98	0.62	85	18.0	5.0	0.25	0.59	2.02		
Meal, dehulled	89.8	50.9	45.8	0.83	0.56	75	0.8	2.8	0.26	0.62	1.71		
Meal, expeller	90.0	43.8	37.3	0.84	0.56	77	4.7	6.0	0.27	0.63	1.97		
Meal, solvent	89.0	45.8	39.0	0.78	0.53	72	0.9	6.0	0.32	0.67			
Urea (45% N)		281.0											
Urea (42% N)		262.0											

TABLE 17-1 (continued)

	Dry matter (%)	Total protein (%)	Dig. protein (%)	NE_m[b] (mcal/lb)	NE_g[c] (mcal/lb)	TDN (%)	Fat (%)	Crude fiber (%)	Calcium (%)	Phosphorus (%)	Potassium (%)	Sodium (%)	Sulfur (%)
Dry roughages													
Alfalfa													
Hay, early bloom	90.0	16.6	11.4	0.55	0.28	51	2.0	26.7	1.13	0.21	1.87	0.14	0.27
Hay, midbloom	89.2	15.3	10.8	0.50	0.24	51	1.8	27.6	1.20	0.20	1.30		
Hay, mature	91.2	12.4	8.7	0.48	0.20	50	1.5	34.2					
Meal, dehyd.	93.0	17.9	14.0	0.55	0.29	58	3.0	24.3	1.33	0.24	2.49	0.09	0.30
Brome-smooth hay													
Early bloom	90.3	10.9	6.7	0.54	0.28	56	2.2	28.2		0.20	2.56		
Mature	92.8	11.0	4.6	0.49	0.20	54	2.8	31.7	0.40				
Clover hay, red	87.7	13.1	7.8	0.50	0.25	52	2.5	26.4	1.41	0.19	1.54	0.15	0.13
Corn cobs, ground	90.4	2.5	0.0	0.43	0.10	42	0.5	32.4	0.11	0.04	0.76		0.42
Corn stover	87.2	5.1	1.9	0.48	0.22	51	1.0	32.4	0.43	0.08			
Cottonseed hulls	90.3	3.9	0.2	0.42	0.08	37	1.4	42.9	0.14	0.10	0.76	0.02	
Fescue hay													
Early bloom	90.0	10.8	6.7	0.51	0.25	52	2.4	28.4	0.36	0.31	3.07	0.14	
Mature	90.0	7.5	3.8	0.50	0.23	51	1.8	30.5	0.34	0.21	1.98	0.13	
Grain sorghum stover	85.1	4.5	1.5	0.47	0.21	49	1.8	27.7	0.34	0.09			
Lespedeza hay													
Early bloom	93.4	14.5	9.7	0.53	0.25	54	3.9	27.6	1.15	0.23	0.93		
Full bloom	93.2	12.5	7.9	0.49	0.20	51	2.9	28.9	0.97	0.21	0.96		
Orchard grass hay													
Full bloom	89.2	11.1	6.6	0.49	0.23	51	3.2	19.5	0.40	0.33	1.85		0.23
Mature	90.4	7.1	4.3	0.49	0.23	51	2.4	34.3					
Soybean straw	87.6	4.8	1.5	0.34	0.00	33			38.6	1.39	0.05	0.46	
Sudan grass hay	88.9	11.3	4.9	0.51	0.26	52	2.0	25.7	0.50	0.28	1.37	0.02	0.05

Timothy hay													
Early bloom	87.7	7.6	4.4	0.50	0.25	52	2.3	29.1	0.53	0.23	0.81	0.13	0.17
Midbloom	88.4	7.5	4.1	0.53	0.28	54	2.4	29.6	0.36	0.17	1.42	0.05	0.11
Late bloom	88.0	7.3	3.6	0.49	0.24	51	2.2	28.5	0.33	0.16			
Wheat straw	90.1	3.2	0.36	0.42	0.08	43	1.5	37.4	0.15	0.07	1.00		
Fresh forage													
Alfalfa	27.2	5.3	4.1	0.16	0.09	17	0.8	7.5	0.47	0.08	0.55	0.03	
Brome													
Immature	32.5	7.2	5.4	0.22	0.14	22	1.4	7.3	0.20	0.18			
Mature	56.1	3.6	1.9	0.36	0.21	36	1.6	18.5	0.17	0.15	0.70		
Fescue tall													
April–May	20.9	3.9	2.8	0.13	0.08	14	0.7	5.5	0.08	0.07	0.71		
July–Aug.	35.3	3.0	1.6	0.22	0.13	23	0.7	12.0	0.13	0.08	0.79		
Sept.–Nov.	50.0	4.2	2.0	0.32	0.18	32	1.1	16.4	0.20	0.10	0.80		
Dec.–March	60.0	4.9	2.6	0.38	0.22	38	1.0	19.4	0.25	0.11	0.30		
Lespedeza													
Early bloom	25.0	4.1	3.0	0.17	0.10	17	2.0	8.0	0.44	0.05	0.28		
Mature	35.5	4.5	3.1	0.26	0.17	26	2.1	16.0	0.36	0.11	0.27		
Orchard grass													
Immature	23.9	4.4	3.2	0.15	0.09	16	1.2	5.6	0.14	0.12	0.63	0.01	0.05
Full bloom	29.9	2.5	1.5	0.19	0.11	19	1.0	9.9	0.07	0.07			
Timothy													
Immature	26.1	4.1	2.9	0.17	0.10	17	0.9	5.9	0.12	0.11	0.66	0.02	0.05
Mature	35.8	2.2	1.3	0.23	0.13	24	1.0	12.0	0.06	0.07	0.56		
Silages													
Alfalfa	28.3	5.3	3.2	0.14	0.05	15	1.1	8.2	0.40	0.09	0.67		
Wilted	38.5	6.8	4.1	0.20	0.07	20	1.2	11.0	0.58	0.12			
Wilted (haylage)	55.0	9.8	5.9	0.28	0.09	29	2.0	17.8	0.88	0.21			
Corn dent													
Well matured	40.0	3.2	1.88	0.28	0.18	28	1.2	9.8	0.11	0.08	0.42		
Early maturity	28.0	2.4	1.40	0.20	0.13	20	0.8	7.4	0.08	0.06	0.27		

TABLE 17-1 (continued)

	Dry matter (%)	Total protein (%)	Dig. protein (%)	NEm[b] (mcal/lb)	NEg[c] (mcal/lb)	TDN (%)	Fat (%)	Crude fiber (%)	Calcium (%)	Phosphorus (%)	Potassium (%)	Sodium (%)	Sulfur (%)
Grass													
Timothy	37.5	3.8	2.10	0.21	0.11	22	1.2	12.7	0.21	0.11	0.63		
Legume	29.3	3.5	1.8	0.16	0.07	16	1.0	9.2	0.23	0.08			
Sorghum													
Grain	29.4	2.1	0.59	0.16	0.08	17	1.0	7.7	0.07	0.05			
Sorgo	26.0	1.6	0.44	0.15	0.07	15	0.8	7.0	0.09	0.05	0.32		
Sudan grass	23.3	2.4	1.30	0.13	0.07	14	0.7	8.0	0.15	0.05	0.72		
Mineral supplements													
Bonemeal, steamed	95.0	12.1	8.2			15			29.00	13.60		0.46	
Dicalcium phosphate	96.0								22.20	17.90			
Limestone	100.0								35.84				
Monosodium phosphate	96.7									21.80		32.30	
Phosphate, defluorinated	99.8								33.00	18.00	0.09	3.95	
Sodium sulfate	96.0											32.40	22.50
Sodium tripolyphosphate	96.0									24.94			

[a]Courtesy of Dr. Homer B. Sewell.
[b]NEm = net energy required for maintenance.
[c]NEg = net energy requirement for gain.

Water performs several functions in the animal body. Water within body cells causes them to be distended and hold their normal shape. Various nutrients are dissolved in water and by this means are transported from one part of the body to another. Water is also a necessary ingredient of the many biochemical reactions within the body. Water helps regulate the body temperature of cattle through the process of evaporation. Water is also a principal constituent of blood, lymph, and materials which lubricate the joints and other body parts. Death will occur more quickly in cattle due to a lack of water than to the lack of any other nutrient.

Water is made available to cattle in several ways: (1) through the water they drink, (2) through that present in the feed they eat, and (3) through that present in the form of metabolic water, which results from the breakdown of carbohydrates, proteins, and fats in the body. In producing cattle, much attention must be paid to supplying drinking water at all seasons of the year. It can be supplied in ample quantities, as a general rule, with little cost.

The amount of water required by cattle depends to a great extent on the surrounding air temperature and the amount of dry matter consumed. Requirements of cattle of all different ages and kinds will be met if a free-choice supply is available at all times. This means that methods must be used in the wintertime to keep water from freezing. Freezing of water can be prevented in several ways. Cattle will usually drink from 3 to 9 pounds of water per day with 100 pounds of body weight, but they will drink more in the summer than in the winter months.

Water is supplied to cattle from wells (Figure 17.1), creeks, rivers, ponds, lakes, etc., in many parts of the United States. Cattle may drink directly from these sources or from more elaborate automatic waterers, especially in feedlots where cattle are being fattened for slaughter. Care should be taken to supply water free from contamination of any kind, such as toxic substances and disease-bearing organisms.

Water in some arid or semiarid regions contains large amounts of salt and other minerals that may cause physiological upsets and even death in some instances. Usually such sources of water have been identified and are not made accessible to cattle.

Supplying water in ponds (or tanks) in the range area often makes it possible to utilize grazing areas that otherwise could not be used. These impoundments sometimes contain water only a few days during the year because they have to be built in areas where leakage cannot be prevented because of the soil conditions or rainfall is so little that there is little or no runoff accumulated.

Figure 17.1.   Beef cattle drinking from a pond on a western range.
(Photo by John F. Lasley)

17.2   Carbohydrates

Carbohydrates are organic compounds present in large amounts in
plants, and they contain carbon, hydrogen, and oxygen. They represent most of the dry matter in feeds for beef cattle.

Carbohydrates vary from simple sugars, such as glucose, which
contain only one molecule of sugar to polysaccharides such as starch
and cellulose which contain many molecules of this compound. As
shown in Chapter 16, carbohydrates are absorbed into the body from
the digestive tract mainly as simple sugars such as glucose. Beef
cattle, as mentioned previously, are able to utilize complex carbohydrates such as cellulose that cannot be digested by the animal's
own digestive system through the action of microorganisms in the
rumen.

Carbohydrates supply most of the energy needs of beef cattle
for maintenance, growth, and reproduction. Much of the feed consumed is used to meet these energy needs. If they are not met, the
animals become very thin and are referred to as being *underfed*.
Any energy consumed above that required for maintenance (where

the animal loses or gains no weight when doing no work or producing no product) is used for growth, fattening, or reproduction.

Feeds containing large amounts of carbohydrates include most of the grains, grasses, hays, silages, etc. The grains contain large amounts of concentrated carbohydrates mainly in the form of starch and are called concentrates. Because carbohydrates are so concentrated in the grains, they are used in fattening rations because they supply energy in excess of that needed for maintenance. This excess energy is used for added growth and fattening. Roughages such as hay are bulky and usually do not contain enough concentrated energy in the form of carbohydrates to cause animals to fatten. They do supply some carbohydrates and other nutrients and add bulk to cattle rations necessary for the proper functioning of the digestive system.

17.3   Fats and Oils

Fats and oils, like carbohydrates, contain carbon, hydrogen, and oxygen and are found in many seeds such as soybeans, flax, and cotton which may contain up to 35 percent of their total weights in oils. Seeds, or grain, such as wheat and corn may contain only 4 to 5 percent of such oils. Fats in beef cattle rations supply energy for the function of the body, and excesses above maintenance requirements may be stored as fat. Fats liberate about 2.25 times more energy per unit weight than carbohydrates when digested because they contain a larger proportion of hydrogen and oxygen.

A small amount of fat in cattle rations is desirable because fats are carriers of the fat-soluble vitamins A, D, E, and K. Most rations fed to beef cattle contain adequate amounts of these nutrients so they do not require attention in designing rations for beef cattle. Cows' milk is high in fat content on a dry matter basis, containing up to 40 percent of fats.

17.4   Protein

Proteins are complex organic compounds made up of long chains of amino acids, and they always contain carbon, hydrogen, oxygen, and nitrogen. Proteins may also contain sulfur and phosphorus. Proteins are especially important in beef cattle feeding because they constitute the living protoplasm of all body cells. Proteins are found in the reproductive and growing parts of plants such as the leaves. In animals, all body parts contain proteins, and they make up a large proportion of the tissue found in muscle, internal organs, skin, hair, wool, horns, hooves, and bone marrow. Proteins in the body are produced through the action of genes, and any observed defects in protein tissues are an indication that a mutation, or heredity, may be the cause.

Proteins perform many functions in the body of animals. These functions include the repair of tissue and growth of new tissue. A source of energy when broken down into amino acids, they are the main constituents of vital substances in the body such as antibodies and some hormones, especially those from the anterior pituitary gland, and are essential parts of enzymes which are necessary for all normal body functions. Thus, proteins are necessary for the life and development of all animals.

The twenty or more amino acids necessary for beef cattle production are supplied in the feed. In some animals, especially those with a single stomach, surplus or simple amino acids in the feed cannot be converted into others that are needed by the body. Amino acids that cannot be changed and which are essential for growth must be supplied in the feed. They are called *essential amino acids.* Fortunately, ruminants such as beef cattle synthesize amino acids from simple nitrogen-bearing compounds for their own bodies through the action of microorganisms. Thus, beef cattle can convert inferior proteins or nonprotein nitrogen compounds such as urea into superior proteins which are found in meat and milk and which are made available to humans when they consume these animal products. Urea is a simple nitrogenous compound which is an end product of metabolism. It can be used to meet a portion of the protein needs of beef cattle when added to the ration. It is now produced on a commercial scale by processes which utilize nitrogen from the air.

17.5   Minerals

Many mineral elements are needed by beef cattle for normal body functions. Those needed in relatively large amounts are referred to as *macrominerals* and include calcium, phosphorus, sodium, chlorine, potassium, magnesium, and sulfur. Others which are required in only trace amounts are referred to as *trace minerals* or *microminerals.* They include cobalt, copper, chromium, fluorine, iodine, iron, molybdenum, manganese, selenium, silicon, and zinc.

Calcium, phosphorus, sodium, and chloride are those most often supplied in extra amounts to beef cattle. In areas where they may be deficient, certain trace minerals are also supplied in the feed or in a mineral supplement.

17.5.1   Sodium and Chloride

Sodium and chloride are needed by beef cattle. They may be supplied in the form of salt in many forms. Animals suffering from a salt deficiency often show a depraved appetite and are in a general rundown or unthrifty condition.

17.5.2   Calcium and Phosphorus

Cattle are more likely to suffer from a calcium or phosphorus deficiency than from any other mineral except salt. These two minerals make up about 70 percent of the mineral content of the bones and 30 to 50 percent of the mineral content of milk. Both calcium and phosphorus are required in many other vital body processes. Cattle consuming liberal amounts of legume pasture or hay usually obtain adequate amounts of calcium and phosphorus, but since these may be supplied at relatively low cost, they are usually offered free-choice to cattle in the form of a mineral supplement. A calcium to phosphorus ratio of 1:1 to 1.5:1 is recommended for beef cattle.

17.5.3   Potassium

Potassium content of roughages is usually high enough to meet beef cattle requirements for this mineral. Experiments indicate, however, that due to leaching some forages such as fescue in the winter months of January through March may not contain an adequate supply of this substance to meet the needs of cattle grazing this kind of pasture. Including potassium in a mineral mixture is recommended under such conditions.

17.5.4   Magnesium

Magnesium is needed in a mineral supplement for beef cattle, especially for beef cows that are lactating. The lactating beef cow requires about 20 grams of dietary magnesium daily to maintain a

normal level of this substance in the blood serum as compared to
9 grams for a pregnant beef cow that is not lactating. Lactating cows
receiving low levels of dietary magnesium may suffer from *grass
tetany*, which will be discussed in detail later. Feeding 57 grams
(2 ounces) of magnesium oxide per head daily is recommended when
there is a high risk of grass tetany. Beef cattle need to have a daily
intake of magnesium because this substance is removed from the
blood within 20 to 30 hours after it is consumed. Older animals
store little magnesium in their bodies.

17.5.5   Trace Minerals

Cobalt and iodine are the trace minerals most likely to be deficient in beef cattle rations. Microorganisms in the rumen require
cobalt for the synthesis of vitamin B_{12}. Cobalt should be supplied
every day, especially to beef cattle who are wintered on low-quality
grass or hay, cereal straw, or cornstalks.

Iodine may be deficient in some parts of the United States.
Most areas where this compound is deficient have been identified,
and iodine is included in a ration in the form of iodized salt.

17.6   Vitamins

Vitamin A is the vitamin most likely to be deficient in beef cattle
rations. Carotene, the compound which beef cattle convert to vitamin A, is a pigment found in green and yellow plants. When cattle
are fed rations high in carotene, they store the surplus vitamin A in
the body, mostly in the liver. This stored supply is used as a reserve
in times when carotene is in low supply such as when pasture is
affected by a long drought or in the winter when the feed consists of
poor-quality hay or roughage. Cows fed rations consisting of good-
quality legume hay, haylage, or silage usually receive adequate
amounts of vitamin A.

17.6.1   Vitamin A

A vitamin deficiency in breeding herds is indicated by lowered
fertility, resulting in a poor calving percentage. Cows deficient in
vitamin A may be difficult to settle or may abort or give birth to
calves who are weak or dead at birth.

Both vitamins D and E are required by beef cattle. They are usually included in a supplement with vitamin A or given by injections. Cattle who are exposed to direct sunlight or sun-cured forage should not need supplements containing this vitamin. Vitamin D aids in the development of teeth and bones. Growing calves have a greater need for vitamin D than older cattle. Rickets in calves is a sign of a vitamin D deficiency.

Vitamin E is usually adequate in most rations but is sometimes recommended for cows on low-quality roughages such as cornstalks. White muscle disease in calves has been prevented by the administration of vitamin E. It has also been shown through research that the trace mineral selenium spares or replaces vitamin E in the prevention of or the curing of this disease. Although a vitamin E deficiency has been shown to lower reproductive efficiency in rats, this has not been confirmed in beef cattle.

17.7   Pastures

Pasture crops are extremely important as a source of nutrients for beef cattle. They contain large amounts of protein, TDN, and certain vitamins on a dry matter basis. In regions such as the southern and midwestern states where rainfall is considerable or irrigation may be practiced, pastures may carry one to two head or even more per acre. In some of the range states such as Arizona and New Mexico 100 or more acres may be required to provide yearlong grazing for a single animal. Pastures where rainfall is plentiful may include many highly productive species, but in range areas native grasses such as the gramma grasses which are adapted to the area are utilized. Considerable research has been conducted in range areas to determine efficient management systems and to test new grass species from all parts of the world.

Pasture crops in the United States are quite varied, depending on conditions in a particular area. Conditions are so varied in the different parts of the United States that space does not allow a detailed discussion of specific pasture crops that would be suitable nationwide. For information on pasture crops for a specific area, it is recommended that those seeking such information contact their local or state extension office.

17.7.1   Kinds of Pasture Crops

Native pasture crops are those found growing in a particular area when the country was first settled and still growing there. These include the bluegrasses of Kentucky, the blue stems of the Midwest, and the gramma grasses of the western ranges as well as others. These native grasses were present in certain areas because they were adapted to the local environmental conditions. Some of these native grasses, such as Kentucky Bluegrass, have been introduced successfully in areas outside their original habitat. Some land originally covered with grasses such as the blue stems of southern Missouri and southeast Kansas has been plowed for grain crop production, or these grasses have been replaced by other species which have proved more productive. On rough, rocky land that is not suitable for cultivation, the native blue stem grasses still remain. A good example of this is the blue stem regions of Kansas and Oklahoma.

Most native grasses are perennials and do not need to be reseeded or resodded each year. Close grazing or cutting for hay each year has resulted in some deterioration in soil fertility on which they are grown. For example, many old-timers in southwest Missouri tell of prairie grass in the early years that was shoulder high to a horse. Today the native prairie grass usually grows little more than knee high. These, of course, are merely observations and are not verified by scientific data.

Annual grasses are those that grow from seed each year. They do make some important contributions for grazing purposes, but they are not always dependable sources of forages, especially in regions of very low annual rainfall. An example is alfilaria, sometimes called filaree, an annual plant thought to have been introduced into the southwest portion of the United States from the Mediterranean region by the Spanish conquistadores and now well established in southern Canada and most of the United States, especially in the semi desert ranges of Arizona. It even extends into Mexico. Alfilaria furnishes choice spring forage for all classes of livestock and makes very vigorous early spring growth. It matures rapidly in most regions and then dries up. Cattle readily eat the dried and discolored stems. Hundreds of other annual plants also furnish grazing for livestock (*Range Plant Handbook*, U.S. Department of Agriculture, Forest Service, U.S. Government Printing Office, Washington, D.C., 1937).

Some grasses make maximum growth in warm weather. Examples are the various Bermuda grasses in the South, the gramma grasses of the Southwest, and the blue stems in the Midwest. Annual grasses such as Sudan grass and some sorghum hybrids are also often utilized for grazing in the summer, as is the legume and lespedeza. Cool-weather grasses make their greatest growth early in the spring but make some growth late in the fall. Examples are orchard grass and tall fescue. Animals readily graze fescue early in the spring and in the fall but show some reluctance to graze it in the summer months of June, July, and August. A good pasture program includes species which will give some grazing during as much of the year as possible. Fall-seeded winter oats in the South and winter wheat and rye in Oklahoma and Kansas as well as some other areas also supply considerable fall, winter, and early spring grazing.

Some grasses are referred to as short grasses because of their growth habits. Some of the gramma grasses in the Southwest are good examples. With ample summer rains they make considerable growth and put out stems which produce seed. The blades of grass are short, unlike the blades of orchard grass, for example, which may be a foot or more in height. The short grasses produce small yields per acre, but in the southwestern ranges they retain a large percentage of their nutrients throughout the winter. Grasses such as these along with trees and shrubs which supply some browsing make it possible to graze cattle yearlong in these areas.

Pure legume stands are seldom used for grazing cattle, although in some areas alfalfa that is near maturity may be grazed with little danger from bloat. Legumes appear to work best in combination with grasses such as orchard grass, which has a tendency to grow in bunches. When legumes are used in mixtures with grass, there is less danger of bloat in cattle grazing such mixtures, and the nitrogen fixing qualities of legumes often stimulate additional growth of grasses in the mixture. Grass-legume mixtures for grazing have received increased attention in recent years because of the high cost of complete fertilizers containing nitrogen. Legumes also add protein and often increase palatability.

17.7.2   Nutrient Content of Forages at Different Ages

Most forages contain less fiber and a higher percentage of digestible protein when immature. As they mature, the amount of fiber increases, and the percentage of protein decreases. Proper grazing favors the maintenance of desirable quality in the grass. Clipping pastures not grazed too closely often favors regrowth of more succulent grass after periods of sufficient rainfall. It is not always possible, however, to obtain the greatest possible production per acre and still maintain the highest-quality grass.

17.7.3 Pasture and Range Management

Efficient pasture or range management has several objectives. One is to maintain a good balance of species of grasses for grazing and to prevent the encroachment of unpalatable and sometimes poisonous weeds. Another objective is to encourage rapid growth for an available supply of good-quality forage. It is necessary to compromise somewhat on a balance between yield and quality because it is not possible to obtain both at the same time.

The *continuous system of grazing* is often used. In using this system the same pasture or range is grazed for long periods. This system is satisfactory if the stocking rate is such that overgrazing does not occur. Under such conditions the more palatable species will be grazed first and the less palatable species last, which may cause a reduction in the amount of palatable forages produced. Overgrazing for long periods of time, especially on the range, often results in the death and disappearance of desirable species, followed by an encroachment of unpalatable weeds. The carrying capacity per unit of land becomes less in some cases. Overgrazing of range land in the southwestern United States over long periods of time has resulted in the encroachment of what was once excellent grazing land by trees and shrubs such as juniper and mesquite which prevent the growth of desirable grasses. Continuous grazing where the proper stocking rate is maintained results in less adverse effects and does not always allow the maturity of forages. To a certain extent it does maintain good quality of forages because enough growth remains to allow storage of nutrients in the roots.

Rotational grazing is another system often used. This system usually involves close grazing of a pasture or range for a period of time and then moving the cattle to another pasture, giving the first a rest from grazing. This is a practical system for maintaining good-quality pasture and range, but several fenced pastures, each with its own water supply, are required. Little will be gained from rotational grazing, however, if the stocking rate is not heavy enough to utilize all the rapidly growing plants. This type of grazing system does allow the stockpiling of quantities of forages for winter grazing or for other seasons when it is needed. Rotational grazing also favors a better balance of forages on the pasture or range. This is especially important when pastures contain grass-legume mixtures.

17.7.4 Pasture Improvement

Pasture improvement usually means that the pastures are renovated by working the soil with implements, fertilizing it, and then reseeding or seeding additional species along with those that previously existed. In the semiarid range country, pasture (or range) improvement usually means that efforts are made to prevent overgrazing, with the expectation that under such mangement systems the native grasses will become more vigorous and increase in density, resulting in greater forage production. In some instances stands of grasses have been improved by seeding a species adapted to that particular area.

In regions of sufficient rainfall the ground is worked, limed, and fertilized and a mixture of desirable forages seeded at the recommended time and level. Sometimes legumes, including clovers and birdfoot trefoil, are seeded in pure grass stands. This practice often improves the yield and quality of the forage produced.

17.7.5   Stockpiling Grass

Stockpiling grass (or deferred grazing) means that the grass is not grazed and is allowed to accumulate during the growing season to be grazed at a later time. Usually grass is stockpiled for the winter months. Stockpiling of fescue for winter grazing is practiced in some states of the Midwest, which allows grazing most of the winter. Hay is fed when the fields are covered with snow, and mineral supplements are always supplied. Sometimes protein supplements are fed along with the stockpiled roughage if needed.

17.7.6   Fattening Cattle on Grass

Tender, palatable grass is an excellent feed, but it is not actually a concentrate. Cattle of all ages will gain weight but will not necessarily fatten on grass alone. Cows, of course, can be maintained throughout their productive life on grass supplemented with minerals and proteins in the winter. Older steers and cows that have not nursed a calf during the grazing season are often grass-fat at the end of the grazing season. Fat cows are often marketed directly for slaughter, but in the United States grass-fat steers are usually fed grain for at least a short period to add to their finish before they are slaughtered. In the trail-drive days in the United States grass-fat steers were driven to market and slaughtered without additional feeding. Later some of them were fed for a short period before being slaughtered. These steers were 2 or more years of age and could fatten to a certain extent on grass alone. In countries such as Australia and Argentina very few cattle are grain-fed today because grain is scarce. For this reason they are marketed as grass-fat individuals.

17.8 Hay

Hay is one of the most important roughages fed to beef cattle in the United States. Much of the hay fed is of poor quality because it was not harvested at the proper stage, was too high in moisture content when harvested, or was not stored properly.

For best-quality hay it should be made from a crop that is palatable and includes desirable species of plants. It is also important that the hay be harvested at the proper stage of maturity when its TDN content is at its highest. The time to cut a hay crop varies somewhat with the kind of hay available. Another requirement of good-quality hay is that it should be cut, cured, and handled in such a way that it is leafy and green in color. In recent years many farmers use hay conditioners which cut and press moisture from the plants in a single trip through the field. This speeds up the curing process to the point where the hay can often be cut and cured in about 1 day. The rapid curing helps prevent losses by leaching and shattering if the crop becomes too dry or is rained on before it is baled and stored. The nutrient content of various hay crops is given in Table 17-1.

17.8.1   Legumes as a Hay Crop for Beef Cattle

Alfalfa is a legume hay crop that is grown almost nationwide in the United States. It is well adapted to irrigated sections of the Southwest and West and grows well in nonirrigated regions. Alfalfa will produce several cuttings per year if there is sufficient moisture available and the soil is properly limed and fertilized. Since conditions vary from one region to another, it is best to contact the local agricultural college or extension service for information on how to grow alfalfa, the variety to plant, the time to cut for hay, and the best procedure for fertilization. Alfalfa is the best hay crop available for feeding beef cattle and contains 18 to 20 percent protein on a dry matter basis. It usually has its higher protein content when cut in early bloom.

Red clover and lespedeza hay are two other legumes available as hay crops in some parts of the country. Both crops make satisfactory hay when properly cut and cured but are not as satisfactory in quality, nor do they produce as great a yield per acre as alfalfa.

17.8.2   Grass Hay for Beef Cattle

Grass hays, in general, are lower in protein content than legumes such as alfalfa. Most grass hays contain 10 to 12 percent protein when cut in an immature stage and properly cured. Widely used crops for producing grass hay include brome, orchard grass, fescue, timothy, and some annual grasses such as Sudan grass. Grasses for hay will usually show a good response in yields when fertilized with complete fertilizers containing high amounts of nitrogen. The composition of some of the most widely used grass hays is given in Table 17-1.

17.8.3   Mixtures of Grass and Legumes as Hay for Beef Cattle

Alfalfa and several varieties of the clovers are sown in mixtures with various grasses for hay crops. Early cut and well-cured grass-legume mixtures are very palatable to cattle and have a higher protein content than grass hay alone. The nitrogen fixed in the roots of legumes by bacteria also stimulates additional growth of the grass in the mixture.

The mixtures of grasses and legumes to seed will depend on climatic conditions and other environmental factors.

17.8.4   Crop Residues for Beef Cattle

Crop residues include the forage left when grain is removed from the crops such as corn, sorghums, and small grains. These crop residues are so low in nutrients that they are not used, as a general rule, for fattening cattle, but they can be used for wintering cows and young stock.

Cornstalks are what is left of the plant after the ears have been removed. Farmers have grazed cornstalks with cattle for many years to glean the ears missed in the picking process and the edible portions of the stalk left after the ear is removed. Cattle may also eat some of the corn cobs left in the field when picker-shellers are used. Stalk fields can furnish most of the nutrients needed by beef cows after the ears are picked and up to about the first of the year. Sometimes they can be grazed even longer. A mineral supplement should be fed free-choice to all cattle on cornstalks, and a protein supplement should be fed to calves or younger stock.

Sorghum stalks can also be utilized in the same manner as cornstalks in the early part of the winter, but a protein supplement and a mineral supplement may be needed, as is the case with cornstalks.

Straw from small grains is a crop residue that may be baled and stored to supply a part of the roughage for wintering beef cows. This straw is low in protein and TDN and is seldom used as the only roughage for wintering cows. Nearly all small grain crops are now harvested by means of the combine, and this requires that the crop be very dry and mature at harvest. This late harvesting lowers the palatability and nutrient content of the straw as compared to what it used to be when cut with a binder, shocked, and later threshed.

Crop residues such as cornstalks and sorghum stalks are sometimes placed in stacks or large round bales and removed from the fields to be fed to cows during the winter. Such a procedure involves the added cost of baling and removing the stalks from the fields as well as feeding them. The removal of the stalks will remove about 40 to 50 pounds of nitrogen, 15 to 18 pounds of phosphorus, and 80 to 90 pounds of potash per acre in many fields. Sorghum stalks would contain similar amounts of these elements as cornstalks. In addition, if the stalks are removed and the ground is fall-plowed, more erosion will result than if the ground is not plowed until spring. When cattle graze the stalk fields, they scatter their manure and urine over the field, and fewer nutrients for the next crop will be lost. All of these factors should be considered when a decision is being contemplated as to whether or not to remove the stocks from the field.

Ground ear corn contains the whole ear, which consists of both the grain and the cob. It has been fed to fattening cattle for many years. The cobs add bulk to a grain ration and supply some nutrients. Cottonseed hulls, soybean hulls, rice hulls, and other crop residues have also been fed to cattle when they are available at a reasonable cost. They add bulk to the ration and supply some nutrients when used as a part of the ration.

Poultry litter can be used as a portion of the ration (up to 25 percent of the total ration). Poultry litter contains some crude fiber as well as nonprotein nitrogen. The nitrogen in chicken litter can be used by microorganisms in the rumen to form good-quality proteins. Cattle manure from feedlots can also be reconstituted to form a portion of the ration for wintering beef cows, but this is not a general practice.

17.8.5   Silage and Haylage for Beef Cows

Feeding corn silage to beef cattle has become a common practice in the Corn Belt and other regions of the United States where irrigation can be used to grow large tonnages of this crop per acre. Other regions sometimes utilize sorghum silages for beef cattle feeding.

Silage refers to the chopping of the entire plant at the stage where it is not completely dry and mature. Because the whole plant is harvested, silages which include grain can producd 1 to 2 pounds of gain per day when used as the only feed.

Corn silage is a very desirable crop and a safe feed. It will furnish 50 to 60 percent more nutrients per acre than when the grain alone is harvested. Corn silage works very well for starting cattle on feed, and cattle on such a ration can be easily changed to a high-concentrate ration. Corn silage can also be used to furnish minimum levels of roughages when fed in a finishing ration. Corn should be ensiled when the grain is well dented but before the leaves turn brown or become dry (*Corn Silage for Beef Cattle*, Guide 2061, University of Missouri, Columbia). Enzymes, yeast cultures, antibiotics, etc., have been added to silage, but there is no good evidence to suggest that they improve its feeding value.

The feeding value of corn silage varies with its grain and dry matter content. These should be taken into account when corn silage is used in fattening rations. Corn silage may be used for wintering cows, but if it contains large amounts of grain, there is some danger of the cows becoming too fat.

Sorghum silage is also a good roughage for beef cows. It usually supplies fewer nutrients than corn silage, but sorghums are more drought tolerant than corn and grow in a wider variety of soils. Sorghum silage may be made from a forage sorghum, or it may be a sorghum grain. Forage sorghums will produce more tonnage per acre than grain sorghums. Grain sorghum silage should be rolled to crush the grain. Whole grains of sorghum are not well utilized by cattle.

Sorghum-Sudan grass hybrids are often used for the production of forage silage. They produce a high tonnage per acre and are suitable for wintering mature beef cows. They do not contain enough energy for young calves or cattle in the feedlot to give maximum gains.

Silage can be made from various crops such as alfalfa, clover, bromegrass, and wheat. An advantage of such silage is that it is not seriously damaged by rains or harvest, and few leaves are lost. Leaves have the highest nutritive value of any part of the plant except the grain. Hay silage is usually more palatable than hay and retains a higher percentage of proteins and carotene, especially if ensiled before maturity. Hay silage does not have as high an energy content as corn silage because it does not contain grain.

Haylage is really the same thing as hay silage except it is left in the mower swath a little longer and is drier (40 to 50 percent moisture) than hay silage (65 to 70 percent moisture). Haylage, of course, is a roughage and cannot be expected to supply enough energy for fattening cattle unless fed along with grain.

17.9 High-Energy Grains for Beef Cattle

Grains fed to fattening cattle contain energy in the concentrated form. Feed grains fed to cattle include corn, oats, sorghum, and barley. Wheat may also be fed as part of the grain ration if its price per bushel is low compared to other grains. It is usually used as a human food, however. Four to 6 billion bushels of feed grains are produced in the United States each year.

17.9.1 Corn

Corn is the grain produced in the largest amounts in the United States. Its production is concentrated in the regions of the Corn Belt. Corn is a low-protein feed containing 7 to 9 percent crude protein, but it contains large amounts of TDN (73 to 80 percent). It is the standard with which other grains are compared. Corn may be fed to cattle as shelled corn mixed with a protein supplement to make a balanced ration, but it is usually ground when fed in this manner. Ground ear corn is sometimes fed and has more bulk than shelled corn because it contains the cob as well as the grain.

17.9.2 Sorghum

Sorghum grain contains 8 to 12 percent crude protein and 70 to 80 percent TDN. When fed to cattle, it has about 90 to 95 percent of the value of corn but may be equal to corn when fed as a part of a grain mixture containing corn. It should be ground when fed to cattle. It is of great importance as a high-energy food in the Midwest and in southern plains states.

17.9.3 Barley

Barley is an excellent feed for cattle. It is specifically adapted for grain production to the northern Corn Belt regions and the irrigated regions of Arizona and California. It is a cool-weather crop and responds well in yield to fertilizers high in nitrogen content. It should be ground or rolled when fed to cattle. It has about 90 percent of the value of corn in fattening rations for cattle.

17.9.4   Oats

Oats is a bulky feed containing about 30 percent hulls. It grows in most states in the United States but grows best and has a higher TDN content when grown in the cool climate in the north central states. It usually yields less per acre than some of the other feed grains and is not so widely used for feeding beef cattle. It should be ground or rolled when fed to cattle and is often used in creep rations for calves in this form. Oats has a crude protein content of 10 to 13 percent and a TDN content of 65 to 70 percent. Its value compared to corn when fed to cattle is about 70 to 90 percent.

17.9.5   Wheat

Wheat contains 9 to 16 percent crude protein and 75 to 80 percent TDN. When fed as a part of the grain ration (up to 50 percent), it has a feeding value of 100 to 105 percent as compared to corn. If fed as the only grain, it is likely to cause digestive upsets. Wheat is usually too expensive to feed to animals, but if it is fed, it should be cracked or coarsely ground.

17.9.6   Other Grains

Grains of lesser importance may be fed to cattle if available and of suitable price. These include rye, rice, millet, emmer, triticale (a hybrid of wheat and rye), and spelt. They are usually more valuable when fed as a portion of the grain ration.

17.10   High-Protein Feeds

High-protein feeds in the ration of beef cattle help balance the grain ration for the needed amount of protein. Most high-protein feeds for beef cattle are from a vegetable source. By-products of the meat packing industry such as tankage and meat scraps are usually not fed to beef cattle.

17.10.1   Soybean Meal

Raw soybeans are seldom fed to cattle. Soybean meal is the protein source for beef cattle rations, especially in or near the Corn Belt where so many soybeans are grown. Soybean meal is the part of the soybean left after the oil has been removed. The oil is used for making human foods such as cooking oil and oleomargarine. The crude protein content of soybean meal ranges between 48 and 52 percent and the TDN content from 80 to 85 percent, depending on how the oil has been extracted. Soybean meal is an excellent source of protein for beef cattle and is the one most often fed.

17.10.2   Cottonseed Meal

Cottonseed meal ranks second to soybean meal as a source of protein for beef cattle rations. It is the portion of the cottonseed left after the oil has been extracted. Cottonseed meal contains 44 to 45 percent protein and 75 to 78 percent TDN. It is available in large quantities where cotton is grown on a large scale.

17.10.3   Other High-Protein Feeds

Other sources of protein include linseed meal, peanut meal, safflower meal, sunflower seed meal, wheat mill feeds, rice mill feed, gluten feed and meal, as well as brewer's dried grains. These are not available in large enough quantities to be of great importance in feeding cattle except in some specific areas of the United States and under certain conditions.

17.10.4   Nonprotein Nitrogen

Urea is often used in cattle rations to replace a portion of the protein. Microorganisms in the rumen can utilize urea to form complete proteins and are then digested farther along the digestive tract. This makes the proteins present in the microorganisms available to the ruminant. Liquid urea supplements have been popular in recent years for self-feeding to grazing cows.

Urea is usually cheaper than protein and can be used to reduce the protein cost of a ration. Two thirds of a pound of urea (281) and 4.5 pounds of corn supply about the same amount of protein equivalent as 5 pounds of soybean meal.

Cattle on a high-roughage ration cannot utilize urea as efficiently to meet their protein needs as cattle on high-grain rations, because the release of energy from the roughage is too slow for microorganisms to make the best use of urea. Products such as Starea and

Biruet slow the release of ammonia in the rumen, which may increase the efficiency of utilization of urea in a high-roughage ration. These slow-release products also appear to lessen the danger of toxicity when cattle consume too much urea (see Section 22.4.1).

STUDY QUESTIONS

1. What are the nutrients required by beef cattle?
2. Why are cattle called herbivores?
3. List some important functions of water in animals.
4. How is water supplied to livestock?
5. What are carbohydrates? What is their chemical composition? What do they supply in feeds for beef cattle?
6. What elements do proteins contain? Why should adequate proteins be supplied to beef cattle?
7. What elements do fats and oils contain? What role do fats play in livestock feeds?
8. Why can simple nitrogen-bearing compounds be fed to beef cattle but not to single-stomach animals such as swine?
9. What is meant by microminerals? Macrominerals? What minerals are required in beef cattle rations?
10. What vitamins are required in beef cattle rations?
11. What is meant by *native* pasture crops? Name some native grasses common to specific areas in the United States.
12. What is the difference between annual and perennial pasture crops?
13. What is the difference between short grasses and long grasses?
14. What usually happens to the nutrient content of grasses when they mature?
15. What are some systems of grazing, and what are their advantages and disadvantages?
16. What is meant by *stockpiling* grass?
17. Is grass a concentrate? What cattle may be fattened on grass? Why are limited numbers of cattle fattened in feedlots in Australia and Argentina?
18. What are some of the procedures to follow in the production of good-quality hay?
19. In general, which makes the best-quality hay, legumes or nonlegumes? Why?
20. What are some crop residues often fed to cows? What are some of their limitations when fed to beef cattle?
21. Why is it possible for beef cattle to utilize a limited amount of chicken litter (manure) in their ration?

22. What is meant by silage? Haylage? What is probably the most important silage crop in the United States?

23. What is the comparative feeding value of corn, oats, sorghum, wheat, and barley?

24. What are some important high-protein feeds for beef cattle? Which one is used to the greatest extent in beef cattle rations?

25. What are some nonprotein nitrogen sources for beef cattle rations? In what kinds of rations are they used to best advantage?

CHAPTER 18

rations
for beef cattle

Beef cattle obtain the nutrients they need for growth, reproduction, milk production, and fattening from the feeds they eat. Most of the nutrients in beef cattle rations come from a vegetable source.

The nutrient requirements of beef cattle have been determined from much research. Information summarized in Table 18-1 shows daily nutrient requirements of beef cattle of different ages and in different stages of production. The nutrient content of various feeds is given in Table 17-1.

18.1 Preparation of Feeds

Feed may be prepared for beef cattle in several ways. Feeds may be rolled, crushed, ground, cracked, heated, pelleted, etc. Feeds are usually prepared in a particular way to make them more palatable and to increase their utilization by livestock. They may also be prepared for the purpose of making them more convenient to feed.

TABLE 18-1 Daily nutrient requirements in beef cattle[a]

	Daily nutrients per animal												Nutrient concentration in diet (dry matter)						
Total protein (lb)	Digestible protein (lb)	NEm[b] (mcal)	NEg[c] (mcal)	TDN (lb)	Ca (lb)	P (lb)	Vitamin A (thousands IU)	Body wt. (lb)	Avg. daily gain (lb)	Dry matter (100%) per animal (lb)	Roughage (%)	Total protein (%)	Digestible protein (%)	NEm (mcal)	NEg (mcal)	TDN (%)	Ca (%)	P (%)
1.61	0.91	6.40	2.35	11.1	0.033	0.033	18	800	1.0	18.4	75–85	8.8	4.9	0.60	0.31	60	0.18	0.18
1.63	0.92	6.40	3.77	12.2	0.033	0.033	18	800	1.5	17.9	55–65	9.1	5.1	0.71	0.42	68	0.18	0.18
1.69	1.01	6.40	5.04	13.4	0.037	0.037	18	800	2.0	18.0	25–35	0.4	5.6	0.80	0.51	74	0.21	0.21
1.80	1.12	6.40	6.69	15.4	0.044	0.040	18	800	2.5	18.4	< 15	9.8	6.1	0.94	0.58	84	0.24	0.22
1.12	0.63	6.99	0.00	7.4	0.024	0.024	13	900	0.0	13.2	100	8.5	4.8	0.53	0.00	56	0.18	0.18
1.57	0.92	6.99	2.57	11.8	0.033	0.033	18	900	1.0	18.5	70–80	8.5	5.0	0.60	0.38	64	0.18	0.18
1.74	0.99	6.99	4.09	13.3	0.035	0.035	19	900	1.5	18.9	50–60	9.2	5.3	0.71	0.45	70	0.19	0.19
1.76	1.04	6.99	5.54	14.1	0.037	0.037	19	900	2.0	18.6	20–25	9.5	5.6	0.86	0.53	76	0.20	0.20
1.81	1.09	6.99	6.99	16.1	0.042	0.040	19	900	2.4	18.4	< 15	9.8	5.9	0.94	0.62	86	0.23	0.22
1.21	0.68	7.57	0.00	7.9	0.026	0.026	14	1,000	0.0	14.2	100	8.5	4.8	0.53	0.00	56	0.18	0.18
1.71	0.98	7.57	2.77	12.6	0.035	0.035	19	1,000	1.0	20.0	70–80	8.6	4.9	0.60	0.37	63	0.18	0.18
1.79	1.03	7.57	4.36	14.2	0.037	0.037	20	1,000	1.5	20.1	55–65	8.9	5.1	0.72	0.45	71	0.18	0.18
1.83	1.06	7.57	5.98	15.7	0.035	0.035	20	1,000	2.0	19.5	25–35	9.4	5.4	0.86	0.56	81	0.18	0.18
1.83	1.06	7.57	6.75	16.3	0.042	0.042	20	1,000	2.2	18.7	< 15	9.8	5.7	0.94	0.62	86	0.22	0.22
Pregnant yearling heifers — last third of pregnancy																		
1.37	0.79	5.79	0.84	8.1	0.035	0.035	20	700	1.0	15.6	100	8.8	5.1	0.49	0.17	52	0.23	0.23
1.75	1.00	5.79	1.86	10.8	0.044	0.044	23	700	1.5	19.6	100	8.8	5.1	0.49	0.17	52	0.21	0.21
1.91	1.10	5.79	2.43	12.2	0.049	0.043	27	700	1.8	21.1	85–100	9.0	5.3	0.56	0.27	58	0.23	0.21
1.46	0.84	6.41	0.91	8.5	0.036	0.036	21	800	1.0	16.6	100	8.7	5.0	0.49	0.17	52	0.22	0.22
1.88	1.08	6.41	2.06	11.6	0.045	0.045	26	800	1.5	21.5	100	8.7	5.0	0.49	0.17	52	0.20	0.20
2.04	1.17	6.41	2.70	13.2	0.049	0.048	30	800	1.8	23.4	85–100	8.7	5.0	0.54	0.23	56	0.20	0.20
1.58	0.91	7.0	0.98	9.3	0.038	0.038	24	900	1.0	18.2	100	8.7	5.0	0.49	0.17	52	0.21	0.21
2.05	1.18	7.0	2.25	12.5	0.049	0.049	33	900	1.5	23.4	100	8.7	5.0	0.49	0.17	52	0.20	0.20
2.24	1.29	7.0	2.94	14.2	0.049	0.049	33	900	1.8	25.8	85–100	8.7	5.0	0.53	0.23	55	0.19	0.19
Dry pregnant mature cows — middle third of pregnancy																		
0.73	0.34	6.4		6.8	0.023	0.023	17	800		12.5	100	5.9	2.8	0.49		52	0.18	0.18
0.79	0.37	7.0		7.4	0.025	0.025	18	900		13.6	100	5.9	2.8	0.49		52	0.18	0.18
0.87	0.40	7.6		8.0	0.027	0.027	19	1,000		14.8	100	5.9	2.8	0.49		52	0.18	0.18
0.94	0.44	8.1		8.6	0.029	0.029	21	1,100		15.9	100	5.9	2.8	0.49		52	0.18	0.18

0.99	0.47	8.7		9.1	0.030	0.030	22	1,200		17.0	100	5.9	2.8	0.49		52	0.18	0.18
1.06	0.51	9.2		9.6	0.032	0.032	23	1,300		18.0	100	5.9	2.8	0.49		52	0.18	0.18
1.13	0.54	9.8		10.2	0.034	0.034	25	1,400		19.1	100	5.9	2.8	0.49		52	0.18	0.18
Dry pregnant mature cows — last third of pregnancy																		
0.94	0.43	8.0		8.2	0.028	0.028	21	800	0.9	14.3	100	5.9	2.8	0.49		52	0.18	0.18
1.08	0.46	8.6		8.8	0.030	0.030	22	900	0.9	15.5	100	5.9	2.8	0.49		52	0.18	0.18
1.10	0.50	9.2		9.4	0.032	0.032	23	1,000	0.9	16.7	100	5.9	2.8	0.49		52	0.18	0.18
1.12	0.53	9.7		10.0	0.034	0.034	25	1,100	0.9	17.9	100	5.9	2.8	0.49		52	0.18	0.18
1.18	0.56	9.9		10.6	0.036	0.036	26	1,200	0.9	19.2	100	5.9	2.8	0.49		52	0.18	0.18
1.24	0.59	10.1		11.2	0.038	0.038	27	1,300	0.9	20.5	100	5.9	2.8	0.49		52	0.18	0.18
1.30	0.63	10.3		11.7	0.040	0.040	29	1,400	0.9	21.9	100	5.9	2.8	0.49		52	0.18	0.18
Cows nursing calves — average milking ability — first 3 to 4 months postpartum[d]																		
1.69	0.99	9.4		9.9	0.054	0.054	34	800		18.4	100	9.2	5.4	0.49		52	0.29	0.29
1.78	1.05	9.9		10.5	0.055	0.056	36	900		19.4	100	9.2	5.4	0.49		52	0.28	0.28
1.88	1.11	10.5		11.1	0.058	0.058	37	1,000		20.5	100	9.2	5.4	0.49		52	0.28	0.28
1.98	1.17	11.1		11.7	0.060	0.060	39	1,100		21.6	100	9.2	5.4	0.49		52	0.28	0.28
2.08	1.23	11.6		12.3	0.062	0.062	41	1,200		22.7	100	9.2	5.4	0.49		52	0.27	0.27
2.18	1.29	12.1		12.9	0.062	0.062	43	1,300		23.8	100	9.2	5.4	0.49		52	0.25	0.25
2.29	1.35	12.7		13.5	0.063	0.063	44	1,400		24.9	100	9.2	5.4	0.49		52	0.25	0.25
Cows nursing calves — superior milking ability — first 3 to 4 months postpartum[c]																		
2.48	1.45	12.5		13.0	0.097	0.087	42	800		22.8	100	10.9	6.4	0.53		55	0.44	0.39
2.60	1.52	13.0		13.6	0.098	0.089	44	900		23.8	100	10.9	6.4	0.53		55	0.42	0.38
2.72	1.59	13.6		14.2	0.099	0.091	45	1,000		24.9	100	10.9	6.4	0.53		55	0.40	0.37
2.84	1.67	14.2		14.8	0.100	0.093	47	1,100		26.0	100	10.9	6.4	0.53		55	0.39	0.36
2.95	1.74	14.8		15.4	0.101	0.095	49	1,200		27.0	100	10.9	6.4	0.53		55	0.37	0.35
3.06	1.81	15.4		16.0	0.101	0.095	50	1,300		28.1	100	10.9	6.4	0.53		55	0.36	0.34
3.18	1.88	16.0		16.6	0.101	0.095	52	1,400		29.2	100	10.9	6.4	0.53		55	0.35	0.33
Bulls, growth and maintenance — moderate activity																		
2.09	1.29	5.6	3.8	13.2	0.064	0.053	37	700	2.2	20.5	70–75	10.2	6.3	0.64	0.32	64	0.31	0.26
2.33	1.39	6.9	4.1	15.8	0.052	0.052	44	900	2.0	24.7	70–75	9.4	5.6	0.64	0.30	64	0.21	0.21
2.36	1.36	8.5	3.7	16.5	0.048	0.048	48	1,100	1.5	26.9	80–85	8.8	5.1	0.60	0.30	61	0.18	0.18
2.29	1.32	9.7	3.0	15.9	0.047	0.047	47	1,300	1.1	26.2	80–85	8.5	5.1	0.60	0.30	61	0.18	0.18
2.35	1.33	10.8	2.0	15.2	0.050	0.050	49	1,500	0.7	27.9	90–100	8.5	4.8	0.53	0.26	55	0.18	0.18
1.90	1.07	11.9	0.0	12.3	0.040	0.040	40	1,700		24.6	100	8.5	4.8	0.53	0.00	55	0.18	0.18
2.15	1.22	13.4	0.0	13.9	0.046	0.046	45	2,000		25.2	100	8.5	4.8	0.53	0.00	55	0.18	0.18
2.32	1.32	14.4	0.0	15.2	0.049	0.049	49	2,200		27.3	100	8.5	4.8	0.53	0.00	55	0.18	0.18

TABLE 18-1 (continued)

		Daily nutrients per animal										Nutrient concentration in diet (dry matter)							
	Digest-ible	Energy					Vitamin		Avg.	Dry matter			Digest-ible	Energy					
Total pro-tein (lb)	pro-tein (lb)	NE m [b] (mcal)	NE g [c] (mcal)	TDN (lb)	Ca (lb)	P (lb)	A (thou-sands IU)	Body wt. (lb)	daily gain (lb)	(100%) per animal (lb)	Rough-age (%)	Total pro-tein (%)	pro-tein (%)	NE m (mcal)	NE g (mcal)	TDN (%)	Ca (%)	P (%)
							Growing-finishing steer calves and yearlings											
0.48	0.26	3.07	0.00	3.2	0.008	0.008	5	300	0.0	5.6	100	8.6	4.7	0.53	0.00	57	0.14	0.14
0.88	0.56	3.07	1.12	4.6	0.031	0.024	7	300	1.0	7.8	70-80	11.3	7.2	0.60	0.42	60	0.39	0.31
1.03	0.70	3.07	1.60	5.5	0.042	0.029	8	300	1.5	7.9	50-60	13.0	8.9	0.71	0.45	70	0.53	0.37
1.14	0.79	3.07	2.15	6.0	0.053	0.035	8	300	2.0	7.8	25-30	14.6	10.0	0.82	0.54	77	0.68	0.45
1.23	0.90	3.07	2.76	6.5	0.062	0.042	8	300	2.5	7.7	<15	16.0	11.7	0.94	0.62	84	0.81	0.55
0.59	0.33	3.81	0.00	3.9	0.011	0.011	7	400	0.0	7.0	100	8.3	4.6	0.53	0.00	55	0.15	0.15
1.19	0.66	3.81	1.38	6.1	0.031	0.026	9	400	1.0	10.9	80-90	10.9	6.1	0.56	0.34	56	0.28	0.24
1.25	0.81	3.81	1.98	7.1	0.040	0.031	11	400	1.5	11.1	70-80	11.3	7.3	0.60	0.42	64	0.36	0.28
1.27	0.85	3.81	2.61	7.6	0.051	0.037	11	400	2.0	9.9	30-40	12.8	8.6	0.80	0.51	77	0.52	0.37
1.35	0.93	3.81	3.33	8.0	0.062	0.045	11	400	2.5	9.7	<15	13.9	9.6	0.94	0.59	83	0.64	0.46
0.72	0.39	4.51	0.00	4.6	0.013	0.013	8	500	0.0	8.7	100	8.3	4.5	0.53	0.00	53	0.15	0.15
1.29	0.73	4.51	1.53	7.2	0.031	0.029	11	500	1.0	11.4	75-85	11.3	6.4	0.56	0.46	63	0.27	0.25
1.34	0.85	4.51	2.32	8.3	0.040	0.035	13	500	1.5	12.7	55-65	10.6	6.7	0.74	0.39	65	0.31	0.28
1.43	0.92	4.51	3.09	9.1	0.051	0.040	13	500	2.0	12.3	45-50	11.6	7.5	0.74	0.50	74	0.41	0.33
1.56	1.00	4.51	4.04	9.6	0.060	0.044	13	500	2.5	12.1	<15	12.9	8.3	0.94	0.55	79	0.49	0.36
0.81	0.46	5.17	0.00	5.4	0.018	0.018	9	600	0.0	10.0	100	8.1	4.6	0.53	0.00	54	0.18	0.18
1.22	0.75	5.17	1.78	8.4	0.033	0.029	12	600	1.0	12.8	65-70	9.5	5.9	0.60	0.42	56	0.26	0.23
1.42	0.89	5.17	2.67	10.2	0.040	0.035	14	600	1.5	15.5	55-65	9.2	5.7	0.71	0.32	66	0.26	0.23
1.63	1.03	5.17	3.54	10.7	0.049	0.042	15	600	2.0	15.6	45-55	10.5	6.6	0.74	0.41	69	0.31	0.27
1.70	1.05	5.17	4.62	11.4	0.057	0.046	15	600	2.5	15.4	20-25	11.0	6.7	0.82	0.51	73	0.37	0.30
1.75	1.15	5.17	5.58	12.6	0.066	0.051	15	600	3.0	14.4	<15	12.2	8.0	0.94	0.62	86	0.46	0.35
0.92	0.52	5.79	0.00	6.0	0.020	0.020	11	700	0.0	11.0	100	8.4	4.7	0.53	0.00	55	0.18	0.17
1.55	0.97	5.79	3.05	10.7	0.041	0.036	15	700	1.5	16.9	55-65	9.2	5.7	0.71	0.35	63	0.24	0.21
1.76	1.12	5.79	3.98	12.2	0.049	0.042	17	700	2.0	18.9	45-55	9.3	5.9	0.71	0.37	65	0.26	0.22
1.81	1.15	5.79	5.24	12.9	0.055	0.049	17	700	2.5	17.1	20-25	10.6	6.7	0.82	0.52	75	0.32	0.28
1.90	1.22	5.79	6.28	14.0	0.064	0.051	17	700	3.0	16.8	<15	11.3	7.3	0.94	0.59	83	0.38	0.30
1.03	0.59	6.41	0.00	6.6	0.022	0.022	12	800	0.0	12.0	100	8.6	4.9	0.53	0.00	55	0.18	0.18
1.78	1.09	6.41	4.43	13.0	0.044	0.040	16	800	2.0	17.9	45-55	9.9	6.1	0.74	0.48	70	0.27	0.24
1.86	1.16	6.41	5.74	14.3	0.051	0.044	17	800	2.5	18.3	20-25	10.2	6.3	0.82	0.54	78	0.28	0.24
1.97	1.24	6.41	6.98	16.0	0.057	0.049	18	800	3.0	18.3	<15	10.8	6.8	0.96	0.60	86	0.31	0.27
1.13	0.63	7.00	0.00	7.3	0.024	0.024	13	900	0.0	13.2	100	8.6	4.8	0.53	0.00	55	0.18	0.18

1.88	1.14	7.00	4.92	13.8	0.046	0.044	18	900	2.0	20.9	45–55	9.2	5.5	0.74	0.43	71	0.22	0.21
1.95	1.21	7.00	6.58	15.6	0.051	0.046	19	900	2.5	19.7	20–25	9.9	6.4	0.82	0.59	79	0.26	0.23
2.04	1.27	7.00	7.48	17.1	0.055	0.049	19	900	3.0	19.7	< 15	10.4	6.4	0.98	0.62	86	0.28	0.25
1.19	0.68	7.58	0.00	7.9	0.026	0.026	14	1,000	0.0	14.2	100	8.4	4.8	0.53	0.00	55	0.18	0.18
1.94	1.13	7.58	5.32	15.5	0.044	0.044	19	1,000	2.0	22.5	45–55	8.6	5.0	0.74	0.52	69	0.20	0.20
2.09	1.25	7.58	6.85	17.1	0.051	0.049	20	1,000	2.5	22.1	20–25	9.4	5.7	0.82	0.53	80	0.23	0.22
2.16	1.32	7.58	8.21	17.9	0.053	0.051	20	1,000	3.0	20.8	< 15	10.4	6.3	0.98	0.62	86	0.25	0.25
1.32	0.75	8.14	0.00	8.4	0.029	0.029	15	1,100	0.0	15.4	100	8.6	4.9	0.53	0.00	55	0.19	0.19
2.09	1.23	8.14	5.60	16.5	0.042	0.042	23	1,100	2.0	23.1	45–55	9.0	5.3	0.74	0.46	72	0.18	0.18
2.12	1.27	8.14	7.38	18.9	0.044	0.044	23	1,100	2.5	22.9	20–25	9.3	5.6	0.82	0.56	83	0.20	0.20
2.15	1.33	8.14	8.87	20.6	0.048	0.049	23	1,100	2.9	22.0	< 15	9.7	6.0	0.98	0.62	86	0.22	0.22
Growing-finishing heifer calves and yearlings																		
0.48	0.28	3.06	0.00	3.2	0.008	0.008	6	300	0.0	5.8	100	8.3	4.8	0.53	0.00	55	0.14	0.14
0.90	0.58	3.06	1.13	4.8	0.031	0.024	8	300	1.0	8.1	70–80	11.1	7.2	0.60	0.37	59	0.38	0.29
1.05	0.70	3.06	1.80	5.7	0.042	0.031	8	300	1.5	8.1	50–60	13.0	8.6	0.71	0.47	70	0.52	0.38
1.14	0.79	3.06	2.41	6.3	0.053	0.037	8	300	2.0	8.2	25–30	13.9	9.6	0.82	0.54	77	0.65	0.45
1.30	0.93	3.06	3.21	6.6	0.064	0.044	8	300	2.5	8.2	< 15	15.9	11.3	0.94	0.62	81	0.78	0.54
0.61	0.35	3.74	0.00	4.2	0.011	0.011	7	400	0.0	7.1	100	8.6	4.9	0.53	0.00	59	0.15	0.15
1.12	0.71	3.74	1.38	6.8	0.029	0.026	12	400	1.0	11.4	75–85	9.8	6.2	0.57	0.29	60	0.25	0.22
1.25	0.81	3.74	2.24	7.6	0.031	0.030	12	400	1.5	11.6	65–75	10.7	7.0	0.67	0.37	66	0.27	0.26
1.30	0.85	3.74	2.99	8.1	0.051	0.037	12	400	2.0	10.6	30–40	12.3	8.0	0.80	0.50	76	0.48	0.35
1.40	0.96	3.74	3.99	9.0	0.062	0.044	12	400	2.5	10.1	< 15	13.9	9.5	0.94	0.62	86	0.61	0.44
0.72	0.41	4.50	0.00	4.7	0.013	0.013	9	500	0.0	8.4	100	8.6	4.9	0.53	0.00	56	0.15	0.15
1.30	0.77	4.50	1.76	8.0	0.029	0.029	14	500	1.0	13.6	80–90	9.6	5.7	0.56	0.31	59	0.21	0.21
1.35	0.83	4.50	2.65	8.8	0.040	0.035	14	500	1.5	13.4	60–70	10.1	6.2	0.69	0.38	66	0.30	0.26
1.45	0.90	4.50	3.54	9.5	0.048	0.037	14	500	2.0	12.3	35–45	11.8	7.3	0.80	0.53	77	0.39	0.30
1.52	1.04	4.50	4.73	10.8	0.055	0.042	14	500	2.5	12.4	< 15	12.3	8.4	0.94	0.62	86	0.44	0.34
0.81	0.46	5.16	0.00	5.4	0.015	0.015	9	600	0.0	9.6	100	8.4	4.8	0.53	0.00	56	0.16	0.16
1.39	0.82	5.16	1.89	8.9	0.031	0.031	14	600	1.0	15.2	80–90	9.1	5.4	0.56	0.32	59	0.20	0.20
1.41	0.85	5.16	3.08	9.7	0.037	0.033	15	600	1.5	13.6	55–65	10.4	6.3	0.74	0.47	71	0.27	0.24
1.48	0.94	5.16	4.05	10.7	0.046	0.037	15	600	2.0	13.9	35–45	10.6	6.8	0.82	0.53	77	0.33	0.27
1.68	1.08	5.16	5.43	12.5	0.055	0.045	15	600	2.5	14.7	< 15	11.4	7.3	0.94	0.59	85	0.37	0.31
0.92	0.52	5.79	0.00	5.9	0.020	0.019	11	700	0.0	10.8	100	8.5	4.8	0.53	0.00	55	0.18	0.18
1.47	0.87	5.79	2.13	9.5	0.028	0.028	16	700	1.0	17.0	80–90	8.6	5.1	0.60	0.29	56	0.16	0.16
1.52	0.90	5.79	3.48	10.9	0.035	0.031	17	700	1.5	15.6	55–65	9.7	5.8	0.73	0.45	70	0.22	0.24
1.58	0.99	5.79	4.55	12.1	0.042	0.037	17	700	2.0	16.0	35–45	9.9	6.1	0.81	0.51	76	0.26	0.23
1.78	1.10	5.79	5.96	13.8	0.051	0.044	17	700	2.5	16.9	< 15	10.5	6.5	0.94	0.55	82	0.30	0.26
1.03	0.57	6.40	0.00	6.6	0.022	0.022	12	800	0.0	12.0	100	8.6	4.8	0.53	0.00	55	0.18	0.18

[a]From *Beef Cow/Calf Manual*, Manual 104, Extension Division, University of Missouri, Columbia, p. 36.
[b]NE_m = Net energy required for maintenance.
[c]NE_g = Net energy requirement for gain.
[d]Average milking ability = 10–12 lb/day.
[e]Superior milking ability = 21–23 lb/day.

18.1.1 Preparation of Grain

Whether or not it will pay to grind grain for beef cattle depends on (1) the age of animals fed, (2) the nature of the feed, and (3) whether or not it is to be mixed with other feeds in the ration.

It may be profitable to grind grain for calves up to a few weeks of age before their teeth are developed. After that time and until they are 6 to 9 months of age, calves chew grain so thoroughly that grinding is not necessary. It usually pays to grind all grains other than corn for fattening beef cattle, but coarse grinding is preferred. Corn does not need to be ground for beef cattle, especially if pigs are placed in the feeding pens to eat the unchewed grains in the manure.

Steam rolling or crushing corn does not appear to have any advantage over coarse grinding for beef cattle. Crushed or rolled grain is sometimes preferred, however, for fattening and fitting cattle for the show ring. Some showpeople even prefer to cook their grain for their animals.

18.1.2 Preparation of Roughages

It usually does not pay to grind or chop good-quality hay for beef cattle if hay and grain are fed separately. If a complete ration containing both roughage and grain is fed to cattle, as is often done in performance-testing bull calves, the hay must be ground so it mixes well in the ration.

Coarsely chopping poor-quality roughage for beef cattle is often profitable. Chopping hay in such cases is preferred to grinding because it is cheaper. Feeding results for both are similar. Finely ground hay may be lower in digestibility.

Shredding corn or sorghum fodder often pays because it decreases the amount of these roughages wasted when fed. It does not appear to increase their digestibility, however.

18.1.3 Pelleted or Cubed Feeds

Protein supplements are usually cubed or pelleted when fed to cows on pasture or the range. The cubes or pellets can even be fed on the ground except in rainy or wet weather. Many commercial protein supplements fed to beef cattle are pelleted and mixed with grain when fed.

Poor-quality hay may be fed in the cubed or wafered form,

especially in commercial feedlots. The main advantages are that it is easier to handle, and there is less wastage.

18.2   Manner of Feeding

Rations for beef cattle may be hand-fed or self-fed. There are some advantages and disadvantages with both methods, and under certain conditions one method may be preferred to the other.

Self-feeding has several advantages over hand feeding. When self-feeders are used which have a large hopper, they may contain enough feed to last several days. This requires less labor than hand feeding. Self-fed cattle tend to gain a little faster than those which are hand-fed, and there is less tendency for them to go off feed. This may be due to the fact that each animal that is self-fed can eat whenever it desires and there is less tendency to overeat at any one time. Self-feeding can even be used to start cattle on feed by feeding a ration very high in bulk content and then gradually reducing the amount of bulk to the desired level until the cattle are full-fed on the more concentrated ration desired.

Self-feeding has some disadvantages as compared to hand feeding. Self-fed cattle usually are not observed as closely as those which are hand-fed twice daily. Any animal that is ill or off feed will not come up to the feed bunk and eat when the feed is offered, and such animals can be easily located and treated. Self-fed cattle are usually not observed very closely. Self-feeders must be carefully adjusted to prevent clogging, especially if certain feeds that are bulky are fed. When mixtures of ground grain and pelleted protein supplements are self-fed, cattle tend to seek out the grain and leave the pellets which accumulate in the trough of the self-feeder, and a balanced ration may not be consumed. Feed in a self-feeder must also be protected from rain and moisture, or it will rot and mold and clog the feeder. The feed in a self-feeder cannot be as fresh as that which is hand-fed unless the self-feeder is closely watched and regulated properly.

Most commercial feedlots do not use a self-feeding system. Fence line bunks are usually used and the complete feed put in the bunk each day by means of trucks which deliver the desired amount of feed by means of an auger system. One truck can feed hundreds of cattle each day by this system.

Hand feeding of rations is usually done twice each day. In large commercial operations which use mechanical equipment, the feed is placed in the feeding bunks only once each day. Some feeders try to regulate the amount of feed offered so it will be eaten in 24 to 36 hours. The main objective is to have a supply of feed before the animals at all times so cattle make the best possible daily gains.

Experimental evidence indicates that beef cows and heifers on winter ranges can be fed protein supplements at intervals of 2 to 6 days as effectively as they can be fed daily. This saves considerable labor. Urea supplements should not be fed in this manner because of the danger of toxicity when too much of a urea-bearing feed is fed.

18.3   Fattening Cattle on Pasture

Most beef cattle are fattened in drylot, although some are fattened on pasture. Cattle producers who feed small numbers of cattle are likely to fatten their animals in order to utilize available pasture and market their grain.

Several systems of pasture feeding may be used. One is to fatten cattle on grass alone. Cattle fattened in this manner should be older and early maturing types. Younger animals tend to grow rather than fatten and will not finish as well as older cattle. A second system is to feed a limited amount of grain on pasture. A third system is to full-feed throughout the grazing season. A fourth system is to graze the cattle on pasture alone during the summer and then fatten them in the feedlot on a full feed for 90 or more days. In all these systems a good supply of water and a mineral supplement should be available at all times.

Fattening cattle on pasture has several advantages. Pasture gains are usually cheaper because less protein is fed and it is not necessary to supply a roughage. Pasture feeding requires less labor because grain can be self-fed and cattle can harvest their own roughage. Cattle on pasture scatter manure over the fields, which eliminates the necessity of hauling and spreading it as is necessary when cattle are fed in drylot. Natural shades in pastures may be utilized without erecting expensive shades or buildings. Finally, fattening cattle on pasture allows the utilization of land that can be used for the production of cultivated crops.

Fattening cattle on pasture also has some disadvantages. Pasture-fattened cattle may sell for less than cattle fed in drylot because they may not carry as much finish when marketed. Cattle to be fattened on pasture may have to be purchased in the spring when feeder prices are very high, and they will be marketed in the fall when fat cattle prices may be low. Cattle fattened on pasture may make slow gains because of heat and flies and because drought may lower the quantity and quality of the pasture. Pasture-fattened cattle tend to shrink more when shipped to market, and their fat may be more yellow in color because of its high carotene content.

18.4   Rations for Beef Cattle

The rations fed to beef cattle depend on the age of the animals, the stage of production they are in at a particular time, the feeds available, and the cost of these feeds. Feed costs will vary from one part of the country to the other. In the West, barley may be the chief grain fed because it is available at a nominal cost and in large quantities. In the Corn Belt, corn may be the preferred grain, and corn silage may be the preferred roughage.

18.4.1   Feeding Replacement Heifers

Replacement heifers weaned in the fall will usually be 7 to 8 months of age and weigh 400 to 500 pounds. Heifers from most British breeds will reach puberty at 12 to 14 months of age, and it is desirable for them to weigh 600 to 700 pounds at that time. Weanling heifers should gain 1 to 1.25 pounds per day if they are to come in heat and settle when bred at about 14 months of age. Sample rations that should produce such gains are given in Table 18-2.

Pregnant heifers need to be fed properly during their second winter in order to be in the proper condition to rebreed on time when they are nursing their first calf. This usually requires that they be fed to gain 0.5 to 1.0 pound per day. A first-calf heifer is still growing in addition to producing milk, and unless she is fed properly, she may be slow in breeding during lactation. Some recommend that heifer calves be bred about 30 days before the older cows in the herd to increase the likelihood the mature cow will rebreed and calve on schedule.

First-calf heifers that are heavy milkers and thinner than average should be given extra feed such as grain to induce them to rebreed on schedule.

18.4.2   Feeding Pregnant Cows

Mature cows in average condition in the fall may lose 10 to 15 percent of their weight and still rebreed on schedule in the spring and summer while nursing a calf. From the practical standpoint cows

TABLE 18-2   Winter rations for growing heifers[a]

	Pounds
A. Rations to give 1 lb/head daily gain	
1. Silage and protein supplement:	
Silage	25–35
Protein (44%)	1
Mineral mix	Free-choice
Salt, 1 part	
Dicalcium phosphate, 1 part	
2. Silage and legume hay:	
Silage	20–30
Hay (legume)	5
Mineral mix	Free-choice
Salt, 1 part	
Dicalcium phosphate, 1 part	
3. Hay ration:	
Hay ($\frac{1}{2}$ legume) — full-fed	12–15
Mineral mix	Free-choice
Salt, 1 part	
Monosodium phosphate, 1 part	
B. Rations to give $1\frac{1}{2}$ lb/head daily gain	
1. Silage and protein supplement:	
Silage — full-fed	30–50
Protein (44%)	$1\frac{1}{2}$
Mineral mix	Free-choice
Salt, 1 part	
Dicalcium phosphate, 1 part	
2. Hay and grain:	
Hay (at least $\frac{1}{2}$ legume)	10
Grain (1 lb/100 lb of body wt.)	4–6
Protein (44%)	$\frac{1}{2}$
Mineral mix	Free-choice
Salt, 1 part	
Dicalcium phosphate, 1 part	

[a]From *Beef Cow/Calf Manual,* Manual 104, Extension Division, University of Missouri, Columbia, p. 24.

should be fed to gain 100 to 130 pounds during the last 3 or 4 months of pregnancy. About one half of the gain in weight of the calf in the uterus occurs during the last 2 months of pregnancy. Most cows will lose 100 to 130 pounds at calving because of the birth weight of the calf and the fetal membranes and fluids lost. It is not recommended that cows be starved during the last one third of pregnancy to produce a smaller calf and less trouble at calving. Gain or loss in weight of the cow during this stage of pregnancy seldom affects the birth weight of the calf anyhow, because the cow draws on the nutrient stores in her own body for the proper nutrition of her developing calf. A depletion of nutrient reserves in her body may cause a delay in rebreeding while nursing her next calf.

Example rations for mature pregnant cows are given in Tables 18-3 and 18-4.

TABLE 18-3   Rations for pregnant cows which make maximum use of good-quality roughage[a]

Constituents	Percent-age of moisture	Rations (lb)					
		1	2	3	4	5	6
Legume hay, good							3
Grass hay, good		16		7			
Corn silage, good	67				35		
Atlas sorghum silage	72					50	40
Grass or grass-legume haylage	60		40	25			
Protein suppl. (40%)						1/4	
Mineral mix		+	+	+	+	+	

[a]From *Beef Cow/Calf Manual*, Manual 104, Extension Division, University of Missouri, Columbia, p. 25.

TABLE 18-4   Rations for pregnant cows using low-grade roughage[a]

Constituents	Percent-age of moisture	Rations (lb)					
		1	2	3	4	5	6
Corn cobs, cottonseed hulls, wheat straw, or cornstalks		14	14	10	14	10	10
Corn silage	67						20
Alfalfa meal, dehyd.			2				
Legume hay, good		6		5			
Grass hay, good						8	
Protein suppl. (40%)					3/4		1/2
Molasses (wet)				3			
Ground shelled corn			2 1/2		2 1/2		
Vitamin A[b]				+	+	+	+
Mineral mix		+	+	+	+	+	+

[a]From *Beef Cow/Calf Manual*, Manual 104, Extension Division, University of Missouri, Columbia, p. 25.
[b]30,000 IU/cow daily.

18.4.3 Feeding Lactating Cows

Lactation increases nutrient requirements by 50 to 100 percent as compared to cows not lactating. Heavy milking cows may have even higher nutrient requirements than this because of the extra milk they produce. Cows on good-quality pasture will usually obtain enough nutrients to meet their nutrient requirements and rebreed.

Cows which lactate during a drought or during the winter may need to be fed more feed than those on good pasture during the spring and summer. Energy, phosphorus, and protein are needed for high milk production. Lactating cows on poor-quality roughages may not receive adequate amounts of required nutrients. If they are fed silage and/or good-quality hay, no other feed may be needed. Cows in average condition should gain ¼ to ½ pound per day the first 3 months of pregnancy in order to rebreed on schedule. Example rations for lactating cows are given in Table 18-5.

18.4.4 Creep Rations for Calves

Creep feeding refers to the feeding of concentrates to beef calves while they are still nursing their mothers. The feed is usually supplied in the pasture in a bunk, which allows the calves to enter and eat but not the cows. Extra feed in addition to their mother's milk will often cause calves to weigh 50 to 70 pounds more at weaning (7+ months of age) than calves not creep-fed. About 500 to

TABLE 18-5   Rations for cows in the final 4 months of lactation[a]

Constituents	Percentage of moisture	Rations (lb)					
		1	2	3	4	5	6
Legume hay, good		6				7	
Grass hay, good		18	24				16
Corn silage	70			60			
Atlas sorghum silage	72					45	
Grass or grass-legume haylage	60				50		
Ground shelled corn					3	3	5
Protein suppl. (40%)			½	1½			1
Vitamin A[b]			+	+			+
Mineral mix			+	+	+	+	+

[a]From *Beef Cow/Calf Manual*, Manual 104, Extension Division, University of Missouri, Columbia, p. 25.
[b]45,000 IU/cow daily.

1,000 pounds of grain are required for an extra 100 pounds of gain in creep-fed calves. For this reason creep feeding may pay under certain conditions but not others.

Creep feeding is most likely to give the greatest returns when grain prices are low in relation to feeder calf prices. The following are some conditions when creep feeding is most likely to be economical: (1) in purebred herds where extra gain, finish, and "bloom" are desired; (2) when calves are born in the fall or early winter; (3) when pastures are so poor because of overgrazing or drought during the nursing period that milk production in the cow is reduced; (4) when the herd includes many first-calf heifers or aged cows; (5) when steer calves are to be sold as weanling feeders; and (6) when calves are to be sold for slaughter between 500 and 700 pounds in weight.

In some purebred herds where heavy weights are desired, calves are kept confined at all times, and feed is made available to them when they desire to eat. The cows are run on pasture and then turned with the calves for a period of time in the morning and again at night. This system of management encourages a greater grain consumption by the calves, and the cows can be observed for estrus twice daily. This is important when cows are hand-mated or bred by artificial insemination. Some experimental evidence suggests that cows nursing calves twice each day will show estrus sooner after calving than those with which the calves run continuously and nurse several times per day and night.

Calves sometimes are slow in beginning to eat creep rations. An older calf used to eating grain may be turned with the calves, which sets an example to the younger calves and starts them eating again. Some cattle producers feed some grain to both cows and calves in a corner of the pasture. The calves eat some grain with their mothers and develop a taste for grain. The cows are later fenced away from the grain, but the calves have access to it.

Several different rations may be offered calves in the feeder. Rations containing fiber in addition to grain may be fed to avoid founder. Ground whole ear corn or oats can be mixed with shelled corn. Ground sorghum grain can be fed in the place of corn.

A mixture of two parts shelled corn to one part whole oats may be fed calves 2 to 4 months of age. Calves can efficiently utilize shelled corn up to 6 to 9 months of age, but it can be coarsely ground if desired. Creep rations for calves 4 to 9 months of age may include (1) eight parts shelled corn to one part of a 40 percent protein supplement, (2) six parts of ground ear corn to one part of 40 percent protein supplement, or (3) six parts shelled corn to three parts oats to one part of a 40 percent protein supplement.

It is often desirable to feed good-quality legume hay along with the creep ration for calves not on pasture during the fall and winter months. Both cows and calves should have access to a mineral mixture at all times.

18.4.5   Finishing Rations

A good ration for finishing cattle should be palatable, economical, and free of toxic substances and should supply all the daily nutrient requirements of the beef animal.

Ration ingredients vary from one region of the country to another because some crops are more widely grown in certain areas. In addition, prices of ration ingredients may vary from season to season and year to year. Some complete rations for cattle are given in Tables 18-6 and 18-7.

High-moisture grains are sometimes fed to beef cattle. When harvested with a moisture content of 25 to 30 percent and ensiled, grain has been fed with satisfactory results. High-moisture storage of milo appears to improve its feeding value more than it does for corn in feeding cattle. Milo usually has a hard seed coat which should be ground except when stored with a high-moisture content. Harvesting high-moisture grain results in an early release of the land for another crop or for fall plowing and reduces losses of grain during the harvesting process. In addition, there may be as much as 10 percent improvement in the feed value of the grain. High-moisture grain, however, has to be fed to livestock and cannot be marketed, as can the drier grain. It may also freeze in the feed bunk in winter and may attract flies in the summer.

18.4.6   Feeding the Herd Bull

Rations for bulls can be the same as those fed to cows. Since bulls are larger than cows, they may need more feed per head per day. Recommendations given here are not for young bulls fed on a performance test but are for bulls retained for breeding after they have completed such a test.

Young bulls are still growing and should be fed to gain at least 2 pounds per day. They should be fed a ration containing 9.5 to 10.5 percent crude protein. One pound of grain per 100 pounds of body weight plus 1 pound of 40 percent protein supplement in addition to good-quality hay fed free-choice will usually give the desired gains.

Mature bulls can be maintained mostly on pasture or good-

TABLE 18-6   Complete mixed rations (mixture for 100 lb — as-fed basis)[a]

Ingredient	Ration (lb)		
	1	2	3
Ration A: Corn, soybean meal, alfalfa hay			
Ground shelled corn	50.42	64.70	68.58
Soybean meal	—	—	4.00
Alfalfa hay	49.00	34.80	27.00
Dicalcium phosphate	0.21	0.13	—
Limestone	—	—	0.02
Salt (trace mineral)	0.37	0.37	0.40
	100.00	100.00	100.00
Composition, dry matter basis:			
TDN	72.50%	77.50%	82.50%
Protein	14.40	13.00	12.80
Ca	0.72	0.52	0.47
P	0.33	0.33	0.33
Salt	0.45	0.45	0.45
Ration B: Corn, soybean meal, corn silage			
Ground shelled corn	4.80	17.46	38.00
Soybean meal (45%)	3.20	4.00	5.23
Corn silage	91.60	78.00	56.00
Dicalcium phosphate	0.12	0.08	—
Limestone	0.12	0.27	0.52
Salt, trace mineral	0.16	0.19	0.25
	100.00	100.00	100.00
Composition, dry matter basis:			
TDN	72.50%	77.50%	82.50%
Protein	12.00	12.50	12.80
Ca	0.44	0.44	0.44
P	0.33	0.33	0.33
Salt	0.45	0.45	0.45
Ration C: Corn, soybean meal, urea, timothy hay			
Ground shelled corn	47.62	63.23	78.80
Soybean meal (45%)	0.38	—	—
Urea (281 protein equiv.)	0.89	1.00	1.00
Timothy hay	50.00	34.54	18.85
Dicalcium phosphate	0.33	0.22	0.85
Limestone	0.38	0.61	0.85
Salt, trace mineral	0.40	0.40	0.40
	100.00	100.00	100.00
Composition, dry matter basis:			
TDN	72.50%	77.50%	82.70%

TABLE 18-6   Continued

| | Ration (lb) | | |
Ingredient	1	2	3
Protein	12.00	12.35	12.60
Ca	0.44	0.44	0.44
P	0.33	0.33	0.33
Salt	0.45	0.45	0.45

[a]From *Guide Sheet 2066*, University of Missouri, Columbia, courtesy of Dr. H. B. Sewell.

TABLE 18-7   Finishing rations for beef cattle[a]

	Pounds
1. Shelled corn-corn silage:	
Ground shelled corn	
(1–$1\frac{1}{2}$ lb/100 lb of body wt.)	8–15
Protein (44%)	$1\frac{1}{2}$
Corn silage	Full-fed
Mineral mix[b]	Free-choice
Salt, 1 part	
Limestone, 1 part	
Dicalcium phosphate, 1 part	
2. Shelled corn-corn silage:	
Ground shelled corn	Full-fed
Protein (44%)	$1\frac{1}{2}$
Corn silage	5–10
Mineral mix[b]	Free-choice
Limestone, 2 parts	
Salt, 1 part	
Dicalcium phosphate, 1 part	
3. Shelled corn-grass hay:	
Shelled corn	Full-fed
Hay (grass)	4–6
Protein (44%)	$1\frac{1}{2}$
Mineral mix[b]	Free-choice
Limestone, 2 parts	
Salt, 1 part	
Dicalcium phosphate, 1 part	
4. Shelled corn-legume hay:	
Ground shelled corn	Full-fed
Hay (legume, good quality)	4–6

TABLE 18-7   Continued

	Pounds
Protein (44%)	1
Mineral mix[b]	Free-choice
Salt, 1 part	
Dicalcium phosphate, 1 part	

[a]From *Guide Sheet 2066*, University of Missouri, Columbia, courtesy of Dr. H. B. Sewell.

[b]Use trace mineral salt. You may substitute bonemeal for dicalcium phosphate and tripolyphosphate for monosodium phosphate.

quality roughage between breeding seasons. If they become thin, a small amount of grain may be needed to bring them to the proper condition for the breeding season. Increasing the feed about 60 days before the breeding season begins is often desirable.

18.5   Formulating Rations

The daily nutrient requirements of beef cattle and the nutrient content of feeds are needed in formulating rations. These are given in Table 17-1 and Table 18-1. In some instances a laboratory analysis of feeds to be used may be obtained if desired.

The *first step* in formulating a ration is to record the kind of animal to be fed, its weight, and the daily gain desired. As an example we shall use a yearling steer weighing about 700 pounds with an expected daily gain of 2.5 pounds.

The *second step* is to itemize the feedstuffs to be used in the ration. Their approximate composition may be obtained from Table 17-1. The feeds used in this example and their percentage composition are as follows:

Percentage composition

	Day matter (D.M.)	Total protein	TDN	Calcium	Phosphorus
No. 2 yellow corn	89.0	8.9	81.0	0.02	0.31
Orchard grass hay	89.2	11.1	51.0	0.40	0.33
Soybean meal	89.0	45.8	72.0	0.32	0.67
Limestone	100.0			35.80	

The *third step* is to list the daily nutrient requirements in pounds for the 700-pound steer to gain about 2.5 pounds per day. These are obtained from Table 18-1 and are as follows:

		Pounds of nutrients			
	D.M.	Total protein	TDN	Calcium	Phosphorus
2.5-lb daily gain	17.1	1.97	12.9	0.055	0.49

The *fourth step* is to decide on the pounds of hay to be fed per head per day. We shall use 5 pounds of hay in this example, but more or less can be included, depending on the costs of the feeds available and the rate of gain desired. The nutrients supplied by 5 pounds of orchard grass hay as taken from Table 17-1 would be as follows:

		Pounds of nutrients supplied			
	D.M.	Total protein	TDN	Calcium	Phosphorus
5 lb of orchard grass hay in full bloom	4.46	0.56	2.55	0.02	0.017

In the *fifth step*, subtract the nutrients in 5 pounds of orchard grass hay from the total amounts required as given in step 3. The remainder would be the amount of nutrients to be supplied by the corn, soybean meal, and limestone in the ration. This would be as follows:

		Pounds of nutrients required			
	D.M.	Total protein	TDN	Calcium	Phosphorus
Nutrients from corn, soybean meal, and limestone	12.64	1.41	10.35	0.035	0.032

Since corn and soybean meal have a TDN content of about 77 percent (average), it will take about 13.44 pounds of the two (10.35 ÷ 0.77 = 13.44) to supply the 10.35 pounds of TDN. The protein to be supplied by the mixture of corn and soybean meal is 1.41 pounds. Thus, the percentage of protein needed would be 10.5 percent (1.41 ÷ 13.44 - 10.5).

The square method can now be used to calculate the ratio of corn and soybean meal required to give a 10.5 percent protein mixture from corn and soybean meal (SBM). This may be done in the following steps:

1. Put the percentage of protein (10.5 percent) needed in the center of the square:

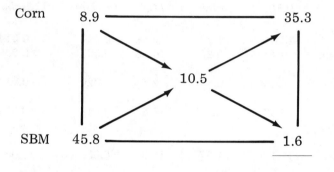

Corn   8.9 ———————————————— 35.3

10.5

SBM   45.8 ———————————————— 1.6

36.9 Total parts

2. Place the corn and soybean meal and their percentages of protein on the left-hand corners of the squares, as shown above.
3. Subtract diagonally the smaller from the larger figures (10.5 – 8.9 = 1.6; 45.8 – 10.5 = 35.3). The required parts of corn to give the 10.5 percent protein mixture would be 35.3, and the parts of soybean meal would be 1.6, giving a total of 36.9 parts.
4. Calculate the percentage of the total parts (36.9) that is corn. This would be 35.3 ÷ 36.9 or 96.0 percent. Next, calculate the percentage of the total parts (36.9) that is soybean meal. This would be 1.6 ÷ 36.9 = 4 percent. Thus, 100 pounds of this mixture would contain 96 pounds of corn and 4 pounds of soybean meal. Since 13.44 pounds of these feeds will be required to supply the daily TDN, 96 percent of this, or 12.9 pounds of corn, 4 percent of 13.44, or 0.54 pound of soybean meal, would be required. This can be rounded off to 13.5 pounds of corn and 0.5 pound of soybean meal.

The balanced ration would contain the following:

Corn	13.5 lb
Soybean meal	0.5 lb
Orchard grass hay	5.0 lb
Limestone	0.02 lb

The daily nutrients supplied by this ration using feed compositions given in Table 17-1 would be as follows:

Pounds of nutrients

	D.M.	Total protein	TDN	Calcium	Phosphorus
13.5 lb of corn	12.0	1.20	12.56	0.027	0.014
0.5 lb of soybean meal	0.45	0.23	0.36	0.001	0.003
5.0 lb of orchard grass hay	4.46	0.56	2.55	0.020	0.017
Nutrients per day supplied by this ration	16.93	1.99	15.47	0.048	0.034
Nutrients required for 2.5-lb daily gain	17.1	1.97	12.9	0.055	0.049

The calculated ration is very close to the requirements, being equal to or exceeding the requirements in each case except for calcium. The deficiency in calcium would be 0.055 – 0.048 = 0.007 pound. Only 0.02 pound of limestone per day (0.007 ÷ 0.3584) added to the above ration would supply the needed calcium.

Various rations can be formulated as in the example shown. In recent years, however, the computer has been used to build nutritionally adequate rations from many feeds at the least cost. Many cattle feeders have access to the computer for this service through their local or state extension division.

18.6   Feed Additives

Many compounds have been added to beef cattle rations in recent years for the purpose of stimulating more rapid and efficient gains. These are called feed additives.

Feed additives are growth-promoting substances added to rations that include antibiotics, hormones, or hormonelike drugs. They improve the health of the individuals and/or stimulate faster and more efficient gains.

Amendments to feed additive laws passed in 1958 and 1962 are of great importance to the livestock industry. The 1958 amendment requires that new feed additives not be used until the safety of the product has been assured. The 1962 amendment requires that new drugs be evaluated for effectiveness as well as for safety. Drug companies have to spend much time and money in developing new products before they meet the rigid requirements. No additive is considered safe if it is shown to induce cancer when ingested by humans or animals. There can be no residue of drugs in animal tissues, milk, or eggs. Even when a drug is approved, the feed manufacturer must be capable of mixing feeds properly, must properly label medicated feeds, must store such feeds properly, must inform customers of precautions in feeding, and must share new information with customers as it becomes available.

Feeders also have certain responsibilities in using approved drugs in feeds. They must use the feeds for the purpose intended, must follow directions on the feed label, and must heed warning statements carried there. They must also store medicated feeds properly and observe withdrawal periods if they are required. The Food and Drug Administration (FDA) of the U.S. government is charged with approving the use of feed additives and the safeguarding of human health.

18.6.1   Hormones and Hormonelike Substances

Diethylstilbestrol (DES) was placed on the market in 1954. When mixed in the feed, it was found to improve rate of gain by 12 to 15 percent and efficiency of gain by 8 to 10 percent. The response to DES was slightly less in heifers than in steers. It should not be fed to breeding animals because it would interfere with their fertility. DES is a synthetic estrogenic compound which produces a physiological response similar to that produced by estrogen, the female sex hormone. Because of possible harmful effects of DES when consumed by humans in animal products, it is no longer used in beef cattle rations. It can still be used as an implant pending further consideration. The discovery of the growth-promoting effects of DES stimulated many pharmaceutical companies to develop other feed additives. Several have been used with good results.

Melengestrol acetate (MGA) is an orally effective progesteronelike compound. It has been approved for use in the rations of fattening heifers. It is not effective in steer or spayed heifer rations. The approved level of MGA in the ration is 0.25 to 0.50 milligrams per head per day. Recent evidence shows that MGA in heifer rations increases gains by 10 to 12 percent and feed efficiency by about 6 to 8 percent. MGA must be withdrawn 48 hours before the heifers are slaughtered.

Rumensin (sodium monensin) alters the metabolic process within the rumen and changes the action of rumen microorganisms so that cattle can use the feeds they consume more efficiently. Many experiments in which 30 grams of Rumensin were added per ton of air-dried ration resulted in a saving of 10 to 11 percent in the pounds of feed required per pound of gain for cattle in the feedlot. The rate of gain was not significantly affected.

Rumensin has also been cleared for cattle on pasture which weigh more than 400 pounds. Rumensin can be fed in a dry supplement carrier or in meals, cubes, cakes, or pellets, or it can be mixed with the grain. The optimum amount of Rumensin for cattle appears to be 200 milligrams per head per day. Results of many experiments indicate that cattle fed this product on pasture gained about 16 percent faster than those not fed Rumensin. It appears to be effective when fed to cattle on all types of grasses.

Rumensin can be fed up to the time of slaughter because it is not a hormone and no withdrawal time is required. It should be kept away from horses because it has resulted in fatalities when horses have eaten it. Rumensin is not recommended at the present for cows or replacement heifers.

18.7   Implants for Beef Cattle

Some drugs and other substances are implanted under the skin behind the ear rather than being added to the feed. DES can be implanted as well as mixed in the ration.

Synovex-S contains 200 milligrams of progesterone and 20 milligrams of estradiol benzoate. *Synovex-H* contains 200 milligrams of testosterone and 20 milligrams of estradiol benzoate. Synovex implants improve daily gains in cattle by 12 to 14 percent and efficiency of gains by 8 to 10 percent. It may be implanted in cattle on pasture or in the feedlot, but it must be removed within 60 days of slaughter.

Ralgro (Zeranol) is a growth-increasing implant cleared with the

FDA in 1970. It may be given to cattle on pasture, but it must be withdrawn 65 days before slaughter. When given at the recommended level (36 milligrams per head per day), it improves the rate of gain in fattening steers and heifers 8 to 12 percent and feed efficiency by 8 to 10 percent. It also improves the rate of gain in suckling calves from 4 to 8 percent in both heifers and steers. Steers implanted with Ralgro on pasture experience 12 to 15 percent faster rates of gain. Implanted heifers on pasture gain about 8 to 10 percent faster than controls.

STUDY QUESTIONS

1. What are the different ways in which feed may be prepared for beef cattle?
2. What determines whether or not it will pay to grind grain for beef cattle?
3. When is it necessary to grind grain for beef cattle?
4. Is it desirable to grind or chop roughages for beef cattle? Explain.
5. What are some advantages and disadvantages of the self-feeding of beef cattle as compared to hand feeding?
6. What system of feeding do most commercial feedlots use? Why?
7. What systems may be used to fatten cattle on pasture? What type of cattle are more likely to fatten on good-quality pasture alone?
8. What are some advantages and disadvantages of fattening cattle on pasture?
9. Outline a method for feeding replacement heifers.
10. Is it desirable to starve cows the last 3 or 4 months of pregnancy to reduce calving difficulties? Explain.
11. When are lactating cows more likely to need feed other than grasses?
12. What are some advantages and disadvantages of the creep feeding of calves? When is creep feeding most economical?
13. Outline some methods that may be used to induce calves to begin eating creep rations.
14. Suggest a good ration for creep-feeding calves.
15. What is high-moisture corn? What are some advantages and disadvantages of feeding it?
16. Assume you have a young herd bull that is quite thin. What and how would you feed him between breeding seasons?
17. How would you feed a mature bull between breeding seasons?
18. Formulate a ration for a 600-pound yearling steer to be fattened in the feedlot using fescue hay, barley, soybean meal, and limestone.
19. Do many cattle feeders calculate their own rations today? Why?

20. What are feed additives? Why are they added to beef cattle rations?

21. Why is it difficult to get new drugs approved for use in beef cattle rations?

22. What is the responsibility of the feed or drug manufacturer in the use of feed additives?

23. What is the responsibility of the feeder in the use of feed additives?

24. What is the present status of DES in feeding beef cattle?

25. What is Rumensin, and what does it do when added to beef cattle rations? Does it have to be withdrawn from the ration before slaughter? Why?

PART SIX

MARKETING
BEEF CATTLE

marketing purebred and commercial beef cattle

The efficiency of beef cattle production is of great importance in a beef cattle operation. Much of the gain due to efficiency of production can be lost, however, if cattle are not sold for what they are worth at market time. Knowing when and how to market one's animals and the efficiency of production determine the success of a beef production enterprise whether it is a commercial or a purebred operation. It is our objective in this chapter to discuss the various aspects of marketing beef cattle as they contribute to profits from the beef enterprise.

19.1 Marketing Purebred Cattle

Seed stock for the commercial production of beef cattle originally comes from purebred herds. The purebred breeder, therefore, should point the way toward producing the most efficient and most desirable individuals for slaughter. In recent years during the change from the small, early maturing type to the larger-framed, meatier, fast-growing individuals, some purebred breeders were slow in making the changes.

The introduction of the large-framed, meaty, fast-growing individuals from the "exotic" breeds provided competition for the older breeds, and there was an increased interest in selecting seed stock within the pure breeds to meet the requirements of the commercial producer. The change in the performance of purebred animals produced is illustrated in Figure 19.1 in changes in average yearling weights of certain pure breeds of animals sold in the annual bull sales held at the Livestock Center on the University of Missouri campus at Columbia. Although this is not necessarily representative of what is happening in all purebred herds in the United States, it does indicate that progress toward the modern type of good-performing individuals within the various breeds is being made each year.

Purebred breeders make their own market to a certain extent. They can do this in a number of ways. They can fit and show some of

Figure 19.1.   Adjusted 365-day weights by year for bulls sold in the Missouri Tested Bull Sales. (Courtesy Agriculture Experimental Station Res. Bulletin 1017, 1976)

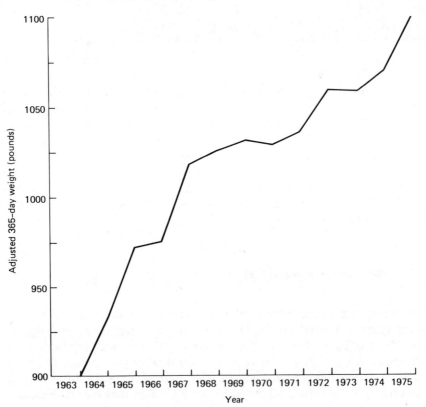

their animals at local and national livestock shows. Even though they may not show a grand champion, and few do, it does place their names and the quality of their cattle before the public. This may bring many prospective buyers to their farms. This is especially true if they show a number of animals that place well up in their classes even though they may not be the winner. Purebred breeders must also advertise their herd in local papers and especially in breed magazines. The building of a reputation as an honest and fair dealer is probably one of the most important bits of advertising a breeder can practice.

Livestock shows are almost always accompanied by a sale of many of the animals shown. This is one way breeders can market their stock.

Many state breed associations hold purebred shows and sales each year. In recent years some states have held sales of performance-tested animals, mostly bulls, once or twice each year. To be accepted for such sales they have to meet rigid requirements for type and performance, which in itself is good advertisement for breeders.

Purebred breeders can sell much of their surplus stock at private treaty and at home. The prospective buyers must be brought to the ranch through some means of advertising, as mentioned previously. Efforts to satisfy the customer help bring purchasers back to the herd for purchases year after year.

19.2   Markets for Commercial Beef Cattle

Most cattle marketed today, except in the purebred industry, sooner or later find their way to slaughter houses to produce beef for human consumption. Most of these are produced in commercial herds except aged, infertile animals or animals of undesirable qualities from purebred herds. Cattle are sold through several kinds of markets.

19.2.1   Terminal Markets

Terminal markets are also referred to as central markets, public stockyards, open markets, and central public markets. They are live-stock trading centers which have complete facilities for receiving, handling, caring for, and selling livestock. The facilities at these markets are owned by a stockyard company, which derives its income from renting the facilities and providing services for livestock sale rather than from purchasing and selling livestock.

Terminal markets are intermediate between the feedlot and the packer. Many years ago most slaughter cattle were marketed in this way, but in recent years most slaughter cattle have been sold directly to the packer. This is known as direct marketing. Because of the increase in direct marketing, several terminal markets have ceased to exist.

The facilities of the terminal market may be used by anyone to buy or sell livestock. In most instances, however, the buyers and sellers are represented by a professional sales agent or buyer.

Two or more commission firms must operate in the terminal market for it to be properly classified as a terminal market. Neither the stockyard company nor the commission firm takes title to the livestock sold. They provide their services for a certain fee. Most livestock sold through such markets are sold at private treaty, but some livestock are sold at auction 1 or more days per week.

19.2.2   Auction Markets

An auction is a form of marketing where many groups of cattle are offered to the public at the same time. They are then sold to the highest bidder. Cattle sold at an auction may be sold by weight on a price per pound basis, or they may be sold by the head.

In some auctions cattle belonging to the same owner may be sold singly, in pairs, or in small lots of uniform size, quality, or condition. In other auctions cattle are consigned to the same sale by a number of owners. These cattle are graded and sorted into large lots containing animals of uniform grade and weight. Feeder cattle are often sold in this manner. Cattle in these lots may be weighed at the time they are sorted, or they may be weighed immediately after they are sold. The time of weighing depends on the procedure customarily followed at a particular auction.

All auction markets charge a commission, or selling fee, which is the primary source of income for those operating the market. Some auction firms also charge fees for yardage, insurance, brand inspections, and health. The market may also deduct other fees such as trucking to the auction barn if the seller so desires. The amount of the charges varies from auction to auction, and the charges may be by the head or by the gross value of the animals sold.

Auction markets have some advantages when cattle from several owners of similar weight, grade, and quality are pooled to make large uniform lots. Prices received may be higher than when cattle are sold singly or in small lots by the same owner. This is particularly true in the sale of feeder cattle. Another advantage of auction sales is that cattle are offered to several buyers at the same time, and the competition at ringside may cause higher prices to be paid. Many times the bidding is quite spirited (Figure 19.2).

Some auctions operate on a regular schedule and are called *regular auctions*. Sales are held on schedule 1 or 2 days per week. This is particularly true of local sale or auction barns. Even these auctions may occasionally hold special sales of feeder and stocker cattle.

Special auctions are usually cooperatively owned and schedule sales of feeder cattle on a seasonal basis. Sale offerings in the spring include some calves and yearlings purchased to go on grass during the grazing season. Some of these cattle move to feedlots to be fattened for slaughter, however. Fall sales include weanling calves and yearlings which have grazed during the summer and will be moved to the feedlot for fattening. Some of these cooperative sales include large enough numbers of cattle that they can be sorted into uniform lots of 50 to 100 or more. Good-quality cattle in such sales command top prices.

In the western range states such as Arizona and New Mexico, cattle from large ranches are sorted into large uniform lots and sold at auction. Where the country is rough and roads for trucks

Figure 19.2.   Selling feeder cattle at public auction. (Courtesy Duane Dailey, University of Missouri, Columbia)

are not available, cattle are herded or driven to the auction pens. In some instances, auction pens are located at a central location where cattle from the range have to be driven only a short distance. After the cattle are sold, they are usually trucked to their final destination.

19.2.3 Local Markets

Local markets are located in the country near areas where livestock are produced. They have facilities such as pens, chutes, and scales for handling livestock. They operate on a schedule of 1 or more days a week and sell both cattle and hogs. Many of these local markets act as a dealer, taking title to the livestock and reselling to a packer. In a few cases the packers own the local market.

19.2.4 Order Buyers and Dealers

Order buyers act as agents for buyers in purchasing livestock. They charge a fee on a per head or weight basis. Livestock dealers operate independently to buy and sell livestock. The dealer usually

Figure 19.3.   Importance of various markets in the United States. (Courtesy U.S.D.A.)

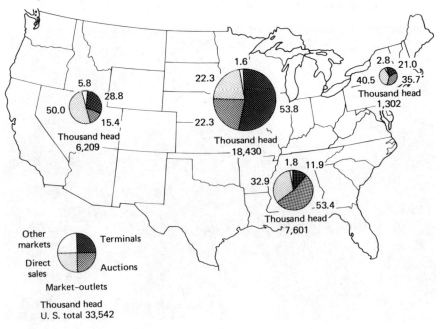

takes title to the livestock and then resells them. Dealers' profits depend on their ability to determine the true value of the cattle and to sell them for more than they paid for them. Many dealers have limited facilities for handling livestock.

Some dealers and order buyers purchase cattle directly from the farm or ranch. Their knowledge of cattle and prices together with a good reputation go a long way toward making their business a profitable one.

The different methods of marketing cattle have their advantages and disadvantages for the seller. The best method to use may depend on the local situation. The producer must be familiar with all the methods and should use the one that seems most advantageous.

The kinds of markets available and most popular depend on the section of the country where the cattle are located. The proportion of cattle sold by the various methods in different regions of the United States is shown in Figure 19.3.

Local marketing associations may be organized through the farm bureau or by commission firms formerly located at terminal market yards.

19.2.5   Slaughter Plants

Many plants in the United States slaughter one or more species of animals. The majority of them sell their meat products wholesale, but some sell direct to consumers. Several slaughter plants have a locker plant combined with the slaughter house. These plants purchase livestock for slaughter in several ways. They may make purchases at a terminal market, at an auction, or through an order buyer. Some slaughter plants purchase livestock at the plant or on the farm. A few operate a local marketplace, and some even feed cattle to be slaughtered in their plant.

19.2.6   Other Methods of Marketing

Two other methods of selling in addition to those mentioned previously include on-the-rail and futures trading.

On-the-rail selling refers to the selling of carcasses rather than live animals. When sold in this manner the price paid is determined for each hundredweight of carcass on the basis of a specific quality and/or yield grade prior to slaughter. After slaughter the packer pays the seller according to the grade and weight of the carcasses produced.

Data in Table 19-1 show that about 23 to 24 percent of cattle purchased by packers are purchased on the basis of carcass grade and weight. The number of cattle and calves purchased on this basis have increased some in the past 10 years. A very small percentage of calves purchased by packers were purchased on a carcass grade and yield basis.

Futures trading refers to the buying and selling of carload lots of beef cattle contracts for future delivery. Only about 2 or 3 percent of the contracts are offset before their future delivery date when the sellers repurchase their contracts or the buyers sell their initial position.

19.2.7 Changes in Importance of Kinds of Livestock Markets

The importance of various kinds of livestock markets has undergone considerable change in recent years. Forty to 50 years ago 85 to 90 percent of all cattle and calves federally inspected were sold through terminal markets. In recent years most of the slaughter cattle have gone directly to the packer, whereas feeder cattle have been sold locally or through other means.

TABLE 19-1   Percentage of cattle purchased on carcass grade and weight from 1967 to 1976[a]

	Percentage of total sales	
Year	*Cattle*	*Calves*
1967 (355)[b]	14.4	3.2
1968 (395)	17.2	3.6
1969 (383)	19.6	3.9
1970 (358)	18.7	4.5
1971 (333)	20.5	5.8
1972 (317)	22.6	6.7
1973 (329)	23.4	5.6
1974 (304)	22.9	6.2
1975 (311)	24.3	9.1
1976 (314)	23.3	8.4

[a]From *Packers and Stockyards Resume,* Vol. 15, No. 3, U.S. Department of Agriculture, Washington, D.C., 1977, p. 26.
[b]Figures in parentheses represent the number of firms purchasing cattle on a carcass grade and weight basis. Total purchases ranged between 30 and 37 million head per year.

19.3 When to Market Beef Cattle

A number of factors may determine the best time to market beef cattle. The ideal time, of course, is to market them when they are ready and when the market price is at its highest. These ideals are often not realized under practical conditions. Livestock prices are generally determined by supply and demand, and these factors vary. They influence prices on a long-term and a seasonal basis.

Seasonal variations in prices for 400- to 500-pound choice feeder steer calves at Kansas City, Missouri, for the years 1967 to 1978 are shown in Figure 19.4. The same information for 900- to 1,100-pound choice slaughter steers at Omaha, Nebraska, is given in Figure 19.5.

Feeder calf prices vary widely from year to year. This great variation may tend to cover up seasonal trends within a given year. Feeder calf prices show some tendency to be highest during May and June. This may be due to an increased demand for calves to graze during the summer months. Prices for feeder cattle during September, October, and November tend to approach the level of spring prices.

Figure 19.4.   Seasonal variation in prices for 400- to 500-pound choice steer calves at Kansas City, Missouri, 1967–1978. (Courtesy Glenn A. Grimes)

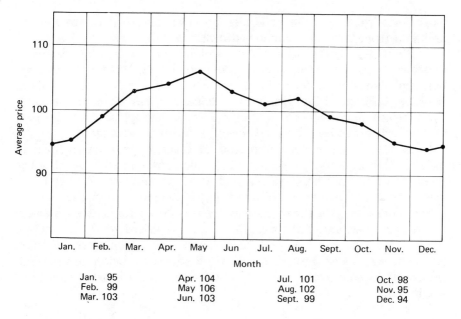

Jan. 95	Apr. 104	Jul. 101	Oct. 98
Feb. 99	May 106	Aug. 102	Nov. 95
Mar. 103	Jun. 103	Sept. 99	Dec. 94

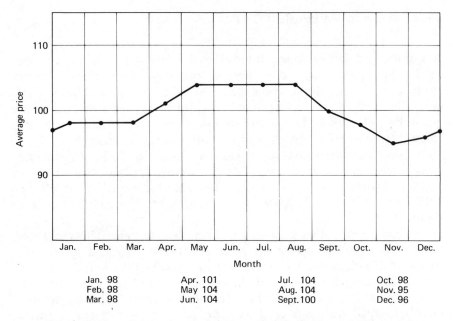

Figure 19.5. Seasonal variation in prices for 900- to 1100-pound choice slaughter steers at Omaha, Nebraska, 1967–1978. (Courtesy Glenn A. Grimes)

This is desirable for the producer because many calves are weaned and sold from farms and ranches during these months.

Slaughter cattle prices also tend to vary from one season to another, with the lowest prices received during November, December, and January. This trend, as is true of prices for feeder calves, may be influenced by rising or declining prices on a year-to-year basis. Seasonal trends do not appear great enough so that the best market prices can be estimated several months in advance.

Cattle numbers over a period of several years are cyclical in nature. This is shown in Figure 19.6. Data since 1925 show that the peaks in cattle numbers are reached at approximately 10-year intervals. The peaks are then followed by a decline in numbers to the point where they begin to increase again. Prices for cattle generally follow the rise and fall in cattle numbers. Prices tend to increase when numbers decrease and to decrease when numbers increase. Dr. V. R. Jacobs of the University of Missouri Extension Division has proposed an equation for estimating changes in the market supply of cattle:

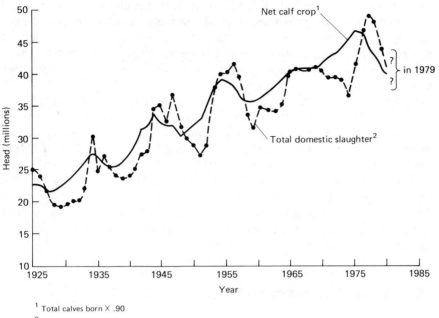

¹ Total calves born X .90

² Total cattle and calf slaughter minus net live imports.

Figure 19.6.   Cyclical nature of beef cattle numbers since 1925.
(Courtesy Dr. Victor Jacobs, University of Missouri, Columbia)

$$\begin{matrix} \text{changes in} \\ \text{market supply} \end{matrix} = \begin{matrix} \text{changes in net} \\ \text{production and} \\ \text{live imports} \end{matrix} - \begin{matrix} \text{change in the rate of} \\ \text{change in the national} \\ \text{herd size} \end{matrix}$$

Other factors which may have an effect on the cycle include grain exports, grain prices, federal price support policies, acres diverted from grain crops to grass, and government price controls and land use regulations.

19.4   Livestock Market News

Prices for various classes of cattle are reported from the U.S.D.A. Market News Service by radio and television on most days of the week except Saturday, Sunday, and national holidays. They are usually reported one or more times per day on farm programs. Many newspapers also report prices for various classes of beef cattle in a market section along with a report on the stock market. Some state departments of agriculture also print and circulate a daily market summary for cattle prices at some of the major markets in the United States in addition to reports of prices at many local markets within the state. By using one or more of these cattle price quotations, beef cattle producers can keep current on prices for the class of livestock they have for sale.

Forecasts of what prices might be for cattle in the future are often referred to as *price outlooks*. Many state extension economists study information on prices, numbers of cattle likely to be available, as well as other information to make an educated estimate of future prices. These outlooks cannot be expected to be 100 percent accurate all the time, but they do give good estimates of what future prices might be. This sort of information is valuable to livestock feeders and producers to help them plan their future programs. The U.S.D.A. also releases information from time to time on cattle numbers and prices in the United States.

19.5   Costs of Marketing Beef Cattle

The major costs of marketing cattle involve transportation, selling costs at the market, and shrinkage. These must be considered by both the buyer and seller.

19.5.1   Transportation

The major means of transportation from the farm or ranch to the market includes shipment by truck.

The costs of transporting cattle to market vary widely and are determined by the distance traveled and the number of cattle shipped per unit hauled. A full load will usually be cheaper to transport per head than a partial load. Another possible transportation expense may be a tariff charged by the state or federal commissions. Sellers should investigate the cost of transportation within their particular area before the final decision is made as to which one to use. If they are selling cattle at a market close to home, transportation charges, of course, will be less than if they are sold at a market some distance away.

Trucking companies and railroads that haul livestock must pay a 3 percent federal transportation tax based on charges for transportation. The seller pays this tax indirectly since the transportation companies include this in their charges.

A very small percentage of cattle shipped may die or be crippled in transit. The death loss is usually only one in several thousand, although more death losses occur in calves than in larger, more mature animals. The number of cattle crippled in transit is slightly higher than the number that die, but this is still a very small fraction of those shipped. Transportation insurance is available and is sometimes required. It will usually cover such losses. These insurance rates are usually low because losses are low.

Slaughter animals shipped to market are damaged more by bruises than by death and crippling. These bruises cannot be detected visually until after the animals are slaughtered. The bruised areas must be trimmed from the carcass and represent losses of 15 to 20 cents per pound of live weight as a general rule. Bruises are caused by crowding, trampling, horns, etc. Every effort possible should be made to reduce such losses by careful handling while loading and unloading the cattle at home and at the slaughter pens, by removal of horns when animals go into the feedlot, and by not overloading on trucks and rail cars when cattle are shipped to market. Even then some bruises may occur.

19.5.2   Selling Costs at the Market

Costs of selling cattle at the market may include commissions, yardage, feed, and other incidental costs.

Cattle shipped to central terminal markets are consigned to a livestock commission firm. Employees of these firms handle the cattle in the stockyards, feed and water them, and show them to prospective buyers. Finally a representative of the commission firm sells them to the highest bidder. Thus, sellers have little or nothing to do in the process of selling their animals once they arrive at the stockyards. Commission charges at auction sales are similar to those at terminal markets. These charges may be made on a per head or gross sales basis and usually vary between 1 and 3 percent of the gross receipts.

Stockyard companies at central markets usually charge $1 to $1.50 per head for the use of their pens, scales, and so on. This is the main source of income for these companies. If cattle are kept in pens for any length of time, they must be fed and watered. The seller is charged for the feed used in these instances.

Commission firms and yards that cooperate with the National Livestock and Meat Board deduct 3 cents per head for all animals sold. This board is a nonprofit organization which represents all segments of the livestock industry. It (1) initiates, sponsors, and encourages scientific research on the place of red meat in the diet and its relation to human health; (2) conducts a continuous and extensive program of education and information about red meat products; and (3) assists the livestock and meat industry in presenting a constantly improving meat product to the public.

19.5.3   Shrinkage

Shrinkage refers to weight losses in cattle from the time they leave the seller's premises until they are weighed to the buyer.

Weight losses may be classed as excretory shrink or tissue shrink. Excretory shrink is a loss of weight due to the loss of digestive tract contents when the animals receive no food or water for a period of time. Shrinkage when cattle are kept off feed for 10 to 12 hours is usually excretory shrink. Shrinkage of this kind is rapidly recovered when the animals are given feed and water and when their digestive tracts are again filled. Tissue shrinkage refers to weight losses from tissues within the carcass. Such shrinkage occurs when cattle are shipped long distances or when they are held without feed and water for extended periods. Recovery from tissue shrinkage takes longer than that from excretory shrinkage. Both may occur simultaneously. Weight losses in cattle are very important to the buyer and seller, because they greatly affect the price per pound of the cattle involved.

The conditions under which cattle are kept, or have been kept, have a great influence on the amount of shrinkage (R. Brownson, "Shrinkage in Beef Cattle," *Great Plains Beef Cow-Calf Handbook*, GPE-4002, Cooperative Extension Service — Great Plains States). Cattle which have been on grass or silage will usually shrink about 4 percent when in an overnight stand for about 12 hours without feed and water. Fat cattle which have been fed concentrates will shrink less (2.5 to 3 percent). When feed and water are available free-choice, shrinkage in fat cattle will be only about 2 percent. Cattle accustomed to running on open range or pasture will lose 5 percent or more of their weight if confined to a pen overnight. Shrinkage of 7 to 9 percent may occur in cattle shipped long distances, with more shrinkage occurring as the length of the shipment increases. Excessive shrinkage may occur in calves that are weaned, vaccinated, and shipped at the same time. Cattle and calves often require 2 weeks or more to regain weight losses suffered in transit.

In recent years some feeder cattle have been preconditioned before being sold to feedlot operators. Preconditioning refers to the preparation of cattle by the producer to reduce as much loss as possible from shrinkage, sickness, and the time required to get cattle on a full feed. It includes vaccinations, worming, treatment for lice and grubs, dehorning, castrating, weaning, and becoming accustomed to eating grain or concentrated rations. Buyers will often pay 2 to 4 cents more per pound for preconditioned yearlings if their precondition can be verified. This is sometimes done by furnishing a preconditioning certificate to the buyer which has been certified by a veterinarian. These certificates can be obtained from the state veterinarian's office in some states.

STUDY QUESTIONS

1. What change has taken place in purebred animals the past few years and why?

2. What markets are available to breeders of purebred cattle, and how can they take advantage of them?

3. What is a terminal market? What are some other names for such a market?

4. What is an auction market? What are some advantages of the auction market?

5. What is a regular auction? Special auction?

6. Describe the role of order buyers and dealers.

7. Where do most slaughter plants obtain their cattle?

8. What is meant by on-the-rail marketing?

9. What changes have occurred in the kinds of livestock markets in recent years?

10. When is the best time to market cattle? Are there variations in cattle prices according to seasons? Years?

11. What is meant by the cattle cycle? How long is it as a general rule? What factors are responsible for the cattle cycle?

12. Describe livestock news reports and who makes them. What is meant by price outlooks?

13. What are some of the major costs involved in selling cattle?

14. What major means are used for transporting cattle to market?

15. Why are cattle shipped to market often crippled and bruised? How can these losses be reduced?

16. What is the National Livestock and Meat Board, and where does it obtain its funds for operation? What are its objectives?

17. What are the two kinds of shrinkage, and what causes them?

18. What percentage of shrinkage is likely to occur in various classes of cattle?

19. Why is shrinkage an important factor to consider in the marketing of cattle?

CHAPTER 20

U.S.D.A. grades
for beef cattle
and beef carcasses

The Agricultural Marketing Act of 1946 (as amended) makes the U.S.D.A. responsible for providing meaningful grade standards for facilitating the marketing of livestock and meat. The act directs the Secretary of Agriculture to develop and improve standards for quality, condition, quantity, and grade of such products and recommend and demonstrate such standards in order to encourage uniformity and consistency in commercial practice. The act also directs the Secretary of Agriculture to inspect, certify, and identify the class, quality, and condition of agricultural products so they may be marketed to best advantage and so that trading consumers may be able to obtain the quality product they desire; however, no person is required to use the service.

In compliance with the act the U.S.D.A. has developed standards for various grades of cattle and beef carcasses. These are made available to the public. The standards are revised from time to time as more research information becomes available and consumer demands change. Changes in the standards are made only after various groups such as producers, slaughterers, wholesale and retail meat dealers, agricultural college workers, and others interested in the marketing of livestock and meat have had the opportunity to make suggestions for improving such standards.

In this chapter the standards developed by the U.S.D.A. for various classes of livestock and meat will be discussed.

20.1   U.S.D.A. Feeder Calf Grades

U.S. standards for grades of feeder cattle have been revised in recent years. This is in keeping with efforts to use the latest information to upgrade the standards used. The latest standards are based on frame size and muscling.

20.1.1   Specifications for Official U.S. Standards for Grades of Thrifty Cattle

Large frame (L).   Feeder cattle which possess typical minimum qualifications for grade L have large frames, are thrifty, and are tall and long bodied for their age. They would be expected to excel in growth rate, but steers and heifers would not be expected to produce U.S. choice beef carcasses until their live weights exceed about 1,200 pounds and 1,000 pounds, respectively.

Medium frame (M).   Feeder cattle which possess typical minimum qualifications for grade M are thrifty, have slightly large frames, and are slightly tall and slightly long bodied for their age. They would be expected to have an average growth rate. Steers and heifers would be expected to produce U.S. choice beef carcasses at live weights of 1,000 to 1,200 pounds and 850 and 1,000 pounds, respectively.

Small frame (S).   Feeder cattle included in grade S have small frames, are thrifty, and are shorter bodied and not as tall as specified as the minimum for the medium frame grade. They would be expected to have a relatively slow growth rate. Steers and heifers would be expected to produce U.S. choice grade carcasses at live weights of less than 1,000 pounds and 850 pounds, respectively.

20.1.2   Specifications for Official U.S. Standards for Grades of Thrifty Feeder Cattle

Number 1.   Feeder cattle with a slightly thin covering of fat which possess typical minimum qualifications for grade 1 are thrifty and are very thick throughout. They are very wide through the chest and through the middle part of the rounds. The forearm and gaskin are very thick and full, and the back and loin appear full and well rounded. The legs are set very wide apart, both front and rear. They usually show no evidence of nonbeef breeding. Cattle with a higher degree of thickness — extremely thick — are also recognized in this grade. Double-muscled cattle are not eligible for this grade.

Number 2. Feeder cattle with a slightly thin covering of fat which possess typical minimum qualifications for grade 2 are thrifty and are slightly thick throughout. They are slightly wide through the chest and through the middle of the rounds. The forearm and gaskin are slightly thick and full, and the back and loin have a rounded appearance. The legs are set slightly wide apart, both front and rear, and show a very high proportion of beef breeding. Cattle with two higher degrees of thickness — moderately thick and thick — are also recognized in this grade.

Number 3. Feeder cattle included in grade 3 include thrifty animals which are inferior in their thickness to the minimum requirements specified for grade 2.

20.1.3 Specifications for Official U.S. Standards for Grades of Unthrifty Feeder Cattle

Inferior. The inferior grade includes those feeder cattle which are unthrifty because of such factors as mismanagement, disease, parasitism, or lack of feed and those that are double-muscled. Cattle in this grade may have any combination of muscling and frame size.

20.2 Veal Calves

A veal calf may be defined as one that is less than 3 months of age when slaughtered. A veal calf has grown and developed largely on milk. Its rumen shows very little development. Most veal calves are produced by dairy cows and are produced from straight dairy breeding or by cows bred to beef bulls. Very few calves of straight beef breeding are sold as vealers. Veal calves represent only a small proportion of the total calves slaughtered. Live veal calves are graded on the basis of conformation, finish, and quality. Grades of veal calves include prime, choice, good, standard, utility, and cull, with the prime grade being the most desirable.

20.3   Slaughter Calves

Slaughter calves are slaughtered at weaning or shortly afterward. They will usually range between 7 and 12 months of age and weigh 350 to 700 pounds at slaughter. Slaughter calves are usually those who have obtained enough milk from their mothers in addition to grass to produce a good finish at weaning or shortly afterward.

Fat slaughter calves are produced in areas such as the South and Southwest where abundant grass from small grain pastures or irrigated pastures is available. Such calves are often creep-fed.

Good milk-producing cows from crossbred dairy × beef breeding or, in some areas, crossbred cows from Brahman crosses bred to beef bulls are used to produce slaughter calves. Calves from such crosses are often in good condition at weaning but might not command top prices as feeder calves.

Slaughter calf grades also include prime, choice, good, standard, utility, and cull.

20.4   Official U.S. Standards for Quality Grades of Slaughter Steers, Heifers, and Cows[1]

The quality grades for slaughter steers, heifers, and cows include prime, choice, good, standard, commercial, utility, cutter, and canner. Quality grades are based on an evaluation of the factors related to the palatability of the lean (or quality) in slaughter cattle. These factors include (1) the amount and distribution of finish, (2) the firmness of muscling, and (3) the physical characteristics of the animal associated with maturity.

Forty-two months is the approximate maximum age limitation for the prime, choice, good, and standard grades of steers, heifers, and cows. The commercial grade of steers, heifers, and cows includes only cattle over approximately 42 months of age. Utility, cutter, and canner grades of steers, heifers, and cows have no age limitations. Twenty-four months is the maximum age limitation for all grades of bullocks.

[1] From U.S.D.A. Agricultural Marketing Service, reprint from *Federal Register*, 1975-40FR, 11535.

20.4.1 Prime

Cows are not eligible for the prime grade. Slaughter steers and heifers between 30 and 42 months of age possess several minimum qualifications for this grade. They must possess a tendency for a thick, fat covering over the crops, back, ribs, loin, and rump. The brisket, flanks, and cod or udder should appear full and distended, and the muscling should be firm. The fat covering should be smooth, with only slight indications of patchiness. Steers and heifers under 30 months of age should have a moderately thick but smooth covering of fat which should extend over the back, ribs, loin, and rump. The brisket, flanks, and cod or udder should show a marked fullness, and the muscling should be firm.

20.4.2 Choice

Slaughter steers, heifers, and cows 30 to 42 months of age possessing the minimum qualifications for choice should have a fat covering over the crops, back, loin, rump, and ribs that tends to be moderately thick. The brisket, flanks, and cod or udder should show a marked fullness, and the muscling should be firm. Cattle under 30 months of age should carry a slightly thick covering of fat over the top. The brisket, flanks, and cod or udder should appear moderately full, and muscling should be firm.

20.4.3 Good

Slaughter steers, heifers, and cows 30 to 42 months of age possessing the minimum qualifications for the good grade should have a fat covering that tends to be slightly thin, with some fullness evident in the brisket, flanks, twist, and cod or udder, and the muscling should be firm. Cattle under 30 months of age have a thin fat covering which is largely restricted to the back and loin. The brisket, flanks, twist, and cod or udder should be slightly full, and the muscling should be slightly firm.

20.4.4 Standard

Slaughter steers, heifers, and cows 30 to 42 months of age and possessing the minimum qualifications for the standard grade should have a fat covering primarily over the back, loin, and ribs which tends to be very thin. Cattle under 30 months of age should have a very thin covering of fat that is largely restricted to the back, loin, and upper ribs.

20.4.5   Commercial

The commercial grade is limited to steers, heifers, and cows over approximately 42 months of age. Slaughter cattle possessing the minimum qualifications for the commercial grade and slightly exceeding the minimum maturity for the commercial grade have a slightly thick fat covering over the back, ribs, loin, and rump, and the muscling is moderately firm. Very mature cattle usually have at least a moderately thick fat covering over the back, ribs, loin, and rump, and frequently considerable patchiness is evident about the tailhead. The brisket, flanks, and cod or udder appear to be moderately full and the muscling firm.

20.4.6   Utility

The minimum degree of finish required for slaughter steers, heifers, and cows to qualify for the utility grade varies throughout the range of maturity permitted in this grade from a very thin covering of fat for cattle under 30 months of age to a slightly thick fat covering, generally restricted to the back, loin, and rump, for the very mature cattle in this grade. In such mature cattle, the crops are slightly thin, and the brisket, flanks, and cod or udder indicate a very slight fullness.

20.4.7   Cutter

In slaughter cattle in the cutter grade, the degree of finish ranges from practically none in cattle under 30 months of age to very mature cattle which have only a very thin covering of fat.

20.4.8   Canner

Canner grade cattle are those which are inferior to the minimum specified for the cutter grade.

Official U.S. standards for quality grades of slaughter bullocks include prime, choice, good, standard, and utility.

20.5   Official U.S. Standards for Yield Grades
of Slaughter Cattle

Yield grades include 1, 2, 3, 4, and 5 and refer to the cutability or the expected yields of boneless retail cuts from live slaughter animals.

20.5.1   Yield Grade 1

Cattle in grade 1 produce carcasses with very high yields of boneless retail cuts. Cattle in this grade are moderately wide, and the width through the shoulders and rounds is greater than through the back. The top is well rounded with no evidence of flatness, and the back and loin are thick and full. The rounds are deep, thick, and full, and the width through the middle part of the rounds is greater than through the back. The shoulders are slightly prominent, and the forearms are thick and full. These cattle have only a thin covering of fat over the back and rump. The flanks are slightly shallow, and the brisket and cod or udder have little evidence of fullness. Slaughter cattle of this description producing 600-pound carcasses usually have about 0.3 inch of fat over the ribeye and about 13.0 square inches of ribeye area.

20.5.2   Yield Grade 2

Slaughter cattle of grade 2 produce carcasses with high yields of boneless retail cuts. Very thickly muscled cattle typical of the minimum of this grade have a high proportion of lean to bone. They are wide through the back and loin and have slightly greater width through the shoulders and rounds than through the back. The top is well rounded with little evidence of flatness, and the back and loin are thick and full. The rounds are thick, full, and deep, and the thickness through the middle part of the rounds is greater than that over the top. The shoulders are slightly prominent, and the forearms are thick and full. There is a slightly thick covering of fat over the back and rump, and the flanks are slightly deep. The brisket and cod or udder are slightly full. Slaughter cattle of this description producing 600-pound carcasses usually have about 0.6 inch of fat over the ribeye and about 12.5 square inches of ribeye area.

20.5.3   Yield Grade 3

Cattle of grade 3 are very thickly muscled and have a high proportion of lean to bone. They are very wide through the back and loin and are uniform in width from front to rear. The back or top is nearly flat with only a slight tendency toward roundness and there is a slight break into the sides. The back and loin are very full and thick. The rounds are deep, thick, and full. The shoulders are smooth, and the forearms are thick and full. They have a moderately thick covering of fat over the back and rump. The flanks are deep and full, and the brisket and cod or udder are full. Slaughter cattle of this description producing about 600-pound carcasses usually have about 0.9 inch of fat over the ribeye and about 12.0 square inches of ribeye area.

20.5.4   Yield Grade 4

Cattle of grade 4 produce carcasses with moderately low yields of boneless retail cuts. Very thickly muscled cattle typical of the minimum of this grade have a high proportion of lean to bone. They appear wider over the top than through the shoulders or rounds. The back and loin are very thick and full, are nearly flat, and break sharply into the sides. The rounds are deep, thick, and full. The shoulders are smooth, and the forearms are thick and full. These cattle have a thick covering of fat over the back and rump. The flanks are very deep and full, and the brisket and cod or udder are very full. Slaughter cattle of this description producing 600-pound carcasses usually have about 1.1 inches of fat over the ribeye and about 11.5 square inches of ribeye area.

20.5.5   Yield Grade 5

Slaughter cattle of grade 5 produce carcasses with low yields of boneless retail cuts. Cattle of this grade consist of those not meeting the minimum requirements for yield grade 4 because of either more fat or less muscle or a combination of these two characteristics. Practically all cattle of this grade will qualify for either prime or choice grade.

20.6   Evaluation of Carcass Beef

Present-day consumers demand more lean and less fat in beef products than consumers of several years ago. They also demand a high-quality product with a minimum of waste in fat or bone. Beef cuts at the retail level must be attractive and possess a desired red color. Thus, beef carcasses must have desirable quality to suit consumers and cutability to suit the needs of retailers. The different parts of the beef carcass that produces wholesale cuts of beef are shown in Figure 20.1.

20.6.1   U.S.D.A. Beef Quality Grades for Beef Carcasses

The main purpose of U.S.D.A. beef quality grades is to identify eating qualities of beef. Several factors contribute to the quality of beef carcasses, including maturity, marbling, texture of the lean, firmness of the lean and fat, and the color of the lean and fat. The most useful part of the carcass in determining quality is the *longissimus dorsi* (ribeye) in the ribbed carcass.

The quality grades for steer, heifer, and cow carcasses are prime, choice, good, standard, commercial, utility, cutter, and canner. Cow carcasses are not eligible for the prime grade.

Carcass grades for bullocks include prime, choice, good, standard, and utility.

Figure 20.1.   Wholesale cuts of beef. (Courtesy Missouri Beef Steer Project)

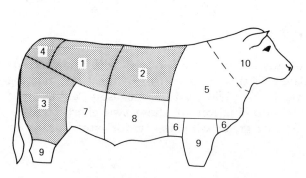
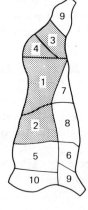

Higher value wholesale cuts	Lower value wholesale cuts
1. Loin	5. Chuck
2. Rib	6. Brisket
3. Round	7. Flank
4. Rump	8. Plate
	9. Shank
	10. Neck

The age of the animal (maturity) at slaughter is closely related to the eating quality of beef. In general the younger animal produces beef that is more desirable. In carcasses from young animals the split vertebrae (especially the thoracic vertebrae) are soft, red, and porous. They are also tipped with large amounts of soft pearly white cartilage. When the animal matures, the vertebrae become harder and whiter, and the cartilaginous parts become progressively ossified, beginning in the sacral vertebrae and ending in very mature animals in the anterior thoracic vertebrae.

The color of the meat becomes a darker red as the animal matures. Veal is usually a pale pink, whereas in yearling cattle the lean is a very bright cherry red. The lean is usually a very dark red color in mature cows.

Marbling refers to the occurrence of fat between the lean fibers in the muscle. The degree of marbling is usually associated with the quality of the lean. U.S.D.A. standards for grading carcass beef include at least ten grades of marbling: (1) very abundant, (2) abundant, (3) moderately abundant, (4) slightly abundant, (5) moderate, (6) modest, (7) small, (8) slight, (9) traces, and (10) practically devoid. Good-quality beef should have abundant marbling.

The texture of the lean is also related to beef quality. The texture of the lean refers to the prominence of muscle fibers in the exposed surface of the lean cuts. A coarse-textured lean shows many visible muscle fibers surrounded by heavy connective tissue. Fine-textured lean has few if any visible muscle fibers separated by connective tissue. Lean cuts of coarse texture tend to be less tender than those of a finer texture. Beef from young animals is usually finer textured than beef from older animals.

The firmness of the lean and fat is an indicator of meat quality. Soft, low-quality muscles showing excessive moisture on the surface are not attractive to the consumer and are associated with lower-quality meat. A firm fat which does not have a greasy or an oily appearance is preferred and indicates better quality.

The color of the fat is an important indicator of meat quality. Fat with a white creamy color is preferred by the consumer. The yellow color is due to an accumulation of carotene in the fat. Dairy cattle, especially those from the Guernsey breed, store carotene in the fat rather than storing it as vitamin A, which is practically colorless. Cattle on green, lush pastures tend to possess larger amounts of carotene in their fat than cattle fed concentrates containing grains. Cattle fattened on pasture also tend to possess a soft, oily fat which is undesirable to the consumer. It is probably as suitable from the nutrient standpoint as the firmer, white fat.

20.6.2 U.S.D.A. Yield Grades (Cutability) for Beef Carcasses

Yield grades are used to identify the proportion of the beef carcass that produces trimmed and salable retail cuts. The yield grade is determined to a great extent by the degree of muscling and the degree of finish in the carcass. Yield grades are designated by 1, 2, 3, 4, and 5, with 1 being the highest grade.

Finish refers to the amount of fat which covers the outside of the carcass and that which is deposited within the body around the heart and kidneys. Excessive fat reduces the carcass value because much of it must be trimmed from the retail cuts offered to the consumer. The amount of finish is related to the length of time the cattle are full-fed and to the breed or kind of animal fed. In recent years a greater emphasis has been placed on breeding animals that will possess more lean and less fat at slaughter even though they are full-fed for varying lengths of time. A certain amount of finish is required, however, in order for live cattle to grade high at slaughter and command good prices. A desirable beef carcass should have no more than 0.3 inch of fat over the ribeye at the twelfth rib. Since the fat is deposited unevenly in different animals, it is important to evaluate the quantity of fat deposited in every part of each carcass. The larger the amount of trimmable fat on the carcass, the lower its cutability.

The amount of muscle in the carcass is another important factor affecting cutability. The degree of fatness in live animals makes it difficult to accurately estimate the degree of muscling in the animal. The trimmed round is a good indication of the yield of salable cuts from the carcass. The desirable round should be long, wide, and plump and free of excessive trimmable fat. The loin should also be long, wide, and full with a minimum of trimmable fat. The ribeye area is also an indication of the yield of salable cuts. The ratio of ribeye area to fat cover at the twelfth rib adds to other factors in estimating the yield of salable cuts. The desirable ribeye is long and wide. The chuck and other cuts should show thickness of muscling and should not show excessive fat. It should be remembered, however, that some beef carcasses can have a thin covering of fat but may not produce a high yield of salable cuts because the muscling is light and poorly developed.

The U.S.D.A. standards for yields of carcass beef ("Official U.S. Standards for Grades of Slaughter Cattle," U.S.D.A. Bulletin, S.R.A. 112, U.S. Government Printing Office, Washington, D.C.) are determined from the following equation: 2.50 + (2.50 × adjusted fat thickness, inches) + (0.20 × percent kidney, pelvic, and heart fat) +

(0.0038 × hot carcass weight, pounds) – (0.32 × area ribeye, square inches). Changes in standards for U.S.D.A. yield grades are being considered at the present time.

Further details concerning the evaluation of beef carcasses are given in the *Meat Evaluation Handbook*, published by the National Livestock and Meat Board, 444 North Michigan Avenue, Chicago, Ill. 60611.

STUDY QUESTIONS

1. Why does the U.S.D.A. publish standards for the classification of beef cattle and beef carcasses?

2. What major points are included in the U.S.D.A. feeder calf grades?

3. What is a veal calf? How are they produced in some parts of the country, and why are they produced?

4. What are slaughter calves? Why are some calves sold for slaughter rather than as feeder individuals?

5. What is meant by quality grades in slaughter steers, heifers, and cows?

6. Describe the quality grades in slaughter cattle.

7. What is a cutter? A canner?

8. What are the U.S.D.A. yield grades for slaughter cattle?

9. What are the official U.S. standards for quality grades of slaughter bullocks? What are bullocks?

10. What are the yield grades for slaughter cattle? What does yield grade refer to?

11. Describe cattle that should be in yield grade 1.

12. What are the quality grades for beef carcasses? What determines quality in beef carcasses?

13. Name the wholesale cuts of the beef carcass. Which ones have the highest value?

14. Why do some slaughter cattle have a yellow fat, whereas others have a white fat?

15. What are the cutability or yield grades for beef carcasses? What factors determine the yield of the carcasses to a great extent?

PART SEVEN

KEEPING BEEF CATTLE HEALTHY

control
of parasites

Parasites cause large economic losses to beef cattle producers. The U.S.D.A. has estimated that more than $1 billion per year is lost in the United States because of various pests that affect cattle. Many of these losses are caused by parasites that produce unnoticed losses, although such infestations may cause death losses in some instances. Losses due to parasitism may be due to direct damage caused by the parasites themselves or to losses which may occur because parasites rob the animal of nutrients and cause damage to vital organs. All of these detrimental effects may make the animal more susceptible to bacterial infections and other disease-causing organisms.

The migration of parasitic worms through the liver, lungs, blood vessels, and the digestive tract can have serious effects in the animal. One of the most important effects may be that the damage results in the poor digestion and utilization of feed, resulting in inefficient gains. This poor performance may be associated with the production of poor-quality meat and meat products. In addition, millions of dollars are spent each year in efforts to control parasitic infestations.

Parasites are of two general kinds. These include external and internal parasites. External parasites include insects and insectlike organisms. Internal parasites include protozoa (coccidia), flukes or trematodes (liver flukes), tapeworms (cestodes), and roundworms (nematodes). Certain types of parasitism occur more often in one part of the country than another. Cattle producers should consult their local veterinarian or their state extension veterinarian or entomologist about those most prevalent in their area and how to control them.

The discussion here will be limited to the most important parasites and how to control them.

21.1   Chemical Control of Parasites

Many chemicals have been developed, especially in recent years, for the control of external and internal parasites. Some of these chemicals are contact poisons which kill insects on contact, whereas others are stomach poisons and have to be eaten by the insects to be effective. Most chemicals used for controlling external parasites in beef cattle are contact poisons.

Forty to 50 years ago coal-tar creosote dips and arsenical dips were used for controlling lice and ticks in cattle (*U.S. Dep. Agric. Farmers Bull.* 909 and 1057). More recently, compounds such as malathion, lindane, toxaphene, and methyoxychlor have been used for lice control. Many chemicals for controlling parasites are fat soluble and may be present in meat and meat by-products at slaughter. For this reason treatment of cattle with some chemical pesticides requires several days between the time they are last used until slaughter to prevent the chemicals from being present in meat at levels high enough to be toxic to humans who eat the meat or meat products. The waiting period, if one is required, varies with the product used, and recommendations for use are usually printed on the container in which the chemical is sold. Some chemicals may be harmful to livestock and humans if not properly used, so directions for their use should be carefully followed. Some chemicals have been banned for use in recent years by the Federal Drug Administration, and it is possible that others may be in the future.

Some products are systemics in that they may be given orally to an animal and then be carried by the blood to various parts of the body where they affect a specific organism. Several compounds may be applied topically, for example, poured on the back of an animal in the treatment process.

21.2 Biological Control of Parasites

Insects prey upon other insects. In this way many insect numbers are kept under control biologically. More than 1,000 viruses, fungi, nematodes, protozoa, and rikettsiae are known to infect and kill insects. Much more needs to be learned in this area, and researchers are working very diligently to gain further information that might be helpful in controlling parasites.

Biological controls may act in various ways. For example, viral and bacterial diseases may infect insects and cause their death. Other insects insert their eggs on or in the larva of other insects, and when the eggs hatch, they attack and destroy the developing larva. In female insects such as the screwworm fly, which mates only once during her lifetime, releasing males sterilized by radiation decreases the number of fertile eggs laid and has been used to eliminate the screwworm flies in some areas. Some female insects secrete chemical substances (sex attractants) known as pheromones which attract males to the females at the time of mating. Some of these sex attractants have been identified as to their chemical composition and have been synthesized in the laboratory. Some of them have been used to lure male insects to their death in traps designed for this purpose.

Another method of biological control is through the use of certain growth regulators, the majority of which interfere with the growth and development of insects in the immature or juvenile stage. They are often referred to as *juvenile hormones.* Some substances such as metroprene prevent the survival of viable adult insects (*Agric. Res.* 26:7, 1978).

21.3 Natural Resistance to Parasites

Very little research has been done in relation to natural resistance (probably inherited) to parasites. It has been observed on a practical basis that some individuals are more susceptible to parasites than others.

Genetic resistance to ticks in cattle has been reported in several countries. R. B. Kelly (*Bull. Coun. Sci. Ind. Res.* 172, 1943) reported that purebred zebus are completely tick resistant, and they pass this characteristic to their offspring almost in direct proportion to the amount of zebu blood the offspring possess. Cattle from the Shorthorn and Hereford breeds carry many ticks under the same conditions. A breed of cattle developed in Australia contained about one-half *Bos indicus* and one-half *Bos taurus* blood and was about ten times more resistant to ticks than the pure *Bos taurus* breeds. Through selection and breeding, tick resistance in this breed was further increased. All of these factors tend to confirm that tick resistance is genetic, although the actual mode of resistance is not fully known. One study showed that the only one of several characteristics studied in cattle correlated with tick resistance within a breed was serum amylase (J. Francis and J. Ashton, *Aust. J. Exp. Biol. Med. Sci.* 45: 131–140, 1967).

Some animals are also genetically resistant to internal parasites. Romney sheep in England have been shown to be resistant to trichostrongyle worms, probably because the breed was developed in an area where there was a great exposure to this parasite during its development. Romney sheep in California were also found to be resistant to these parasites (P. W. Gregory, *Proc. Am. Soc. Anim. Prod.* p. 316, 1937). Experiments in Texas showed that resistance to stomach worms (*Haemonchus contortus*) could be increased through selection in both sheep and goats (B. L. Warwick et al., *J. Anim. Sci.* 8:609, 1949). Undoubtedly genetic resistance to both internal and external parasites exists in many animals and could be increased by proper selection methods.

21.4   Some External Parasites and Their Control

The most common external parasites which affect beef cattle include cattle grubs, cattle lice, cattle mange, ticks, flies, and screwworms. Several effective methods of controlling such parasites have been developed in recent years.

Several precautions should be followed in using insecticides. Instructions on the labels of the product should be strictly followed. Any insecticide whose trade name is preceded by PRU ("Pending Restricted Use") means that all or some uses of this product have or will be restricted by the Environmental Protection Agency (EPA). The person doing the application must be certified and licensed before purchasing the product.

Insecticides should not be used on calves less than 3 months of age or on animals that are ill or under stress. Light applications may be used on calves 3 to 6 months of age. Insecticides should not be used in conjunction with oral drenches and other internal medica-

tions such as phenothyazine, natural or synthetic pyrethroids, or
their synergists or with other organic phosphates. Be careful not to
contaminate feed and drinking water; also, swine should be kept
away from runoff areas. Animals should not be sprayed in non-
ventilated areas.

21.4.1   Cattle Grubs

Other names given to cattle grubs include the cattle warble, the
warble fly, or the heel fly.

The heel fly is hairy and is black with yellow stripes. It resembles
a small bumblebee in appearance. It is about three times larger than
the ordinary housefly and has no mouth parts and does not bite or
sting. Nevertheless, they annoy cattle a great deal by depositing
their eggs on the hairs of the heels or legs when cattle are standing.
If cattle are lying down, the eggs are deposited in the flanks. The
common heel fly appears during the first warm days of spring and
lays its eggs. Cattle attempt to escape from heel flies by running with
their tails held high over their backs.

The eggs of the heel fly hatch in 3 to 5 days, and the larvae then
migrate to the back of the animal where they cause painful swellings
called *warbles*. The larvae then cut a hole through the skin through
which they breathe. The larvae develop under the skin for about 7
weeks and then drop to the ground where they pupate, producing
adult winged flies in 2 to 11 weeks.

The U.S.D.A. estimates that losses of about $120 million per
year occur because of this insect in the United States. The losses are
due to many tons of meat that must be discarded because it is infested with grubs and to damaged hides which must be discarded.

Several systemic insecticides may be used for controlling
cattle grubs. They may be applied by the pour-on method or by
spraying or as a feed or mineral additive. The instructions on the
insecticide label should always be followed.

The insecticide should be applied before the grubs appear on
the backs of the animals. Cattle from Texas, Oklahoma, Kansas,
Louisiana, Mississippi, Arkansas, the eastern one third of New
Mexico, Colorado, and the southwestern half of Missouri should
not be treated for grubs after October 1, because of the stage of
development of the grubs and possible adverse effects on host animals. Cattle from the rest of the United States should not be treated
after November 1.

21.4.2 Cattle Lice

Five species of cattle lice are found in North America. The short-nosed cattle louse (*Haematopinus eurysternus*), the long-nosed cattle louse (*Linognathus vituli*), the little blue cattle louse (*Solenopotes capillatus*), and the cattle tail louse (*Haematopinus quadripertusus*) all survive by sucking blood from their cattle hosts. Another species, the cattle biting louse (*Bovicola bovis*), does not suck blood but feeds on the skin tissue.

All species of cattle lice spend their entire lives on cattle. The females lay eggs which are glued to individual hairs close to the skin. Their life cycle is completed in a few days (25 to 28 days). These species vary in their importance from one region of the United States to another. Two or more species may be present on cattle at the same time.

Heavy lice infestations are more likely to occur in the winter months. The hair coat is usually lighter in spring and summer, and the sunshine and rain encountered help keep numbers low during these seasons. The first signs that cattle are lousy is observed when cattle scratch and rub themselves against fences, trees, feed bunks, and other objects. This is not a sure indication of lousiness, however. The animals should be checked for lice to make certain by parting the hair and looking for the lice or their eggs (nits) attached to the hairs. If present, control measures can then be used.

Heavy infestations of blood-sucking lice in the wintertime may cause adult animals to become anemic, with their red blood cell numbers reduced as much as one fourth or one half. Heavy infestations cause animals to become unthrifty and more susceptible to disease. They may grow weak, and the skin around the eyes, muzzle, and udder may become very pale. When heavily infested cattle are shipped or handled, they may become exhausted and may even die if they are strenuously exercised. Ridding infested cattle of lice usually results in rapid improvement, and they may recover completely in 3 or 4 weeks. When such cattle are treated for lice, they should be handled as gently as possible. Animals in the weakened condition may also be more easily poisoned by insecticides, especially systemics.

Lice may be controlled on beef cattle by means of cable-type back rubbers or dust bags charged with insecticides. These should be made available to cattle throughout the year. Systemic insecticides may be applied by means of pouring them on the animal, starting behind the shoulders and along the back, for a distance of 18 to 24 inches. Insecticides may also be applied by means of a spray, using a sprayer with a pressure of at least 200 pounds per square inch to wet both the skin and hair. No dry spots should be left as they may serve as a source of a new infestation. The first spraying of cattle should be about November 1, with a second spraying in 14 to 18 days to kill lice that were in the egg stage at the first spraying. Such a spraying procedure should control lice throughout the winter without the need for further spraying during very cold weather.

Dust bags may be charged with 1 percent coumaphos (Co-Ral) or 3 percent Stirofos (Rabon) livestock dusting powder. Back rubbers may be charged with 1 percent ronnel or 5 percent toxaphene (livestock formulation). Pour-ons may include various mixtures containing crufomate (Roulene) or fenthion (Lysoff). Insecticide sprays that may be used include coumaphos (Co-Ral), malathion, ronnel (Korlan), as well as others. Cattle producers should contact their extension workers for up-to-date recommendations on what insecticide to use and how to use it.

21.4.3   Cattle Mange

The terms *mange* and *scabies* are often used to describe the same condition. They are somewhat different, however. Mange is any skin condition caused by a mite. Scabies is a particularly serious, debilitating mange condition.

Four species of mites are found on cattle in North America that are regarded as mange mites. Those of most importance include the cattle itch mite (*Psorergates bos*) and the hair follicle mite (*Demodex bovis*). The latter is found everywhere in the world.

Three species of scabies mites are of importance. Psoroptic scabies is caused by a parasite known as *Psoroptes bovis*, which spreads quickly among cattle of all classes and ages. This form of scabies is by far the most injurious form of cattle scabies. Wherever found it requires immediate quarantine and control measures. Sarcoptic scabies is a second form caused by the sarcoptic scab mite, *Sarcoptes scabiei*. Cattle with this form of scabies are also subject to a quarantine and control measures wherever found. Chorioptic scabies is a third form of scabies caused by the mite *Chorioptes bovis*. In some states this form of scabies is reported to the state veterinarian. Whether or not quarantine and control measures are taken is decided, as a general rule, by the state animal health agencies.

Scabies mites produce the most severe skin lesions during the fall, winter, and spring months. When cattle shed their long hair in the spring and the skin is exposed to the hot summer sun, the infestation often clears up. Some mites, however, in parts of the body protected from direct sunlight survive during the summer and reinfest the animals when cool weather arrives once more.

Three chemicals are presently recognized by the U.S.D.A. as suitable for the treatment of scabies. These include water solutions of lime-sulfur; water emulsions of toxaphene, a chlorinated hydrocarbon; and water suspensions of coumaphos, an organic phosphate compound. When coumaphos is used, all animals which are infested or even just exposed must be treated twice. When treated with coumaphos, cattle may be immediately slaughtered with no holding period required.

21.4.4   Cattle Ticks

Tick fever in cattle was a very serious problem in the United States in the early 1900s, especially in southern and southwestern states which included about one fourth of the United States. Conservative estimates indicated that direct and indirect losses to this disease amounted to $40 million annually. Cattle tick fever was originally known by several different names including *Texas fever*. This name was not correct because it did not originate in Texas, nor was it confined to this state. The best name for the disease was probably *cattle tick fever*, since it was transmitted only by the cattle fever tick (*Boophilus annulatus*). An effort to eradicate this tick was begun officially in 1906 and was very successful, with what were once major losses being practically eliminated ("Keeping Livestock Healthy," *U.S.D.A. Yearbook of Agriculture*, U.S. Government Printing Office, Washington, D.C., 1942).

Three different species of ticks are of importance in the United States at the present time. These are the Lone Star tick, the Rocky Mountain wood tick, and the American dog tick.

The Lone Star tick (*Amblyomma americanum*) was given this name because of a conspicuous white spot on the rear angle of the scutum of the female. This spot sometimes shows a red or green tinge. The male is marked by two horseshoe-shaped areas at the rear side of the body. These whitish spots which form the horseshoe outline may also be tinged with red or green. The distinguishing marks in both sexes are very white in color, which makes a contrast with the reddish-brown background of the body. The Lone Star tick is known to be present primarily in eastern Texas and Oklahoma, but it has also been observed in Kansas and Montana. The damage it causes is due to irritation, blood loss, and loss of weight, especially in calves.

The actual economic loss in cattle caused by this tick has not been determined. The Lone Star tick feeds on three different host animals. Cattle and deer host large numbers of females ready to lay eggs. This tick may be controlled by appropriate dips and sprays.

The Rocky Mountain wood tick (*Dermacentor andersoni*) is a common parasite of mammals in the western United States in part or all of Oregon, Washington, Idaho, Montana, North Dakota, South Dakota, Wyoming, Nebraska, Colorado, New Mexico, Arizona, Utah, Nevada, and California. It does its damage to cattle through the sucking of blood and the transmission of diseases such as Colorado tick fever, Q fever, and tularemia. It also may cause a condition known as tick paralysis. Adult male and female ticks may feed on many animals such as cattle, horses, and sheep. They may also feed on porcupines and jack rabbits. After feeding on the host for about 1 week, the female wood tick drops to the ground, lays 5,000 to 10,000 eggs, and then dies. The eggs hatch in about 1 month and the six-legged larvae attach themselves to rodents such as mice and squirrels and feed for 3 to 6 days. Then they drop to the ground and molt into eight-legged nymphs, which again seek out rodent hosts, feed for about 1 week and then drop to the ground again, seek shelter, and form adults which attach themselves to large animal hosts, completing the life cycle.

The American dog tick (*Dermacentor veriabilis*) is distributed widely in North America from the eastern Rocky Mountains to the Atlantic Coast. It is also found along the coast of California, northern Washington, and northern Idaho. Its life cycle is similar to that of the Rocky Mountain wood tick. Its favorite host is the dog, as the name implies, but it will feed on humans, cattle, deer, and others.

All species of ticks can be controlled by spraying with recommended sprays as soon as the ticks are noticed in the spring. Spraying should be continued as long as the cattle are reinfested. Spraying is usually done at intervals of 2 to 4 weeks.

21.4.5   Face Flies (Musca autumnalis — DeGreer)

The presence of the face fly was first reported in the United States in 1952. Since then it has spread to the northern half of the United States and southern Canada and now ranges as far south as Alabama. It is a pest that is usually found in regions of high rainfall, although it is found near irrigated pastures of the semiarid regions also. The flies breed in the manure, and in open arid ranges the manure apparently dries out before the larvae are completely developed.

The U.S.D.A. has estimated that losses from damage inflicted on cattle by the face fly amount to about $68 million per year. The face fly also appears to be involved in the transmission of pinkeye. It does not suck blood but feeds on various secretions such as tears, saliva, nasal mucus, etc. Adult face flies resemble houseflies but are about twice as large and darker.

Face flies are difficult to control. One of the reasons for this is that most face flies on cattle are females which spend considerable time away from the cattle. Another reason is that it is more difficult to get insecticides on the face and around the eyes than on the body. Dust bags and oilers are used to control face flies, but this is usually on a free-choice basis and often is not too successful. Forced treatment is necessary for good control, and treatment must be applied every 2 or 3 days to be effective.

Face flies lay their eggs in fresh cow manure. The eggs hatch in 1 to 2 days, the larval stage is 3 to 6 days, and the transition from larvae to adults (pupal stage) is 7 to 10 days. Face flies overwinter as adults in barns and other farm buildings. Insecticides are sometimes incorporated into salt, minerals, and protein supplements fed to livestock. The insecticide is then present in the manure and destroys the larvae. This is not always successful since animals vary in their intake and sometimes do not eat enough insecticide for it to be present in large enough quantities to control the larvae in the manure.

21.4.6   The Horn Fly (Haematobia irritans [L])

The horn fly is of European origin. It was first reported in New Jersey in 1887 and now is found in Mexico, the United States, and the southern half of Canada. Losses from this pest have been estimated at $130 million per year by the U.S.D.A. Horn flies are blood-sucking parasites and cause considerable losses in weight gain and possibly milk production.

Horn flies are smaller than the housefly, being about one half their size. In addition to being present around the horns, these flies are also found on the back and underside of the body, which cannot be reached by the head or tail of the animal. They often cause sores around the naval where their feeding is heavy.

Horn flies spend their entire life on cattle. The females leave the animal only long enough to lay their eggs in fresh cow manure. About 10 days are required to pass from the egg to the adult.

Dust bags and oilers (back rubbers) containing certain insecticides are widely used for horn fly control. They may be used free-
choice or by forced use, which gives good control. Forced use is
attained by making cattle go under the insecticide carriers to get to
water, minerals, or feed. Power sprayers or dusters are sometimes
used for horn fly control. In addition, insecticides such as phenothiazine and ronnel may be supplied as a feed additive, but the control obtained in this manner is erratic since some animals do not eat
enough of the carrier to control the larvae development in the
manure.

21.4.7   The Screwworm Fly

The screwworm fly is found widely dispersed in Mexico, Central
America, and the southern United States. In the past, losses from
screwworm infestations have amounted to millions of dollars in the
United States alone. The screwworm fly is about twice the size of the
common housefly. The female deposits masses of eggs (200 to 400)
at the margins of wounds, and the eggs hatch in 12 to 24 hours. The
maggots have a tapered head and possess rasping mouth parts. They
feed only on living tissues. Wounds infested with screwworm maggots
emit a bloody, foul-smelling discharge which attracts more screwworm flies.

The female screwworm fly mates only once during her lifetime.
The male, however, may mate several times. The normal life cycle of
the screwworm fly is about 21 days, but the female seeks a mate
when only about 2 days of age.

Screwworm eradication has been successful in some areas of the
United States. This is done by releasing millions of sterile male flies
at weekly intervals which compete with normal males in mating with
females. The male flies are sterilized when in the pupal stage by exposing them to radiation. Since the female mates only once, she produces all infertile eggs when mated with a sterilized male. Sterile
males are continuously released and finally overwhelm the native
normal male population, with a resultant reduction and finally the
annihilation of the entire population. This method has worked very
well when sterile male flies are distributed carefully and often in
numbers large enough to dominate the native male population. If
the population is not completely wiped out, however, only a few
fertile females can start a new infestation. Reinfestations by transport
of immature or adult flies into a depopulated area can occur. For this

reason cattle should be observed often in screwworm season and treated to keep more flies from being produced by infected wounds.

Screwworm infestations occur in the late spring and summer months. The dehorning and castration of cattle should be delayed to the winter months when the wounds are not likely to be subject to screwworm infestations. When certain surgery must be done during the fly season, the wounds produced should be treated with a suitable fly repellent.

A number of insecticides such as coumaphos, ronnel, lindane, and toxaphene are effective in wound infestations and fly control. They should always be used according to the directions on the label of the container.

21.5   Internal Parasites

Internal parasites work inside the animal and are not easily observed. When heavy infestations occur, clinical symptoms appear, but these are often ascribed to other conditions. Internal parasites probably affect young animals (yearlings) more than mature animals. More research is needed to gain further knowledge on internal parasites in cattle. The important internal parasites of cattle include roundworms, tapeworms, flukes, and protozoa. The U.S.D.A. estimates that these internal parasites cost the cattle industry more than $160 million per year.

21.5.1   Roundworms (nematodes)

Twenty-five species of nematodes infect ruminants. Ten of these species are of economic importance. The different species vary greatly in size and occur as adults or larvae in many parts of the body. Each species has its own preferred location in the host. Roundworms which occur in the highest incidence and have the ability to produce disease usually inhabit the abomasum or the small intestines of cattle.

Animals become infested with roundworms by grazing contaminated pastures. Female roundworms in the digestive tract produce eggs which are eliminated from the body in the manure. Under favorable conditions the eggs hatch into larvae in the manure. These develop into the infective stage of larvae in 2 weeks or less. These larvae migrate onto the grass and are picked up by grazing cattle and grow into adults in the digestive tract, completing the life cycle (Figure 21.1).

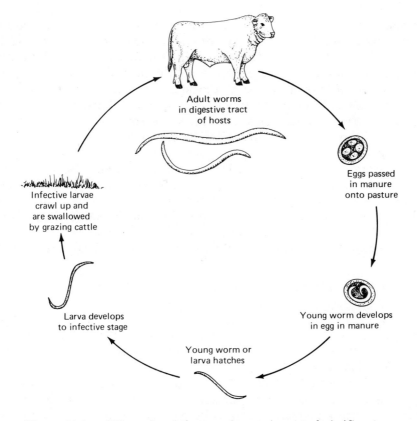

Figure 21.1.   Life cycle of the roundworm (nematodes). (Courtesy U.S.D.A.)

Roundworm infestations seldom kill cattle, but they may produce harmful effects. This is done by robbing the host of nutrients, damaging vital organs, and causing infested animals to be more susceptible to bacteria and other disease-causing organisms. These infestations lower the performance of affected individuals and result in more costly gains.

Probably the best method of prevention of infestations with roundworms is to prevent overgrazing of pastures. Overgrazing causes the pastures to be shorter, and the cattle will graze more closely, apparently picking up a greater number of parasitic larvae. Other factors may also be involved, however. Overgrazing can be prevented by rotating pastures and not grazing too large a number of cattle on the same pasture for long periods of time. Supplemental feeding of calves on pasture with grain appears to reduce infestations. Grazing animals of all ages together results in a higher probability of calves becoming infested. Unhealthy or diseased cattle are more susceptible to parasitic infestations. The accumulation of manure in barns and sheds and around water places increases the possibility of infestation. Some evidence indicates that some cattle are genetically resistant to these parasites, but such lines have not been developed on a commercial scale. Evidence also suggests that some animals seem to develop some resistance to these parasites if they are not heavily infested initially.

Several drugs may be administered as a means of treatment. Some of the drugs used include phenothiazine, organophosphate compounds, and many others. *L*-tetramisole is a new drug that can be administered either orally or by subcutaneous injection. Regardless of the drug administered, the directions on the label should be followed carefully. Treatment should be repeated periodically as needed.

21.5.2   The Common Tapeworm

Cattle in the United States may harbor both adult and larval tapeworms. The two species of tapeworms known to occur in the small intestines of cattle in the United States are *Moniezia expansa* and *Moniezia benedeni.*

Adult tapeworms have a head that possesses four suckers and a body made up of a number of segments joined together to form a chain. Mature tapeworms produce large numbers of eggs in the small intestines, and segments and eggs are expelled from the body in the manure which falls to the ground in the pastures. These eggs are eaten by beetle mites and develop within the body cavity of the beetles. When the weather is warm, the beetle mites crawl up on the grass and are taken into the digestive tract of the cattle where they mature and start another life cycle (Figure 21.2).

Many of the drugs used for the treatment of roundworms are also effective against tapeworms.

21.5.3   Flukes

Cattle in the United States may be infested with one or more species of flukes. Flukes are flat, leaflike parasites that may occur in the liver or in the rumen. Mature flukes produce eggs that pass out of the body with the manure and hatch in water, releasing ciliated embryos known as miracidiae. The miracidiae enter the body of certain species of snails where they develop into circariae which encyst

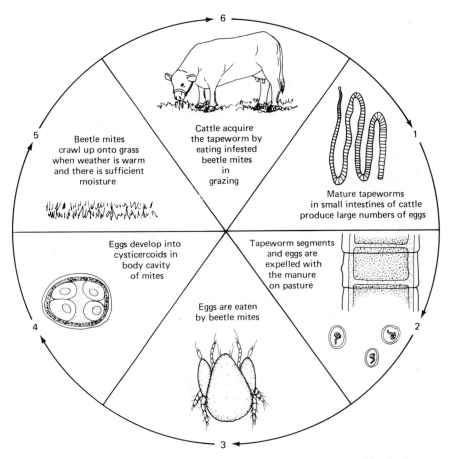

Figure 21.2.   Life cycle of the common tapeworm (*Moniezia expansa*). (Courtesy U.S.D.A. Circular 614)

on vegetation. The encysted circariae are eaten by cattle and develop into mature flukes (Figure 21.3).

The common liver fluke, *Fasciola hepatica*, is a flattened leaf-like brown worm about 1 inch long. Flukes of this species are found in the liver, bile ducts, and the gall bladder. In heavy infestations it has been found in abscesses in the lungs.

This species of fluke occurs worldwide. In the United States it occurs in the Pacific Coast states, the Rocky Mountain states, the Southwest, and the Southeast. Fluke infestations cause unthriftiness and anemia similar to those conditions found in worm infestations.

The recommended way to control flukes is to drain wet areas where snails propagate and to treat areas with copper sulfate.

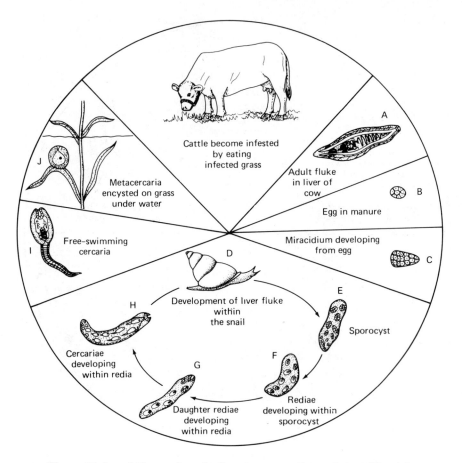

Figure 21.3.   Life cycle of the common liver fluke (*Fasciola hepatica*). (Courtesy U.S.D.A. Circular 614)

Copper sulfate is very effective in killing snails even in very dilute concentrations.

21.5.4 Protozoa

Protozoa are minute parasites of microscopic size. Many protozoa are reported to be parasitic in cattle, but only a few are known to be harmful. The most harmful belong to the two species *Eimeria zurni* and *Eimeria bovis*. These parasites occur in the wall of the intestine of cattle, especially calves. They cause a disease in calves known as coccidiosis or bloody diarrhea. This disease appears to be widespread throughout the United States.

The spore form of this parasite is usually the form found in microscopic examinations of fecal specimens. The spore is ovoid in shape and is four or five times larger than a red blood cell.

Coccidiosis is accompanied by diarrhea, with fecal matter smeared over the rump as far around as the tail will reach. It may or may not contain blood. An acute case of this disease may result in the death of the calf. Sometimes death occurs at a later period because of secondary complications.

Good management and feeding practices accompanied by sanitation practices help prevent the disease. Sulfonamides have been used for treating this disease for many years. If sulfonamides are given before signs of the disease appear, it may be a good preventative measure.

21.5.5   Trichomonads

The trichomonad, another parasite of cattle, is a one-celled organism which is microscopic in size, with three anterior flagella, each one about as long as the body, and a posterior flagellum (see Figure 21.4). When studied under the microscope in a saline solution, the organism shows rapid movement. In a more viscous medium its movement is restricted.

Two species have been reported as parasites of cattle: *Trichomonas foetus*, which is found in the reproductive organs of cattle suffering from genital trichomoniasis, and *Trichomonas ruminantium*, which is a common parasite of the digestive tract. The latter is associated with cases of diarrhea in cattle but is not always the cause.

Trichomonas foetus is of the greatest economic importance to beef cattle producers. It is widely distributed throughout the United States. It reproduces by each organism dividing to form two new individuals. It is usually transmitted from one animal to another by the act of mating. In cows an infection of this organism may result in failure to conceive. The cow may also develop a uterine infection and a vaginal discharge, which is more noticeable during estrual periods. Estrous periods may be irregular. In other cases conception may take place but will end in abortion. Abortion may occur at any time during gestation, but it usually occurs from 8 to 16 weeks after conception. In other cases the fetus may die and not be expelled from the uterus. In such cases the fetus becomes macerated, and the uterus becomes filled with a grayish white, almost odorless fluid. In such cases the cow appears to be pregnant, but there are no signs of illness and no signs of approaching parturition. At the end of gestation no calf is produced. In other instances in infected animals a normal gestation and parturition may occur.

Bulls may also be infected. The prepuce may become inflamed, and there may be a discharge of pus from the sheath. The penis is also often inflamed and may show many nodules when examined. The ampullae, vas deferens, and epididymides may also be infected. An infection in the bull is usually chronic.

The prevention of the occurrence of the disease is very important. Any new, mature animals introduced into the herd should be examined carefully to make certain they are not infected. Such animals should come from a herd whose past reproductive history shows no signs of the disease. Cows known to be free of the disease should not be bred to a bull away from home. If the disease is suspected, a veterinarian should be contacted. Trichomoniasis is often difficult to cure.

21.5.6   Anaplasmosis

Anaplasmosis is a disease of the blood of cattle produced by a microscopic parasite which destroys the red blood cells. If the blood of infected cattle is placed on a glass slide, stained, and observed under the microscope, certain deep staining bodies called *Anaplasma marginale* may be seen near the margins of the red blood cells. These are small bodies many times smaller than the red blood cells themselves and appear as dark round dots.

Anaplasmosis occurs in cattle in many states in the southern half of the United States but it is not necessarily limited to warm climates. It also occurs in such states as Wyoming and Ohio, which are far enough north to have very low winter temperatures. Anaplasmosis causes millions of dollars of losses each year and seems to be increasing in incidence. Practically all clinical cases occur in mature cattle, but calves and yearlings may have the disease, although they may show only mild symptoms.

Signs of anaplasmosis include rapid breathing, rapid weight loss, and sunken eyes, and some infected animals become constipated. The animals may have very pale lips, nostrils, and mouth linings. The body temperature is high, ranging between 104°F and 106°F. The pulse rate is increased, and the respiration rate is short and rapid. The death rate varies, but it has been as high as 50 percent in some herds where the infection occurred. Recovery from the disease is very slow and may require several months. Animals that recover may be carriers of the disease the rest of their lives unless treated with large doses of antibiotics.

Anaplasmosis may be transmitted by blood-sucking insects such as horseflies and ticks. It can also be transmitted mechanically by unclean instruments. It is possible to spread the disease by dehorning, tattooing, and other surgical procedures.

Treatment of infected animals is often beneficial if done in the early stages of the disease. Blood transfusions and antibiotics are often used. Treatment to clear the carrier state includes the injection of tetracycline at the rate of 5 milligrams per pound of body weight for 10 to 14 days. An alternative is to include 5 milligrams of this antibiotic per pound of body weight in the feed for a 45-day period. During the insect vector season, a method of control is to supply 0.5 milligram of the antibiotic per pound of body weight per day in the feed. The problem with this method is that some animals may not consume enough feed to consume this level of antibiotic.

A vaccine has been developed to control anaplasmosis. Initially two injections of the vaccine are given 4 to 6 weeks apart, followed by a booster shot given annually. Vaccinated and nonvaccinated animals can be run together with no danger. A small percentage of vaccinated cows develop antibodies in their blood which destroy the red blood cells of their calf when it obtains the milk through the nursing process. This causes the death of the calf in many instances. The vaccination program protects the treated animal from an acute attack of the disease, but it does not stop an infection in progress or prevent the carrier state.

Once the herd is infected, tests may be conducted on each animal to determine which react positively or negatively. Once this is determined, those that test positive may be kept apart in different herds, may be sold, or may be treated with antibiotics to clear the carrier state.

STUDY QUESTIONS

1. Although infestations with parasites seldom cause the death of individuals, why are losses of billions of dollars encountered each year?

2. What are the two general kinds of parasites that affect cattle?

3. Why is it necessary for livestock producers to keep abreast of the kinds and methods of treatments of parasites when chemicals are used?

4. What is meant by the biological control of parasites? How may biological controls work?

5. What are juvenile hormones? What are pheromones?

6. Are some cattle naturally resistant to parasites? Discuss.

7. What are the most common external parasites that affect beef cattle?

8. What precautions, in general, should be taken in treating animals with chemical insecticides?

9. What are cattle grubs? How do they affect cattle?

10. When are cattle lice most likely to have undesirable effects on cattle? How may cattle lice be controlled?

11. Distinguish between cattle mange and scabies in cattle. If an infestation of scabies in cattle is recognized, what must be done?

12. What is Texas fever in cattle? What insect was responsible for its spread? How was it controlled?

13. What are three different species of ticks that affect cattle in different parts of the United States? How may they be controlled?

14. What are face flies? How do they feed on cattle? Why are they difficult to control?

15. Describe the life cycle of the face fly.

16. How does the horn fly differ from the face fly in its appearance and in its effect on cattle?

17. What is the best way to control horn flies on cattle?

18. In what countries of the Americas are screwworm flies the most important?

19. What method was used to reduce and eliminate the screwworm fly in some parts of the United States? What was the main reason this method could be used?

20. When a screwworm population is eliminated in a region, what might be responsible for it reinfesting animals in that region?

21. List the internal parasites of most importance in cattle in the United States.

22. Discuss methods that may be used to control roundworm infestation in cattle.

23. Describe the life cycle of roundworms that affect cattle.

24. What two species of tapeworms may affect cattle in the United States? Describe an adult tapeworm.

25. Describe the life cycle of the common tapeworm *Moniezia expansa.*

26. What are flukes? What parts of the body do they usually affect in cattle?

27. What methods may be used to control liver flukes?

28. What two species of protozoa are important parasites of cattle in the United States?

29. Describe a trichomonad. What species is the most important parasite in cattle in the United States? How may trichomoniasis be prevented and controlled?

30. What causes anaplasmosis in cattle? Describe the symptoms of this disease in cattle.

31. How may anaplasmosis be transmitted from one animal to another?

32. Describe in detail methods that may be used for treating and controlling anaplasmosis.

some common diseases of cattle

Diseases cost beef cattle producers in the United States millions of dollars each year. These costs include actual losses of cattle and their poorer performance as well as the costs of treatment and prevention. Some diseases in cattle are so common and widespread that it is a normal practice to routinely vaccinate against them year after year. Others are not so common, but when cattle are infected, losses may be severe, not only from losses due to death but from losses due to unthriftiness and poor performance.

Diseases are of two general kinds: infectious and noninfectious. Infectious diseases are those caused by bacteria, other organisms, and viruses. Noninfectious diseases are those caused by mechanical ailments, ingestion of toxic materials, digestive or nutritional upsets, and other causes. Some genetic defects may also be classified as diseases.

Infectious diseases may spread from one animal to another in a number of ways. These include the following: contact with other animals that are diseased; drinking water from polluted streams or other sources; contact with organisms in vehicles used to transport cattle; contact with contaminated facilities and equipment; contact with disease organisms transported by dogs, birds, and other animals which feed upon infected carcasses; and infestations with insects. So many different means of spreading disease are possible that not all of them can be discussed in detail in this chapter.

22.1   Body Defenses against Infectious Diseases

The body has many defenses against infectious disease. The first line of defense is the skin, which provides a wall against many disease-producing organisms that are always present in the surrounding environment. The skin may also contain certain chemical substances such as enzymes that fight against organisms, and in some instances it may secrete and carry chemicals that repel disease-bearing insects such as flies and ticks. The importance of the skin is very evident to everyone because when the skin is cut or its surface broken, infection often results.

The mucous membranes which line the openings of the body such as the digestive tract are also an important means of defense against disease. The secretions of these membranes help control infections. Some mucous membranes contain cilia (whiplike projections) that help propel foreign objects such as microorganisms from the body openings. Secretions from some of the mucous membranes such as nasal discharges may expel foreign objects from the body. Still other secretions may contain antibodies that help destroy disease-causing organisms.

Infectious organisms which penetrate the skin and mucous membranes come face to face with other body defense mechanisms. Some of the white blood cells are phagocytic in nature. When a foreign microorganism enters the bloodstream, the phagocytes are attracted to it and destroy it. This is called *phagocytosis*. If the microorganism is not destroyed, it may be carried by these cells to other parts of the body, which spreads the disease.

Some white blood cells such as the lymphocytes release antibodies when microorganisms enter the bloodstream. The antibodies also aid in the destruction of these disease-causing organisms. In a sense, antibodies are the body's last line of defense. Vaccines containing antibodies or which cause the body to produce them are widely used in the control and prevention of many diseases. Even fever (a higher body temperature than normal) aids in the fight against infectious diseases.

22.2 Resistance of Cattle to Infectious Organisms

The ability of an animal to resist infectious diseases is either natural or acquired.

22.2.1 Natural Resistance or Immunity

Natural immunity is present in the individual without stimulation of any kind. The natural resistance may be due to mechanical barriers such as the thickness of the skin and hair, or it may be physiological, such as the presence of inherited antibodies in the blood. Natural resistance can be present on a species or an individual basis. Species resistance means that a disease which affects one species may not affect another. For example, cattle have Texas fever, but humans do not. Individual resistance refers to the fact that even within the same herd when a disease strikes, some individuals may die, others may become ill but recover, and still others may not show any signs of being ill. Individual differences in resistance may be due to inherent differences, but noninherited differences such as age, retention of antibodies from colostrum, or a slight unobserved illness in some individuals may occur. It is also possible that some individuals may not be exposed enough to the disease to contract it.

Some cattle probably possess inherited resistance to almost any infectious disease. Lines probably could be developed that would be resistant as a line, but this has not been attempted on a practical basis for several reasons. Developing resistant lines would be a costly and long time process because animals would have to be exposed to the disease to locate those that were resistant. Exposure would have to be made for several generations, and some animals would become sick and even die. Even if such a line were developed, it would probably be susceptible to other diseases for which selection was not practiced. Furthermore, it may be cheaper and more practical to vaccinate against diseases in the entire herd.

22.2.2 Active Immunity

Active immunity occurs when an animal has a disease and recovers or is vaccinated against it. Exposure to the disease causes antibodies against the disease to be produced. Following vaccination or recovery from the disease the animal is resistant to the disease

for varying lengths of time because the animal builds its own antibodies against that specific disease. Vaccinations for some diseases
require booster shots at periodic intervals to keep the antibody level
in the blood high enough to be effective.

22.2.3   Passive Immunity

Passive immunity develops when antiserums or antitoxins are
given to an animal. In other words, this type of immunity results
from borrowing antibodies from another animal. The protection is
immediate but lasts only a short time (15 to 25 days). Antibodies received by the calf through the fetal membranes and the colostrum
give a passive immunity. The maternal antibodies received by the calf
through the colostrum give the calf protection from diseases until it
develops the ability to produce its own. In rare cases, calves may fail
to produce their own antibodies and will usually die of disease when
the supply received from their mother through the colostrum is
depleted.

22.3   Disease Due to Ingestion of Foreign Objects

Hardware disease which develops from cattle ingesting foreign objects in their feed is often encountered. The reticulum, which is one
compartment of the multiple stomach of cattle, catches nearly all the
heavy objects ingested with the feed. Lighter objects or material pass
through the reticulum into the rumen. Sharp objects such as nails
and bits of wire may perforate the wall of the reticulum and go into
the heart sac when muscular contractions occur. Hardware disease is
the term often given to traumatic gastritis and traumatic reticulitis.

The rumen and the reticulum are forced forward during pregnancy and as the calf increases in size. At parturition and during
labor, the muscle contractions may be responsible for forcing the
sharp objects through the wall of the reticulum into the heart sac.
Sometimes the object may be digested if it is present for a long
period of time and may not be present when the animal is slaughtered.

Symptoms of hardware disease are many. The cow may have a
poor appetite and may be reluctant to move about. Signs of indiges-
tion sometimes occur, and the animal may show signs of pain when
defecating. Infections may occur and produce considerable fluid
around the heart. Sometimes large amounts of fluid accumulate in
the brisket. Sometimes affected animals will bloat.

If an affected animal is not valuable, it should be sold for slaughter. If the animal is valuable, a veterinarian can perform an operation in the interior portion of the reticulum, and any foreign object found can be removed. If the condition is diagnosed early, about two thirds of the affected animals can be treated and returned to normal. Sometimes magnets are inserted into the reticulum with a balling gun or stomach tube. Foreign objects will adhere to the magnet, thus reducing the chances that they will penetrate the reticulum and go into the heart sac. The best prevention, however, is to keep feed bunks, pastures, and lots free of any objects that may be ingested and cause the disease.

22.4   Ingestion of Toxic Materials

The ingestion of toxic materials such as lead paint sometimes causes sickness and death in animals. This can be prevented by making certain no chemicals which might cause sickness and death are placed where cattle can get to them. Other toxic materials may be present in the feed or water.

22.4.1   Nitrate Poisoning

Nitrates are not particularly poisonous to cattle because most forages contain them. Nitrates eaten by cattle are normally converted to protein in the rumen by microorganisms. Nitrite is the cause of nitrate poisoning. It is one of the intermediate products of nitrate metabolism.

Nitrite is absorbed into the bloodstream where it changes the red pigment of hemoglobin into methemoglobin. Hemoglobin normally carries oxygen from the lungs to other tissues of the body and is a bright red color. Methemoglobin does not carry oxygen. When enough nitrite is absorbed into the blood to form high amounts of hemoglobin, the oxygen-carrying capacity of the blood is reduced so much that the animal may die. The development of a toxic level of nitrite in the blood depends on how much and how fast the nitrate is consumed.

Methemoglobin causes the blood to be brownish in color, and the disease can be detected from the appearance of this color in the blood. Death of the individual and fetal abortion in cows can result. Deaths are due to asphyxiation because of the lack of oxygen in the tissues.

High nitrate levels in forages may cause nitrate poisoning. Many factors can affect the nitrate level of feeds. These include hot, dry weather, the use of herbicides, plant diseases, stage of growth, plant parts, the plant species, and nitrogen fertilization. Grain usually does not contain large amounts of nitrates. This compound is more likely to be found in the lower third of the stalks of crops such as corn, sorghum, and Sudan grass. Nitrates may be present in water in ponds, road ditches, etc., where there is drainage from feedlots, heavily fertilized fields, silos, septic tanks, or manure disposal legumes. Well water seldom contains toxic levels of nitrates.

The amount of nitrates in forages can be determined by chemical tests. Harvesting a forage crop as silage reduces the nitrate content by fermentation by 40 to 60 percent. Crops harvested as nearly as possible to maturity will usually reduce their nitrate content.

Forages containing high levels of nitrates can be safely fed. They can be mixed with other forages low in nitrates, or they can be fed along with grain in a balanced ration.

22.5   Some Noninfectious Diseases

Many diseases of cattle are not due to infections caused by bacteria, viruses, and other organisms but rather due to other causes such as stances found in certain feeds that are eaten or to an imbalance of substances within a ration.

22.5.1   Acorn Poisoning

In some areas of the United States periodic outbreaks of acorn poisoning occur. Calves and yearlings appear to be affected more often than older animals. When a large crop is produced, white oak acorns may be eaten in large enough quantities in the fall of the year to cause poisoning. Where pastures contain large quantities of acorns in the fall, cattle should be kept out or at least closely observed during the acorn season.

Signs of acorn poisoning are a poor appetite and constipation followed by diarrhea in which the feces are dark brown in color and have an offensive odor. Poisoned animals also appear to be "tucked up" in the middle and show signs of pain. Poisoned animals also appear depressed, drink water often, and urinate frequently.

22.5.2 Bloat

Bloat is a condition in cattle which is usually due to an accumulation of gas in the rumen which causes its inflation or expansion. Normally these gases are expelled quite freely by eructation (belching) and to a lesser extent by absorption into the blood and elimination through air exhaled from the lungs. Bloat may be due to overeating, the consumption of spoiled feeds or certain feeds, or an occlusion of the opening of the forestomach by hair balls or other foreign bodies. Bloat may also be caused by anything that interferes with normal rumination and intestinal movements. The tendency to bloat appears to be inherited. Extreme bloat can cause the death of the affected individual.

Two general kinds of bloat are recognized, pasture bloat and feedlot bloat. Pasture bloat usually occurs in animals grazing immature lush alfalfa and certain clovers or when these are fed as a green chop. Feedlot bloat usually occurs in cattle fed high-grain rations which may, or may not, contain legume hays.

Several methods are used for preventing pasture bloat. One is to make certain that a legume-grass pasture contains no more than 50 percent legumes. It is also helpful if cattle are filled with hay or grass pasture before they are turned on legume-grass pastures. An antifoaming agent, Poloxalene, will prevent pasture bloat if fed in proper amounts and at the proper intervals. Poloxalene can be fed as a top dressing on feed, in a grain mixture fed free-choice, or on certain kinds of salt blocks. The main problem is to assure that cattle ingest the proper amount of this substance each day to prevent bloat. Directions which come with this compound should be followed carefully.

Feedlot bloat usually occurs repeatedly in only a few individuals. Poloxalene does not appear effective in preventing this kind of bloat in cattle. Feedlot bloat can often be reduced in incidence by adding 15 percent coarsely ground roughage to the grain mixture. Feeding grain that is coarsely ground or rolled also helps prevent bloat.

Severe acute bloat may result in the death of the individual unless the animal is treated promptly. In severe cases an emergency treatment may be applied by inserting a rubber hose $3/4$ to 1 inch in diameter down the mouth into the rumen. This tube often releases the accumulation of gas. In foamy bloat this treatment may not be effective. An antifoaming agent may be administered through the tube or by direct injection into the rumen. This tends to break up the foam and allows the gas to pass from the rumen through the tube.

In extreme cases gas may be released from the rumen by means of a trocar or a sharp knife. In using the trocar, it should be inserted into the rumen at a point halfway between the last rib and the hook bone on the left side about 3 to 4 inches from the edge of the loin. If the trocar does not give relief, a sharp knife may be used to open a slit about 2 inches long. This should be spread apart by the fingers to allow the gas to escape. The best way to solve the chronic bloater problem is to send the affected animal to slaughter.

22.5.3   Fescue Foot

Fescue foot is a condition of lameness, the sloughing off of the end of the tail and hooves in some cattle which graze tall fescue. Affected animals usually show swelling or pain in one or more of their feet. One of the hind feet is usually the first to be affected. As the condition progresses, an indented line appears somewhere between the hock and the dew claws as though a wire had been tied around the leg. The affected foot or tail is deprived of blood and will drop off as the condition progresses. Cattle grazing fescue, especially in the winter months, should be checked for signs of stiffness in the morning when they arise from their overnight bed. Affected animals should be placed inside a building with straw bedding and fed storage other than fescue hay. Some grain and vitamin A may help the animal recover.

The actual cause of fescue foot is not known. Some researchers believe that a fungus or mold growing on the plant produces a poison or toxin that causes the disease. Others believe that the fescue plant itself produces the toxic substance.

Fescue foot may occur in some fields and not in others. Some animals appear to be more susceptible to the disease than others, and there may be an inherited susceptibility involved. Turning cattle on fescue fields which have not been pastured or mowed for several months appears to favor the development of the disease. This is especially true if the fescue has made a thick growth and lodged. This condition may favor the growth of molds and fungi. More cases appear to occur in the winter months when it is very cold and snow is on the ground. Cattle on low-quality rations before being turned on fescue appear to be more susceptible to the disease than those who have been on good-quality rations. Fescue hay can be fed with little or no danger of producing the disease.

Preventing fescue foot is more important than treating it. Fescue pasture should be grazed closely to prevent excessive growth and lodging. A rotation of pastures where one or more contains grass other than fescue may be advantageous when possible and practical. Feeding some hay while cattle are on fescue pasture may reduce fescue grass intake and reduce the possibility of the disease occurring. The animals should be closely checked each day, and those showing early symptoms of the disease should be placed on another feed or pasture.

Fescue pasture is an excellent winter feed in many midwestern states. Losses from fescue foot are less than losses due to malnutrition, parasite infestations, and such. With proper management the gains from fescue pasture are much more than occasional losses from fescue foot.

22.5.4   Grass Tetany

Grass tetany has occurred in many states of the United States and throughout the world. Other names for this condition are winter tetany, grass staggers, wheat poisoning, magnesium tetany, and hypomagnesemia. Its incidence appears to be increasing where grass alone supplies most of the nutrients consumed by cattle.

Clinical signs of grass tetany vary. Affected animals may become excitable, have a wild stare, or stumble and lack coordination. Muscles of the animal may tremble; it may grind its teeth; it may have violent convulsions followed by a deep coma and death. In some cases no symptoms may have been observed, but the cow is found dead. Symptoms of grass tetany are similar to those seen in a number of other diseases.

Grass tetany may occur in the fall but probably occurs mostly in the spring when grass is green and lush. Tetany occurs frequently in cows 6 years of age and older when they are nursing calves under 2 months of age. Most cases occur when cows are grazing grass pasture or when they consume grass hay. Cattle grazing legumes or grass-legume mixtures are less likely to suffer from this disease.

Grass tetany is caused by a deficiency of magnesium in the blood serum. Treatment in early stages of the disease is usually successful. Two hundred cubic centimeters of a saturated solution of magnesium sulfate (epsom salts) injected under the skin will produce a high level of magnesium in the blood in 15 minutes. Four injections of 50 cubic centimeters each at separate locations are recommended. Cows in the early stages of the disease should be handled as gently as possible to avoid excitement.

Animals recovering from an attack of grass tetany appear to be quite susceptible to recurring attacks. The recovered animal should be removed from pasture and fed hay and grain. The animal should also be given 30 grams of magnesium daily by mixing magnesium oxide in the grain fed.

Grass tetany may be prevented in several ways. A mineral mixture containing magnesium should be made available in mineral boxes at several locations in the pasture. Grass pasture should not be grazed too early in the spring because there is less magnesium in young grass than in that which is more mature. Low-risk animals such as heifers, dry cows, cows whose calves are 4 months of age or older, and stocker cattle should be grazed on high-risk pastures. Likewise high-risk animals should be placed on low-risk pasture. High-risk animals are those with very young calves or those that have had grass tetany previously. Soils low in magnesium should be treated to increase the level of this mineral in the soil. Dolomitic limestone is a good source of magnesium when spread on the soil.

22.5.5 Urinary Calculi

Urinary calculi are due to the deposition of certain minerals in the urinary organs. The mineral deposits may be of various shapes and sizes and may be varied in composition. Urinary calculi in range-reared calves are likely to contain calcium, magnesium, and ammonium phosphate. Females usually have urinary calculi less frequently than males. Older animals are less likely to be affected because the deposits are usually present in the bladder, which is less likely to cause a problem. If the calculi become lodged in the urethral tube, however, the tube becomes blocked. Continued production of urine may cause the distension of the urethra and rupture of the bladder or urethral tube. When this occurs, the urine flows into the body cavity and tissue around the penis in males, causing what is commonly called *water belly*.

Urinary calculi may be caused by rations high in phosphorus. Fattening rations are usually higher in phosphorus than high-roughage rations and may cause more calculi to be produced. Rations such as prairie hay are high in silica content, which may also contribute to calculi formation.

Early signs of urinary calculi formation include restlessness and vigorous tail switching. The affected animal attempts frequent urination which may be limited to a small dribble. Sometimes rectal prolapse occurs because of frequent straining. In the advanced stages affected animals may die of uremic poisoning.

Treatment of urinary calculi is difficult and costly. Urinary tract relaxants may be given; they help keep the urethra open so the mineral deposits may pass from the body in the urine. Surgery in early stages of the disease may be effective enough to get the animal to market.

Prevention is more important than treatment for urinary calculi. The ration of fattening animals should contain a two-to-one calcium to phosphorus ratio. When one anticipates that the problem may be encountered, the ration should contain low amounts of high-silica grasses such as sorghum, wheat straw, cottonseed meal, and sugar beet pulp. Feeding common salt at 3 to 5 percent of the total ration will increase water consumption and urination, which helps prevent the disease. Ammonium chloride fed at a level of 0.3 percent of the total ration increases water consumption and may prevent urinary calculi from forming.

22.5.6   Cancer Eye

Other names for cancer eye are epithelioma or carcinoma. This is a malignant type of tumor which affects the eye and related tissues. It appears to be more prevalent in Hereford cattle in the West and Southwest than in other regions of the United States. The specific cause of cancer eye is not known, but exposure to intense sunlight, dust, and other eye irritants appears to be related to a higher incidence of the disease. Hereford cattle lacking pigment around the eyes appear to be more susceptible to the disease. Heredity also seems to be involved, with a heritability estimate for cancer eye in cattle being about 25 to 30 percent.

Cancer eye usually occurs in older animals 5 to 6 years of age or more. Treatment often includes surgical removal of the eye if only the eye is involved. Selection against breeding animals which have the disease may also be effective.

22.6   Bacterial Diseases

Cattle are subject to a number of bacterial diseases. Some diseases occur more frequently than others and are of greater economic importance. Only those of greatest economic importance will be discussed in this section.

22.6.1 Brucellosis

Brucellosis is caused by an infection with the bacterium *Brucella melitensis*, discovered in 1887 by a British army surgeon named Sir David Bruce. This disease in cattle is also called *Bang's disease* after a Danish veterinarian, Dr. Bernard Bang, who isolated *Brucella abortus* in 1897. Dr. Bang showed that an infection with this organism caused abortion in pregnant cows, because the *Brucella* organism becomes localized in the placental tissues of the pregnant cow, causing inflammation which may decrease blood circulation to the fetus. The aborted membranes and fluids contain many organisms which may contaminate food, bedding, and other materials and may be spread by dogs, birds, and other animals. When the contaminated materials are eaten by cattle, a brucellosis infection may result. Even water containing the organisms may cause brucellosis to occur.

A nationwide brucellosis eradication program in the bovine has been successful in the United States in recent years. Thirty states were certified as brucellosis-free areas in 1974. It has been greatly reduced in incidence in many other states.

22.6.2 Blackleg

Blackleg is a disease caused by the bacterium *Clostridium chauvoei*. It usually affects cattle under 6 to 18 months of age. The organism enters the body through the mouth or small wounds in the skin.

The first symptoms of the disease include lameness and depression. The disease is accompanied by a high temperature, rapidly swelling tumors, a great depression, and violent convulsions. The animal usually dies within 36 hours, and the carcass is rapidly bloated with gas. Swelling in affected animals when felt with the hand expresses a crackling sensation.

Large doses of penicillin may be effective if given early. If not diagnosed and treated in the early stages, chances for survival are poor.

Vaccination with blackleg bacterins is a very effective preventative. Calves vaccinated at less than 4 months of age should be revaccinated at 5 to 6 months of age because immunity is not permanent when calves are vaccinated too young.

22.6.3 Malignant Edema

Malignant edema is a disease in cattle caused by the bacterium *Clostridium septicum*. It may affect cattle of any age. *Clostridium septicum* bacteria are present in the feces of most domestic animals. They are also present in large numbers in the soil where livestock have been kept in large numbers.

This bacterium enters the body through wounds and can gain entrance into the body of the cow through vaginal or uterine wounds resulting from difficult calving.

Symptoms of the disease include depression, a high body temperature, and a poor appetite. Treatment with massive doses of penicillin in early stages of the disease may be effective.

Vaccinations with *Clostridium septicum* bacterins usually produced in combination with other bacterins prevent the disease from occurring.

22.6.4   Other Clostridial Diseases

Diseases such as black disease caused by *Clostridium novyi*, enterotoxemia caused by *Clostridium perfringens*, and botulism caused by *Clostridium botulinum* are other clostridial diseases which produce fatal infections. These organisms produce spores which may remain in the soil for long periods of time and are potentially infective to livestock. Many biological products are available for immunizing cattle against these diseases. The beef cattle producer should consult with his or her local veterinarian to determine the products to use in a particular herd.

22.6.5   Leptospirosis

Six lepto serotypes have been diagnosed in cattle in the United States: pomona, hardjo, grippotyphosa, ecterohemorrhagiae, canicola, and szwajezak. Hardjo and pomona are the serotypes most frequently diagnosed in cattle. Leptospirosis is a disease produced by the leptospira bacterium (*lepto* from the Latin word meaning "thin" and *spira* meaning "spiral"). Both wild and domesticated animals are affected by this disease.

The lepto organism may be shed in the urine, which is sometimes a red "port wine" in color. Renal dysfunctions also occur in infected animals, and abortions may occur, especially late in pregnancy. It may also cause infertility in cows and bulls and mastitis in affected cows. Leptospirosis can be transmitted in stagnant water, by rodents, and through the semen of infected bulls. Leptospirae may be transmitted from cattle to humans. It is one of the "big five" diseases that can be transmitted through the semen of the bull.

These five are tuberculosis, brucellosis, leptospirosis, vibriosis, and trichomoniasis.

Vaccination at 6- to 8-month intervals with *L. Pomona* bacterins has been widely used as a means of control. More recently, many beef cattle producers have been vaccinating their cattle with the newer multiple serotypes (multivalent) lepto bacterins (vaccines).

22.6.6   Calf Scours

Calf scours is also called calf diarrhea. It causes great losses to the cow-calf producer. Scours in calves is a clinical sign of diseases which can have many causes. The intestines of calves with scours fail to absorb fluids from the intestinal contents, and the amount of fluids in the feces is increased, resulting in diarrhea. The loss of fluids from the body causes a dehydration and loss of electrolytes from the body, which changes the body chemistry. This is the primary cause of death in calves rather than the infectious agents, but the infectious agent triggers the trouble. Prevention of the infection, of course, is important from this standpoint. The younger the calf when it is infected, the less its chances to survive.

Hundreds of types of salmonella bacteria are known which cause diseases in livestock. The salmonella organism produces a potent toxin (poison) within its own cells. Following treatment with antibiotics the cells may release their toxin, causing shock in the calf. This type of scours usually affects calves 6 days of age or older. The affected calves have diarrhea, and blood and fibrin are seen in the feces. The calves become greatly depressed and have an elevated temperature. The source of infection can be birds, rodents, other cattle, the water supply, or even a human carrier. Antibiotic therapy is effective when given early, but treatment with fluids orally or intravenously may be the most effective.

E. coli can cause scours in calves and it is usually accompanied by dehydration. The severity of the disease is greater in young calves and with certain types of the organism. Oral or intravenous injections of fluids and treatment with antibiotics are recommended. To prevent infections, calves should always receive an adequate supply of colostrum, and those calves affected with scours should be isolated and treated to prevent the spread of the disease.

Enterotoxemia is caused by toxins produced by six different types of the bacterium *Clostridium perfringen*. It can be fatal to calves, and the onset may be sudden. The disease is associated with conditions which prevent the calf from nursing for a longer than usual period of time. When the calf does nurse, it is likely to consume too much milk, which establishes conditions in the gut favorable to rapid growth and toxin production by the clostridial organisms.

Symptoms of the disease in the calf are listlessness, straining, and kicking of the abdomen. Bloody diarrhea may or may not occur. The only treatment for affected calves is injections with antitoxins and antibiotics. Enterotexemia may be controlled by vaccinating cows with the toxoid 60 days and again 30 days before parturition. A single booster shot is recommended in treated cows each year before calving.

22.6.7   Foot Rot

Foot rot is also known by the term *foul foot* and results in the animal showing varying degrees of lameness in one foot — from slight lameness to that which is so great that the affected animal is reluctant to move. Many organisms may be involved, but some people believe that it is largely caused by *Sphaerophous necrophorus*. The disease may arise due to injuries from sharp objects. The skin between the toes, the soft part of the hooves, or the skin around the coronary band may be sites of infection. Exposure to manure or mud around feed bunks or watering troughs appears to contribute to foot infections. Means of prevention include putting cattle in clean, dry pens free of sharp objects that might cause injuries of the feet. Hoof care and proper trimming of the hooves are also important. Sulfonamides and antibiotics are effective in the treatment of infected animals.

22.6.8   Pinkeye

Pinkeye is also known as infectious bovine keratoconjunctivitis, or IBK for short. It affects large numbers of cattle each year in the United States. It may cause temporary blindness in one or both eyes, and sometimes the damage is so severe that permanent blindness results. The greatest economic loss probably occurs from weight losses and decreased production of cows nursing calves.

The bacterium *Moraxella bovis* is thought to be the main cause of the disease. Other infectious agents may also be involved, however. The virus which causes infectious bovine rhinotracheitis (IBR) (red nose) also produces this condition in some animals. Face flies which feed on fluids from the eyes may spread the disease.

The first sign of pinkeye is a watery discharge from the eye. This becomes progressively worse with time, and the discharge may become thick and yellow. The infected eye appears to be very painful. One or both eyes may be affected. The condition usually lasts 3 or 4 weeks, and some animals may be blinded temporarily. Usually vision is regained after the disease runs its course.

Prevention is difficult. Some breeds appear to be more susceptible than others, although some individuals in any breed may have the disease. Controlling face flies may be helpful in preventing pinkeye.

Early treatment with antibiotics and certain steroid compounds may lessen the severity of the disease. After treatment, covering the eye with a cloth patch cemented to the skin around the eye keeps out the bright sunlight and may prevent further infection. It appears to be beneficial to the animal's recovery from the disease.

22.6.9 Vibriosis

Vibriosis is caused by a bacterium known as *Vibrio fetus*, which infects the reproductive tract of cattle. It is spread in the mating process and produces infertility and occasional abortions as well as the death of the fetus in pregnant cows. The most usual result is infertility rather than abortion.

Infected cows usually recover and become normal breeders once a normal pregnancy occurs. Some cows, however, may become carriers of the organism and may be a source of infection for bulls during the next breeding season. Infected bulls vary in the length of time they can transmit the disease.

Effective vaccines for the control of the disease are available. If the disease is suspected in a herd, a veterinarian should be consulted.

22.7 Viral Diseases

Viruses are very minute agents which may contain genetic material such as RNA or DNA but not both. They cannot reproduce unless they invade a normal, healthy cell and take over the cell's metabolic processes. Many serious diseases are caused by viruses in both humans and farm animals. As a general rule they cannot be controlled by antibiotics and other drugs, but vaccines against them are sometimes developed and are effective. Some of the most economically important viral diseases will be discussed in this section.

22.7.1 Calf Scours

Viruses as well as bacteria can cause certain forms of scours (diarrhea) in calves.

A *riolike* virus may cause scours in calves, usually within a few hours after birth. The feces of affected calves are watery and yellow to green in color. Calves lose their appetite when infected, and about 50 percent of them die. Treatment includes the oral administration or intravenous injection of fluids containing electrolytes and the administration of antibiotics to control or prevent secondary bacterial infections. A vaccine is available for this type of scours.

Scours in calves may also be caused by the *corona* virus. It usually affects calves over 5 days and up to 6 weeks of age. If calves scour for several hours, the feces may contain a mucus resembling the white of an egg. Treatment is the same as for *rio* scours, although an effective vaccine is not available to prevent it. Calves in a herd may be affected by both the rio and corona viruses.

The virus which causes bovine virus diarrhea (BVD) in cattle can also cause scours in young calves which have been exposed to it. The diarrhea begins in calves 1 to 3 days after exposure and may persist for several days. The affected calf may have ulcers on the tongue and lips and in the mouth. Treatment is similar to that for scours caused by other viruses. The virus may be controlled by vaccinating all replacement heifers 1 to 2 months before breeding.

22.7.2 Blue Tongue

Blue tongue is a viral disease which affects cattle in many western states of the United States. The virus is spread from animal to animal by blood-sucking insects, especially gnats. Although many cattle in a herd may be infected, usually less than 10 percent show visible signs of the disease.

In the early stages of the disease cattle may have a high temperature of up to 106°F. They may appear depressed and are reluctant to move. Infected cattle slobber profusely, and the muzzle may become crusty. Ulcers are present in foreparts of the mouth such as the dental pads and the lips. Sometimes the tongue may become swollen and bluish in color and may protrude from the mouth. Some affected animals become lame because the coronary band of the hoof becomes reddened and swollen. The virus infects cells lining the blood vessels in such parts of the body as the mouth, tongue, and skin. Affected tissue may die because of an improper blood supply.

Blue tongue viral infections in early gestation may destroy the brain tissue of unborn calves, producing what is known as "dummy" calf syndrome. The course of the disease is usually several weeks, and growth and production of infected animals may be lessened. Death in infected animals is usually due to secondary pneumonia.

The control of blood-sucking insects is helpful in preventing the disease. A vaccine is available, but pregnant cows should not be vaccinated because of possible damage to the brain of the fetal calf. A veterinarian should be consulted if the disease is suspected.

22.7.3   Bovine Virus Diarrhea (BVD)

BVD is a viral disease of the epithelial cells of the digestive and respiratory tracts and their associated lymphoid tissues. This virus contains the genetic material RNA. It is eliminated from the body of infected animals in the feces and in the discharges from the nose, the eyes, and the respiratory system. The virus can cross the placental membranes of pregnant cows and infect the fetus. It infects cattle of all ages and is apparently widespread in its occurrence in cattle in the United States.

Mild, acute, and chronic forms of the disease occur. Clinical signs of acute forms of BVD include fever, diarrhea, and excessive salivation. It may also be associated with dehydration, lameness, and diarrhea. Severe conjunctivitis similar to pinkeye and congestion and ulceration of the mucous membranes of the mouth also occur. In pregnant cows abortions may occur, and calves exposed to the BVD virus before they are born may suffer brain damage.

Treatment of the disease has not been too successful, but a modified viral vaccine has been developed that may be administered by a veterinarian.

22.7.4   Infectious Bovine Rhinotracheitis (IBR)

IBR is also known as red nose and is caused by a virus of the herpes group. The virus is widespread. Three forms of the disease have been described.

Respiratory IBR produces a high fever, open mouth breathing, salivation, and a nasal discharge. Membranes which line the nose often become red and inflamed, which is the reason the disease is often called red nose.

Genital IBR is also called infectious pustular vaginitis (IPV) because the vagina may develop blisterlike nodules. Pus may be discharged from the vagina. The IBR virus may also cause the sheath and penis of the bull to become inflamed. IBR may cause abortions in pregnant cows.

Conjunctival IBR affects the eyes and causes a condition resembling pinkeye.

All forms of IBR may be prevented by vaccination, but vaccinations should be under the supervision of a veterinarian.

22.7.5   Warts

Warts may be caused by several different strains of viruses and are contagious. They may occur on many parts of the body, but in calves under 1 year of age they are usually found on the head, neck, and/or shoulders. The main damage from warts is their unsightly appearance and their damage to hides used for tanning. Vaccines may be used to treat warts, but since several viral strains are involved, several different vaccines may be used before one is found that is effective.

22.8   Lumpy Jaw

Other names for lumpy jaw include big jaw, wooden tongue, and actinomycosis. This disease is characterized by tumorlike swellings or enlargements of the throat and head in cattle. Two forms of the disease occur, and possibly a fungus or bacterium may be the causative factor in each form.

One form of lumpy jaw produces a chronic disease of the jawbone and other tissues of the head. This form is thought to be due to the ray fungus actinomyce. The bone becomes abscessed and swollen and may be greatly damaged. Complete recovery seldom occurs, but cattle may recover enough to be marketed.

The second form of the disease is similar to the first but affects mostly the soft tissues, rather than the bones, of the head and jaw. This form sometimes affects the tongue—hence the name woody tongue. This form is caused by a bacteria and is called actinobacillosis. This form of the disease often responds to treatment with iodides or antibiotics. Early treatment is important, and a veterinarian should be consulted.

STUDY QUESTIONS

1. What factors are involved in economic losses from disease in beef cattle?

2. What are the two general kinds of disease? Describe each.

3. How are many infectious diseases spread from animal to animal?

4. What is the body's first line of defense against disease? How does it defend the body against disease?

5. What is phagocytosis? What part does this play in the body's defense against disease?

6. What are antibodies? Where are they found, and what do they do to defend the body against disease?

7. What is natural resistance to disease?

8. Why haven't lines of cattle been developed that are resistant to disease?

9. What is active immunity? Passive immunity?

10. Define what is meant by *hardware* disease and how it may be controlled or treated.

11. Discuss nitrate poisoning and how it may be detected and treated.

12. What is acorn poisoning? What are the symptoms of this disease?

13. How can one tell if an animal is bloated? What two general kinds of bloat are encountered, and what is the treatment for them?

14. Describe symptoms of fescue foot and how it may be prevented.

15. What is the major cause of grass tetany? What are the symptoms of this disease? How may it be prevented and treated?

16. Discuss urinary calculi and its cause, control, and treatment.

17. Is cancer eye heritable? Describe the symptoms and occurrence of this disease.

18. Has Bang's disease increased or decreased in incidence in the United States in recent years? Explain.

19. Name three clostridial diseases and how to control them.

20. What does the name leptospirosis mean? How may it be treated and controlled?

21. Discuss the symptoms, control, and treatment of calf scours caused by bacteria.

22. Foot rot and pinkeye probably seldom cause the death of an affected animal. Why are they so important economically?

23. What are the symptoms, control, and treatment of vibriosis in cattle?

24. Why can't calf scours caused by viruses be controlled by treatment with antibiotics? Describe the recommended treatment for this kind of scours.

25. What is blue tongue? In what part of the United States is it most important?

26. Blue tongue and some other diseases may infect cattle, but some may show extreme symptoms, whereas others may not. Why this difference?

27. What are BVD and IBR? Describe the three different kinds of IBR.

28. What causes warts in cattle? How may they be treated?

29. Why is lumpy jaw sometimes called woody tongue? What two different kinds of lumpy jaw affect cattle?

some plants poisonous to cattle in the United States

Poisonous plants are those that contain or produce toxic substances themselves or when decomposed cause harm or even death to live-stock that eat them. Poisonous plants cause large losses to livestock all over the world. Losses of this kind have always been large in the United States, especially in the range country. Much is known about poisonous plants and their effects on animals, but much more needs to be learned. Our discussion here will be limited to those poisonous plants of greatest economic importance in the United States. Refer-ences for a more comprehensive study of the subject are given at the end of the chapter. Photographs of some important poisonous plants are given in Table 23-1.

23.1   Why Cattle Eat Poisonous Plants

Many poisonous plants grow in pastures and ranges in association with desirable forages. Most of them are not palatable to cattle and are not consumed in large enough quantities, as a general rule, to be toxic when cattle graze free-choice. Overgrazing of pastures and ranges for any reason increases the chances that enough of the toxic plants will be consumed to be toxic to animals. Overgrazing associated with drought conditions may result in considerable death losses.

Poisonous plants vary greatly in their toxicity and the amount that must be eaten to cause injury or death. Some are acute poisons and require only small amounts to have their effects. Others, however, must be eaten in considerable quantities over a long period of time to be toxic. Fortunately, many poisonous plants can be eaten for long periods without being toxic and will have little or no effect on animals eating them.

23.2 Why Plants May Be Toxic

Plants may be toxic for several reasons. The plants themselves may contain toxic substances. Others may contain substances which are not toxic but which when eaten may form certain toxic substances. For example, amygdalin, a nontoxic glucoside in the wild cherry, may hydrolyze to highly toxic prussic acid when eaten. Some plants may not contain toxic substances, but under certain conditions they may be acted upon by molds which form substances toxic to animals. Some plants may also absorb toxic substances from the soil when they are present in large quantities and store them in large enough quantities to be toxic. An example is selenium (page 445).

23.3 Avoiding Losses from Poisonous Plants

The identification and eradication of poisonous plants where they cause losses in livestock comprise an effective way of reducing losses. Eradication may require a long time and may be very costly in some instances. In some pastures that can be plowed and farmed, poisonous plants may be eradicated in this way. On many ranges in range country, however, this cannot be done, and removing plants by hand or spraying them with an herbicide may be necessary. This is a slow and costly procedure and often is not practical, except in some limited areas.

Some poisonous plants are the first to grow in the spring, especially in high mountain ranges. Losses from such plants may be reduced by a delay in grazing such ranges until other grasses and nonharmful plants make enough growth to supply sufficient palatable plants to grazing livestock.

TABLE 23-1   Photographs of some important poisonous plants in the U.S.A.[a]

Arrow Grass-
Triglochin spp.

Death Camas-
Zygadenus spp.

Johnson Grass-
Sorghum halepense

Larkspur-
Delphinium spp.

Locoweed-
Astragalus spp.

[a]These photographs are from the collection of Dr. A. A. Case, Dept. of Vet. Med. and Surgery, University of Missouri, Columbia.

Lupines-
Lupinus spp.

Rayless Goldenrod-
Aplopappus hetero-
phyllus

White Snakeroot-
Eupatorium rugosum

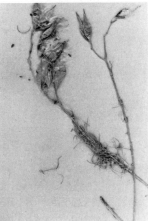

Whorled Milkweed-
Asclepias spp.

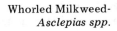

Wild Black Cherry-
Prunus serotina

439

Overgrazing should be avoided on pastures or ranges where poisonous plants are known to occur. In periods of drought livestock should be removed from such pastures, or other forages should be supplied to meet livestock needs. The rotation of pastures also may help to reduce losses due to poisonous plants.

Species of animals differ in their susceptibility to poisonous plants. The plant may not be as palatable to one species as it is to another, or one species may be less susceptible to the toxic substance the plant contains. In some high mountain ranges of the West, cattle cannot be grazed safely because of the presence of certain poisonous plants. These ranges are sometimes stocked with sheep who are not susceptible to those plants.

Annual forages such as some of the sorghums have been poisonous to cattle in the past. These are very palatable to cattle but can be harvested and fed as a dry winter roughage when they mature. Sorghums should not be allowed to grow in fence rows or other places where they are accessible to grazing cattle. In recent years sorghums such as Sudan grass and various sorghum hybrids which are not toxic to animals have been developed by plant breeders. Such plants should be grazed with care, however, and recommendations for grazing them should be closely followed.

The ability to identify potential poisonous plants is important. Information which helps identify poisonous plants may be obtained from extension workers in many states. If a pasture or range is known to contain toxic plants, proper management of the pasture may reduce or eliminate losses.

23.4   Some Plants which Contain Toxic Substances

Toxic substances in plants include alkaloids, glucosides, saponins, resinoids, oxalic acid, tremetol, as well as other known and unknown substances. The toxic substance may be present in the leaves and stems, but it sometimes occurs in seeds. Some plants are poisonous to humans as well as animals.

A knowledge of the kind of poisonous substance contained by toxic plants is important. A knowledge of the chemical nature of the toxic substance is helpful in identifying symptoms of poisoning expressed by affected animals. It may also be helpful in preventing the poisoning of animals eating such plants and sometimes is helpful in treating affected individuals. A knowledge of the chemical poison involved may help identify toxicity symptoms not seen in mature individuals but seen in the young when they are born. The poisonous substance may enter the blood of the pregnant cow and interfere with the proper development of the fetus in early gestation.

Some toxic substances affect different parts of the body such as the central nervous system and the gastrointestinal tract. Some effects of toxic plants may mimic those of infectious diseases.

23.4.1   Alkaloids

Plants containing alkaloids may have an effect on the nervous system, causing nervous disorders such as convulsions and paralysis. Some of these plants are palatable to livestock and in some instances may even attract them.

Some species of larkspurs contain alkaloids of complex chemical composition which are highly poisonous to animals. Some of the species of larkspurs are nontoxic and are often mistaken for those which are poisonous. The main poisonous species are tall larkspur (*Delphinium barbeyi*) and low larkspur (*Delphinium menziesii*). These may cause death losses in grazing animals when the plants are small or eaten in large quantities.

Western ranges may contain lupines in large quantities. Lupines are commonly known as bluebonnets, Quaker bonnets, etc. Lupines belong to the legume family and are very palatable to cattle. Some species are harmless and are excellent feed for livestock. Others are dangerous only at certain times, while still others are poisonous at any stage of growth. The five species known to be poisonous are *Lupinus sericeus*, *L. leukophyllus*, *L. argenteus*, *L. caudatus*, and *L. perrenis*. *L. sericeus* has been shown to produce a syndrome in calves known as the *crooked calf disease* which occurs in certain western states of the United States and Canada. The skeletal systems of affected newborn calves are primarily involved. The bones are twisted and the joints malaligned. In severe cases the forelegs may be useless, and the neck and back may be affected. A cleft palate sometimes occurs. Cows which eat lupines between 30 and 90 days of gestation may give birth to calves affected by this toxic substance.

Death camas is the common name for several plants belonging to the genus *Zygadenus*. They are found in the United States mostly west of the Mississippi river. Death camas has grasslike leaves which grow from an onionlike bulb. They sprout early in the spring and are eaten when livestock are hungry for green feed. Any part of the plant may be toxic. The main poisonous species are *Zygadenus gramineus*, *Z. venenosus*, *Z. paniculatus*, and *Z. nuttallii*.

23.4.2 Glucosides

Glucosides are a diverse group of chemical compounds found in plants. They get their name from the fact that they contain various sugars in their molecules along with other substances. Some are called saponins because they form a soapy solution with water. Others are called cyanogenetic glucosides because they produce hydrocyanic or prussic acid under certain conditions. Some other glucosides do not fit into either of these two groups.

Cyanogenetic glucosides contain cyanide, which is a potent and rapidly acting chemical asphyxiant. It deprives tissues of required oxygen through the inhibition of oxidative enzymes such as cyto-chrome oxidase. Cyanide acts slowly with the hemoglobin of the blood to form cyanmethemoglobin, but the poisonous effects are due to the rapid depletion of oxygen in the tissues. Large amounts of prussic acid cause almost instantaneous death with spasms and respiratory paralysis. Smaller amounts may cause only excitement and convulsion followed by a depression. The more important cyano-genetic plants include sorghum (*Sorghum vulgare*), Sudan grass (*Sorghum vulgare var. sudanensis*), Johnson grass (*Sorghum halepensi*), flax (*Linum vsitatissimum*), arrowgrass (*Triglochin maritima* and *T. palustris*), and the wild choke cherry (*Prunis virginiana, P. melanocarpa*, and *P. demissa*).

Leaves of the wild cherry contain a cyanogenetic glucoside called *prunasin*, whereas the seeds may contain another cyanogenetic glucoside called *amygdalin*. Wild cherry leaves are most dangerous in the spring when cattle crave green plants, but they can be eaten in small amounts without a poisonous effect.

Johnson grass and Sudan grass owe their poisonous properties to a cyanogenetic glucoside known as *dhurrin*. The amount of this glucoside these sorghums may contain varies with different varieties and with different conditions. The sorghums may contain larger amounts of the glucoside when growth is interrupted, for example, by drought, frost, or trampling, and the danger decreases as the plants mature. Second growth and green sprouts on the mature plant may be toxic.

Arrowgrass contains little poison when growing normally, but in droughts it may become toxic.

In most cases cyanogenetic glucosides in hay are in quantities too small to be poisonous because it is thought that they evaporate when the hay is dried. In some cases, however, enough may be left in the hay to cause injury when large amounts of such hay are eaten.

Little is known about the poisonous action of saponins. They appear to produce irritation of the digestive tract, with accompanying abdominal pain and diarrhea. Plants containing saponins include rubberweed (*Actinea richardsoni*), seeds of the corncockle (*Agrostemma githago*), bullnettle (*Solanum carolinense*), bittersweet (*S. dulcamara*), and black nightshade (*S. nigrum*).

23.4.3   Resinoids

Poisonous principals of many plants are resinoids which are very complex. Many species of milkweed contain a poisonous resinoid. Symptoms of milkweed poisoning consist of a lack of muscular coordination, rapid and noisy breathing due to swelling, and the presence of fluids in the lungs. In extreme cases violent spasms and struggling occur and finally death due to respiratory failure.

Several species of milkweed are poisonous, but some are more poisonous than others. The whorled milkweed (*Ascelipes galioides*) causes considerable trouble in the southwestern United States. Other species which are poisonous include the broadleaf milkweed (*A. eriocarpa*) and the whorled milkweed species (*A. mexicana, A. pumila*, and *A. verticilata var. geyeri*).

23.4.4   Oxalic Acid

Oxalic acid occurs in several plants but usually is not present in large enough quantities to be poisonous. Most of the sorrels and docks are poisonous because of their oxalic acid content. Greasewood (*Sarcobatus vermiculatus*) frequently causes large losses in sheep in the United States.

23.4.5   Tremetol

Tremetol is one of the higher forms of alcohols found in the leaves and stems of some plants. Symptoms in poisoned animals include a period of depression followed by trembling of the muscles about the nose and in the legs. Labored breathing, weakness, and inability to stand may also occur.

White snakeroot (*Eupatorium rugosum*) contains tremetol and is poisonous to livestock. Trembles are also caused by another plant in the southwestern United States: the rayless goldenrod or jimmyweed (*Aplopappus heterophyllus*). Tremetol is soluble in the fat of milk and may be transmitted through the milk to other animals and humans, causing what is known as milk sickness. Symptoms of the disease in humans are many, including vomiting, dizziness, loss of appetite, stomach pain, swollen and coated tongue, subnormal temperature, slow respiration, a weak pulse, weakness, and sometimes collapse. The disease in humans has been recognized for 150 years or more and in early days terrorized people and baffled physicians before the actual cause was determined.

23.4.6   Other Poisonous Plants

Many plants are poisonous to livestock, but the poisonous substances involved are not always known. These include several species of locoweeds, the copperweed (*Oxytenia acerosa*), horse brush (*Tetradymia glabrata* and *T. canescens*), and bracken (*Pteridium aquilinim*). Still other plants not identified as poisonous may exist.

23.5   Some Plants which Cause Photosensitivity

Photosensitivity looks like sunburn but is usually severe. The skin becomes reddened and itches. The capillaries just beneath the skin become leaky, and fluid from the injured capillaries often accumulates just beneath the skin, producing a watery swelling. In extreme cases the skin may die and slough off.

Photosensitivity occurs when sunlight reacts with certain light-receiving substances (pigments) in the skin. In cattle the skin is burned in white areas where the skin contains little or no pigment. Affected cattle may have burned or swollen udders or muzzles. Inherited conditions such as porphyria may cause the condition in cattle, but certain plants may contain substances which cause photosensitivity.

Plants may cause photosensitivity in two ways. Some plants contain a photosensitizing agent which enters the capillaries of the skin and is not excreted or destroyed before reaching these skin sites. This causes the skin to be sensitive to sunlight. A second way plants may cause photosensitivity is to contain a substance which poisons the liver. A normal liver changes many potentially dangerous compounds into harmless forms or excretes them into the bile.

Chlorophyll in plants is a major source of pigments reaching the liver of animals. Phylloerythrin is one of the products resulting from liver action on chlorophyll. If the liver is injured and does not remove phylloerythrin from the blood, it enters the capillaries of the skin and causes photosensitivity.

Buckwheat (*Fagophrum escutentum*), introduced into the United States from Europe, may cause photosensitivity in both humans and animals. The seeds as well as the plants appear to contain the photosensitization property.

St. Johnswort (*Hypericum perforatum*) is a weed that is commonly found throughout the United States which may cause photosensitivity in animals.

23.6   Toxic Materials from the Soil that May Be Concentrated in Plants

Although plants may contain poisonous substances they produce themselves, under certain conditions plants of certain species may concentrate substances that are toxic to animals that eat them.

23.6.1   Selenium

Selenium poisoning is a worldwide problem and has been known to be a toxic element that may affect many species of animals. Some species of plants concentrate selenium in amounts 1,000 times that found in the soil in which they grow. These plants have the potential of being highly toxic to animals which eat them. Selenium poisoning is sometimes referred to as *alkali* poisoning. It was first mistakenly given this name because it was thought the disease was due to animals drinking alkali water. This was later proved to be incorrect. Selenium poisoning is a problem mostly in the states of South Dakota, Montana, Wyoming, and Kansas, although soils in some other Great Plains and Rocky Mountain states grow plants which possess increased amounts of this element.

Selenium poisoning in animals can occur in an acute or in a chronic form. The acute form usually occurs in animals, and especially cattle, first introduced to an area where the selenium-concentrating plants grow. Affected cattle first become depressed and unaware of their surroundings. Affected animals may have impaired vision and abdominal distress and may grind their teeth and show excessive salivation. Some animals wander about and stumble because of impaired vision. This has led to the term *blind staggers* to describe the condition.

Chronic poisoning is often called *alkali disease.* It occurs in animals forced to consume plants containing low levels of selenium for several weeks. Affected animals become emaciated, the long hair in the tail is lost, and the hoof wall begins to separate just below the coronary band, which is followed by the sloughing off of the old walls. In other cases the old wall is not shed but combines with the new wall, forming a hoof several times longer than normal with a rocker shape. Of course, such animals show varying degrees of lameness.

Treatment of selenium poisoning has not been too effective. Prevention is the best solution to the problem, which includes good pasture management to eliminate plants that concentrate selenium or treatment to promote the growth of plants that do not concentrate this element.

Some species of plants known to concentrate large amounts of selenium include the woody aster (*Xylorrhiza parryi*), the golden weed (*Oonopsis condensata*), certain vetches including *Astragalus biculcatus* and *Astragalus convallarius* as well as others, and prince's plume (*Stanleya bipinnata* and *Stanleya pinnata*). These species appear to require selenium for growth, while other plants do not require it for growth but will concentrate it if it is present in the soil in which they grow. Most of the selenium-concentrating plants are not very palatable to livestock.

In some areas of the world, including some areas in the United States, selenium levels in the soil are too low to support optimum performance. In recent years trace amounts of selenium have been added to diets of poultry and swine with success.

23.6.2 Potassium Nitrate

Some plants are poisonous because they store potassium nitrate. Plants storing nitrates are many, some of which normally do not concentrate this compound. These plants include the leafy parts of plants such as oats, barley, wheat, some sorghums, and corn. Nitrate poisoning was discussed in more detail in Section 22.4.1.

23.7 Toxic Substances in Decaying Plants

Organisms growing on plants or seeds sometimes produce substances which are toxic to humans as well as animals. Some of these toxic substances have been known to exist for many years, whereas others have been identified only recently.

23.7.1 Ergot

Ergot is a fungus (*Claviceps purpurea*) which infects rye, wheat, and other cereal grains. Other species of this fungus may infect other species of grasses.

In one stage of the life cycle the ergot fungus invades the fruiting head of the plant and forms a hard pinkish to purplish mass (sclerotium) similar in shape to the normal grain but somewhat larger in size than the grain it replaces. The sclerotia are the source of many alkaloids with medicinal and drug properties. In recent years the lysergic acid group (LSD) has been mentioned in connection with drug addictions in many publications. It is produced by this particular fungus. Outbreaks of ergotism in humans reached epidemic proportions prior to 1800 before its cause was determined. Isolated outbreaks still occur where the control of cereal grain purity is not effective or closely supervised. Cases of ergotism in the United States sometimes occur in cattle in the South where they infest Dallis grass or Argentine bahia grass.

Two forms of ergotism are recognized: gangrenous and nervous or convulsive. The gangrenous form is probably the most important economically and results in disturbances in the vasomotor system and often is followed by the sloughing off of the hooves, ears, and tail. Abortion may also occur in pregnant females. The nervous form includes a variety of signs such as hyperexcitability, belligerency, ataxia, and convulsions. Clinical signs usually occur from 2 to 7 days after the ingestion of ergot. The most common species causing this form of ergotism is *Claviceps paspal*, although it may be caused by *Claviceps purpurea* also.

Removing cattle from contaminated grain will usually be followed by recovery in 3 to 10 days.

23.7.2 Mycotoxins

Mycotoxins are chemical substances manufactured by molds that may be poisonous or produce toxic symptoms when they are present in food eaten by humans and livestock. Mycotoxins may remain in food long after the mold which produces them has died. Mycotoxins (or other metabolic products) in livestock feeds can remain in residues in meat, milk, or eggs which are consumed by humans. Hundreds of cases of toxicity syndromes in animals have been reported. In some cases the fungi and toxic agents involved are known, but in many cases they are not.

Aflatoxins, one specific group of mycotoxins, have been studied in detail. They are produced by a few strains of fungi belonging to the two species *Aspergillus flavus* and *Aspergillus parasiticus.* These fungi produce spores which are widely disseminated in the soil. When these toxins are eaten by cattle, they are passed on to humans in animal products such as meat and milk. Aflatoxin B_1 is one of the most potent carcinogens (cancer-producing substances) known. Aflatoxins interact with DNA (the genetic material of the cell) and have large effects on the transcription of genetic information in animal cells as well as the cells of microorganisms.

Mycotoxins have caused heavy death losses of steers in the feed-lot and have been traced to moldy feeds in the ration. In some cases death rates are not high, but the rate of gain is decreased, and the feed per pound of gain is increased. Calves appear to be more susceptible to toxicity than older animals.

Treatments are helpful in speeding animals along the road to recovery. Treatment of the affected animals with high levels of a broad spectrum antibiotic helps prevent secondary infections. A high-energy diet with a high level of protein may also decrease mortality and improve the rate and efficiency of gain. Affected animals should be handled in a way to avoid stress as much as possible. Proper handling of feed between harvest and the time it is fed to prevent mold growth is a good means of preventing toxic symptoms. Moldy feeds should not be fed if at all possible.

Mycotoxins have been present in this world for many centuries. Their importance has only been recognized in recent years. Undoubtedly, more of these toxic substances will be identified in the years to come.

Some evidence suggests that fescue foot in cattle may be due to the presence of toxic materials produced on fescue grass by molds. Fescue foot was discussed in more detail in Section 22.5.3.

23.8   References for Further Study

Buck, W. B., G. D. Osweiler, and G. A. Van Gelder, *Clinical and Diagnostic Veterinary Toxicology,* 2nd ed., Kendall/Hunt, Dubuque, Iowa, 1976.

Kingsbury, J. M., *Deadly Harvest,* Holt, Rinehart and Winston, New York, 1975.

Muenscher, W. C., *Poisonous Plants of the United States,* rev. ed., Collier Books, a Division of Macmillan, New York, 1975.

Radeleff, R. D., *Veterinary Toxicology*, 2nd ed., Lea & Febiger, Philadelphia, 1970.

STUDY QUESTIONS

1. Cattle may not eat poisonous plants even though they are present in the pastures they graze. Why?

2. In what ways can plants be poisonous to animals?

3. Why is it important to know how to identify poisonous species or plants?

4. What are some of the poisonous substances in plants?

5. Name two species of plants that may be poisonous because they contain alkaloids?

6. What is *crooked calf disease*? What causes it?

7. How may cyanogenetic glucosides in plants be lethal to animals that eat them?

8. What are some of the symptoms of plant poisoning due to tremetol? What is milk fever?

9. What is meant by photosensitivity? What may cause it in animals grazing certain plants?

10. What are two ways that certain plants may cause photosensitivity in animals?

11. What is selenium poisoning in animals? What are some symptoms of this disease? Where may it occur?

12. What is *blind staggers*?

13. What is *alkali disease*?

14. What causes ergotism? Can it also occur in humans? How?

15. What are the two forms of ergotism?

16. What are mycotoxins? Aflatoxins? Why are aflatoxins in cattle feed so important to humans who eat products from this species?

CHAPTER 24

behavior as related to the management of beef cattle

Behavior refers to the reaction of animals to stimuli within their environment. Individuals vary greatly in their behavior, and in some instances breeds of cattle also differ in a general way. Behavior responses are due to heredity (internal responses) and learning experiences (external responses). Much evidence is available to show that heredity is involved in the behavior of animals. For example, the bulldog is inherently a fighter; a greyhound, on the other hand, is less a fighter and more a runner. Some breeds of cattle have a gentle disposition, while others are very temperamental.

The reaction of animals to certain stimuli is an inborn characteristic. This is referred to as instinctive behavior. Other behavioral responses are due to learning and experience. These responses are of several kinds. Animals sometimes respond to stimuli because of habit or without thinking. This is known as *habituation*. Animals also learn to respond to certain stimuli because of *conditioning*. An animal learns to associate a response with a certain stimulus usually because of reinforcement in which its response is rewarded when it responds successfully or it is punished when it responds unsuccessfully. Because of conditioning, milk cows "let down" their milk when they enter the milking stalls.

450

Animals may also learn from trial and error. For example, cattle learn to stay away from an electric fence by receiving a shock when they touch it.

Cattle, as a general rule, do not show a high degree of reasoning (insight learning). Individual animals appear to show some degree of reasoning, however, under certain conditions.

Knowledge of the behavior of beef cattle is an important aspect of their production and management. Behavior is probably more important in beef cattle than in dairy cattle because dairy cattle are milked at least twice daily and are closely associated with humans every day. Beef cattle, on the other hand, may be in close association with humans only a few times per year. On the range and in many small herds this is usually during roundup time in the spring and fall when calves are branded, vaccinated, and dehorned and the males castrated. If range cattle have a good memory, they would be likely to associate their discomforts at this time with the presence of humans. On small farms in the midwestern United States beef cattle are in closer association with humans than those on the range. This may be more conducive to their good behavior in the presence of humans.

24.1 Social Behavior

Beef cattle are gregarious in nature. Gregarious means that they tend to flock together in herds on pastures or on the range. They appear to be sociable and fond of each other's company. This is particularly true in small herds in good grazing country. In range country, however, where the stocking rate is only eight to twenty head per section of land, cattle tend to be scattered and are in small family bands, possibly because so much land area is needed to obtain enough food for maintenance and production. Even in range country, beef cattle will congregate around a water hole during the day. They will scatter out again after mid-afternoon as a general rule.

Cattle in a herd show a well-organized social rank as a general rule. Since mature bulls are seldom run in isolated groups, this is of little importance from the behavior standpoint, as it would be in a herd of cows. Cows in a herd quickly establish a dominant-subordinate social rank when first placed together in a group. A "boss cow" emerges, with other cows in a declining rank after her from the next dominant to the most timid cow, which may be subordinate to all other cows in the herd. Social rank in cattle on pasture or range may be of little importance, but it may become very important when cattle are confined to a small space and are fed a limited ration. In such cases any subordinate females may receive only a small amount of feed and become very thin and emaciated.

A dominant-subordinate order is soon established when two or more bulls are placed in the same lot or pasture. Almost invariably the bulls will fight to establish a social order. Usually the larger, more aggressive bull assumes the dominant role, and in some instances the initial fighting may result in injury to one or the other. Under range conditions where several bulls run with the cow herd, the social order is established early, and this is especially apparent when one or more cows are in heat. The larger, stronger, more aggressive bull is more likely to leave more progeny than those who are weaker and less aggressive.

Social order in cattle is related to weight, size, aggressiveness, and timidity. For an animal to be at the top of the social order it must be large, vigorous, and aggressive and have the will, desire, and ability to dominate others. Young animals have a lower social rank and seldom try to establish a higher rank over their elders when all run together in a herd for long periods of time. Castration of males lowers their social rank. Social rank in cattle may also be influenced by heredity.

When a number of steers are placed in a feedlot, one, or a few, will be repeatedly mounted and ridden by other steers. The steer that is ridden is referred to as a *buller steer* and may even be injured or ridden to death. A Kansas study (G. R. Brower and G. H. Kiracote, *J. Anim. Sci.* 35:165, 1972) found that 2 percent of the steers in Kansas feedlots were buller steers. Young range bulls between 1 and 2 years of age run together in a group will show the same phenomenon, and one or more young bulls may be ridden to death unless isolated from the rest of the group.

24.2   Agonistic Behavior

Agonistic behavior refers to combat or fighting activities. It is more prevalent among bulls than cows, but cows may fight under certain conditions. Bulls, as mentioned previously, will fight and establish a social order when placed together in the same enclosure. The fight may result in injuries but usually is not a fight to the death, as may be true in the case of some wild animals.

Bulls who run together from the time they are young seldom fight. Most fights occur when sexually mature bulls are placed together for the first time. Under range conditions where several bulls are run with the cow herd, they seldom fight once the social order is established. In fact, the author has observed that under range conditions in Arizona in the fall of the year a large group of mature bulls may group together with no cow or calf in sight. This may be a carry-over from their wild ancestry. It appears to be a temporary arrangement.

Bulls in Spain and Mexico are bred for a strong instinct to fight matadors in the bullring. Bulls and cows in herds which produce these fighting bulls are selected and bred to concentrate their fighting instincts. The fighting instinct appears to be medium to low in heritability.

Some cows are belligerent and will attack a human who bothers their calves. This agonistic attitude appears to be hereditary and is related to the protective efforts of the cow. Since cows in some breeds appear to be more belligerent than in others, this is probably a heritable trait. The protective instinct of cows toward their calves can be of practical importance in range areas or even in small pastures in the Midwest where wild dogs or wolves may attack and kill young calves.

Dairy bulls are more likely to attack humans than are beef bulls. Exceptions are sometimes noted, however. Some dairy bulls are gentle in the presence of humans, whereas some beef bulls are very belligerent. Most cattle producers warn not to trust any mature bull, because in some circumstances even a gentle bull may turn on his master. Humans have been seriously injured or even killed by enraged bulls.

Cattle of all ages may be provoked to attack humans if roped, stressed, or aggravated. Even range steers with long, sharp horns have been known to attack a human on horseback and gored and killed the mount.

Cattle may be gentled to a considerable extent by proper handling and management. Moving quietly among cattle without yelling or making sudden moves helps in this respect. Petting, scratching backs, and offering feed from a container also improve an animal's trust. Some individuals are much more easily gentled than others, and gentleness is often an individual characteristic.

24.3   Sexual Behavior

Proper sexual behavior is necessary for the propagation of the species. The number of young born and weaned is probably the most important single trait in beef cattle production.

Bulls detect cows in heat by sight or by smell. Often a bull will follow close to a cow for several hours before and after she accepts him in the act of mating.

Sexual behavior in bulls is generally related to the influence of gonadotropic hormones secreted by the pituitary gland and of testosterone, the male hormone, which is secreted by the testes. If bull calves are castrated while very young, they will show little or no sexual activity. Bulls castrated after sexual maturity (called stags) may retain sexual activity for several weeks or months after the operation. This suggests that psychological (or learned) as well as hormonal factors may be involved.

Some bulls are more active sexually than others. Some bulls are seldom, if ever, seen mating with cows, but they apparently do because the cows they run with become pregnant and produce calves. Bulls which run with many cows in pastures or on the range during the breeding season soon learn to pace themselves sexually and limit the number of matings with a cow while she is in heat. If two or more cows in a pasture with a single bull are in heat at the same time, the bull may prefer the company of one more than the others. In some cases the bull will mate with one and not the others. This is one reason it is good management to limit the number of cows per bull during the breeding season.

Most young bulls are very active sexually when first turned with a group of cows. Under some conditions they may not spend enough time grazing and may be thin and in poor condition at the end of the breeding season. This may result in lowered fertility. In such cases it may be necessary to give young, active bulls some grain in addition to pasture during the breeding season.

Old bulls which are fat may be lacking in sexual activity. Older bulls should be given just enough feed to keep them in good breeding condition. They should be exercised freely, and it may be necessary to force them to exercise if they are confined to a small pen when used for hand mating or artificial insemination. Some artificial insemination bull studs use an exercise machine to lead the bulls around in a circle to force them to take needed exercise.

Cows in estrus are usually very restless and show a strong mating desire. This is often referred to as the cow being *in heat.* It is controlled by the endocrine system and helps synchronize the release of the ripened egg from the ovary with the introduction of spermatozoa into the female reproductive tract by the bull during the act of mating. Some cows show signs of estrus much more strongly than others. Estrus may last for 1 to 2 days in some cows and heifers and only a few hours in others. Bulls running with a herd of cows will find and breed all cows, but when cows are being observed for estrus without the presence of a bull, estrus in some cows may not be detected. This problem arises when cows are hand-mated or when artificial insemination is practiced. Bulls who have been altered by an operation to prevent them from mating with the cows may be used for detecting cows in estrus. They are sometimes referred to as *gomer* bulls.

Cows in heat may be ridden by other cows. In rare cases cows in early pregnancy will show signs of estrus, and sometimes cows within a few days of parturition will stand when ridden by other cows. Cows in heat will often be followed by several large bull calves if any are present in the same pasture. In very large pastures on the range, bunching cows together early in the morning and again late in the afternoon is helpful in finding cows in heat. The close proximity of many other cows may have a psychological effect on a cow in heat so that she is more likely to show signs of estrus.

Cows kept away from other cows may show signs of estrus by trying to ride their own calves. They may also show signs of increased restlessness and increased activity.

24.4   Maternal Behavior

Prenatal behavioral development has not been studied in beef cattle. In mice it has been found that subjecting pregnant females to emotional trauma caused an increase in emotionality in their offspring (W. R. Thompson, *Science* 125:698, 1957). This may or may not be true in beef cattle.

The beef cow or heifer seeks seclusion from the rest of the herd when parturition draws near. When stockpersons closely observe pregnant cows and heifers in their herd and find one missing, they know that one is probably in the process of calving or has already given birth to a calf. It is also possible that the one that is missing is sick or has strayed from the pasture or has been stolen.

Seeking seclusion at parturition appears to be a carry-over from the wild ancestors of the cow. The secluded spot chosen will be in the timber, in brush, or in a valley or depression in the land where she cannot be seen by other individuals in the herd. Sometimes in winter when cows are fed in muddy lots, an occasional cow may not seek seclusion but will give birth to her calf in the muddy lot where conditions are very unfavorable. Unless the herd is observed closely and cared for, a calf born under such conditions has less chance of survival than one born under more normal conditions.

Most beef cows lick and clean their newborn calf immediately after it is born, and this process may last from 40 to 50 minutes (I. E. Selman et al., *Anim. Behav.* 18:276, 1970). Licking serves to help dry the calf, especially in cold weather, and may be helpful to the cow to become accustomed to the smell of her calf so she can identify it when it mingles with others. The beef mother may also identify her calf by sight and sound.

Some beef cows do not lick and clean their calves following their birth but may still claim them. In rare cases, however, a cow or heifer may not claim her calf, and under range conditions the calf dies. First-calf heifers which have had difficulty calving sometimes will not claim their calves or let them nurse. When twins are born, a cow may claim one and not the other. Sometimes a cow or heifer will not allow her calf to nurse because her teats and udder are swollen and sore, making nursing painful. In such cases, the female may have to be tied and the calf assisted in nursing. The calf must receive colostrum from the mother within a few hours after birth, or it must be given some from another cow; otherwise it may die of an infection. Some beef cattle producers keep a supply of frozen colostrum available in a deep freeze to give to newborn calves when a supply is not available from their mothers.

The newborn calf will normally struggle to its feet shortly after birth. The amount of time required for the calf to stand depends on its vigor and may vary from a few minutes to 1 or more hours. Calves delivered by pulling the front legs may be unable to stand if the delivery was very difficult and the pulling extremely hard. Such calves should be given colostrum by means of a bottle if necessary. Usually such calves will be able to stand and nurse in a day or two if one or both of the legs are not broken.

Failure to claim her calf may be the permanent nature of an occasional cow or heifer. This problem may be solved in several ways. The best way is to sell the cow if her calf can be raised without her. Another way is to confine the cow and calf to a small pen and assist the calf in nursing two or more times per day. Often the cow will eventually accept her calf. Products are on the market which will often induce a cow to claim her calf. Veterinarians usually have a supply on hand to treat such cases.

A cow will often hide her calf and return to the herd after it has been dried and nursed. The calf will hide by closely hugging the ground, and often one can walk within a few feet of a hidden calf without seeing it. When with the herd the cow will cast occasional glances in the general direction of her calf and will return to it from time to time to allow it to nurse. When the cattle producer is trying to find the calf, a cow will go in the general direction of her calf but often not directly to it. If the human takes a dog along and it is in close proximity to the calf, the cow will often go directly to it. After 2 or 3 days the cow will bring her calf into the herd, but this varies with different cows. Some cows will stay close to a calf who is born dead for a day or two and then will leave it and join the herd.

If a cow and calf are separated for some reason while the calf is still nursing, they will locate each other by vocal sounds. The cow and calf appear to recognize the voice of the other. When many calves are weaned at one time, the cows and calves bawl almost continuously. After 2 or 3 days the bawling ceases, and they appear to accept the separation.

The expression of maternal traits is affected by a number of factors (M. M. Lischko, "Maternal/Neonatal Nursing Behavior," M.S. thesis, University of Missouri, Columbia, 1974). These include prior maternal experiences, general weather conditions, and the sex of the offspring. Older mothers appear to take better care of their young than younger mothers. Cool to cold weather was associated with a maximum amount of nursing by the calf. Bull calves generally received more maternal care than heifers in this study.

Cows give maternal care to their young from birth to weaning. They attempt to protect their young from both humans and animals. The intensity of caretaking varies with different individuals. Some cows are so intense in the care of their young that they will attack a human who comes within close proximity of their calf. Other cows will merely show concern but will not attack a human when their calves are molested.

On an Arizona range where the pasture included several thousand acres with only one source of water, cows appeared to take turns caring for a group of baby calves while the mothers traveled to water. One or two cows would often be seen with a dozen or more calves under their care.

24.5   Ingestive Behavior

Ingestive behavior includes eating and drinking and is demonstrated by animals of all ages. Nursing by the calf is the first ingestive behavior trait expressed in cattle.

Within a few minutes after birth the normal calf will get to its feet and stand. The standing and movement are wobbly at first, and the first few efforts to walk may end in the calf falling to the ground again. The strength gradually increases in the muscles of the calf, and it begins to move toward the back of the cow. If the cow stands still, the calf will nose the udder and finally grasp a teat in its mouth and begin to suck. As mentioned previously, it is extremely important for the calf to obtain a good supply of colostrum as soon as possible after birth. The colostrum contains antibodies which protect the calf from infection until it can build its own supply. The blood of the calf contains few, if any, antibodies before it nurses its mother.

The nursing behavior of calves has been studied by several workers. Purebred Hereford calves nursed 3.5 times in 24 hours with a total nursing time of 28.7 minutes. Hereford × Brahman calves nursed 4.2 times in 24 hours for a total nursing time of 38.0 minutes (T. J. Cartwright and J. A. Carpenter, Jr., *Tex. Agric. Exp. Stn. Bull.* 904, 1960). In another study Charolais calves nursed an average of four times in 24 hours, with each nursing lasting an average of 12 minutes (L. A. Gary et al., *J. Anim. Sci.* 29:203, 1966). Calves running with their mother appeared to nurse about four times per day: from 4 to 6 A.M., 9 to 12 A.M., 3 to 6 P.M., and 10:30 to 1:30 P.M.

The peak in milk production in beef cows is reached about 4 weeks postpartum (L. J. Cole and I. Johannsson, *J. Dairy Sci.* 16: 565, 1933). The total milk production in beef cows appears to be related to the capacity of the beef calf to consume milk (W. Gifford, *Arkansas Agric. Exp. Stn. Bull.* 531, 1953). Calves suckling lighter-producing dams appear to spend more time nursing and nurse more frequently. Cows nursing bull calves appear to give more milk than cows nursing heifer calves (A. A. Melton et al., *Texas Agric. Exp. Stn. Bull.* 26, 1965). Gains made by the calf during the nursing period are closely related to the amount of milk produced by its dam.

Cattle have no upper incisors so they ingest food into the mouth by wrapping their tongue around the grass or other material and jerking the material forward so it is cut off by the lower teeth. Very short grass is grasped between the lower teeth and the upper dental pad. It is cut off by a jerking motion of the head. Badly worn or broken teeth in older animals interfere with proper grazing, and affected animals may become thin and emaciated. This is particularly true where grazing material is very short and is often the case when pastures are overgrazed.

Once the food is taken into the mouth, it is swallowed and passed into the rumen. Later it is regurgitated, chewed more thoroughly, and then swallowed again. This process is known as rumination or chewing the cud. Often when cattle are lying down they can be seen quietly chewing their cud. When an animal is lying down, its ears are drooping, and it is not ruminating, it should be examined more closely to determine if it is ill.

Cattle on the range will spend one third to one half of their time grazing. The time spent grazing will depend on the abundance of grazing material available. They will spend less time grazing when forage is abundant. Much of the remaining time will be spent in lying down and ruminating.

Cattle may graze at all hours of the day and night. The peak grazing activity, however, is just after daybreak, in the afternoon, and again just before dark.

Feeding habits of beef cattle in the feedlot have also been studied (D. E. Ray and C. B. Roubicek, *J. Anim. Sci.* 3:72, 1971) in both winter and summer months in Arizona. Two major peaks of feeding were noted in both seasons. The first peak of feeding occurred about sunrise and a more intense peak period during the afternoon. Hot summer conditions resulted in less frequent eating during midday and in the afternoon and an increased feeding activity during the evening hours. In the winter months an increased feeding activity occurred when feed was introduced into the feed bunks. It did not occur in the hot summer months. In average temperature conditions cattle usually drink only in daylight hours, but in this study hot temperatures increased drinking activities during the afternoon and evening hours and sometimes during the night. A similar study in Iowa (M. P. Hoffman and H. L. Self, *J. Anim. Sci.* 37:1438, 1973) showed a total feeding time of approximately 2.5 hours during a 24-hour period. The greatest feeding activity occurred in winter in the late afternoon (3 P.M. to 6 P.M.). Steers spent more time drinking and drank more water in summer than in winter and spent approximately 12 hours per day lying down. This was not affected by shelter or season.

Cattle, unlike some other animals, will not eat snow as a substitute for water. They may suffer from a lack of water with snow and ice all around them. Cattle should be supplied with water at all seasons either in all-season water troughs or fountains heated by oil, gas, or electricity. In recent years, however, heating water in the winter has become very expensive because of the energy shortage.

Grazing habits of beef cows and heifers on the range have also been studied. Nonlactating Hereford, Hereford × Holstein, and Holstein heifers were observed while grazing on a native tall grass range in Oklahoma (J. R. Kropp et al., *J. Anim. Sci.* 36:797, 1973). Grazing occupied 41 to 45 percent of the heifers' time and tended to be greatest from 6 A.M. to 6 P.M. Rumination required 31 to 32 percent of the time, with 83 percent of the ruminating done in a lying-down position. In another study the feeding behavior of Angus and Charolais × Angus cows was studied in Pennsylvania (W. A. Stricklin et al., *J. Anim. Sci.* 43:721, 1976). The crossbred Angus × Charolais cows rested less and grazed more than Angus cows (57.4 versus 52.9 percent). Grazing and resting activities were not correlated with cow weight or gain, however.

24.6   Eliminative Behavior

Eliminative behavior refers to the methods whereby cattle deposit urine and feces. Feces production by cattle being fattened in the feedlot is of great importance. The cost of cleaning the lots is considerable, especially when thousands of animals are fed in a single feedlot. The production of manure has become an important pollution problem in some areas. Manure is useful in many ways, however. Manure has been spread on the land as a fertilizer for many years. In recent years manure has been processed to form methane (natural gas), which could serve as a source of energy for homes and other purposes. This does not appear to be practical at the present time but may be in the future as the energy shortage grows and more is learned about converting wastes to gas. Manure has been processed and fed as a part of a maintenance ration for cattle. This is possible because cattle are not 100 percent efficient in utilizing nutrients in the feed they eat or those produced by microorganisms in the digestive tract. In addition, microorganisms in the rumen can synthesize proteins from the simple nitrogen compounds found in the manure and urine when they are recycled in the feed.

Cattle tend to deposit their feces in a random manner, but they are likely to deposit them at the spot where they arise after lying down for a period of time. When a group of cattle lie down together for a considerable length of time, they will leave many piles of manure when they arise.

Cattle may defecate and scatter feces while walking, but usually they are deposited in a pile. Because feces are deposited in this way by cattle, in many parts of the world such as India the dried manure piles may be gathered and used for fuel. Cattle do not appear to avoid contact with their feces or urine. Cattle also appear to deposit urine in a random manner but often urinate when rising from a period of time spent lying down.

Cattle appear to defecate and urinate more often during the day than at night (L. M. Schake and J. K. Riggs, *J. Anim. Sci.* 31:414,

1970). A high level of feed intake appears to be related to more frequent defecation and urination.

24.7 Learning Behavior

Cattle possess little or no powers of reasoning, but they can learn to do certain things by trial and error combined with reward or punishment. Dairy cows, for example, soon learn which is their stall in the milking barn and will go to that stall immediately after entering the barn. The reward they receive is a certain amount of a grain ration.

Cows soon learn to associate rewards with certain stimuli. When cattle are fed hay or silage in the winter months by hauling the feed pulled by a tractor, most of them will come running to the wagon as soon as they hear the motor of the tractor. This association may last from one winter to the next.

Many calves must be trained to eat creep feed. The best training method is to feed their dams some grain in the area where the creep feeder will be located at a later date. Then when the calves learn to eat with their mothers, the calves are allowed access to the ration through a creep where the cows cannot enter. Sometimes an older calf which has been accustomed to eating grain will lead other calves to the ration in the creep feeder.

Cattle can be trained as draft animals as easily as horses. In early days when settlers moved from the eastern to the western United States, many of them preferred oxen to mules for pulling their covered wagons. Oxen learned to obey the commands of their drivers just as did horses and mules.

Cattle can learn bad as well as good habits. For example, one of the main sports in present-day rodeos is riding bucking bulls. Some bulls are noted for their bucking ability, which they have "learned" through trial and error. Some bulls become so adept at throwing their riders that only the best can ride the required time in a rodeo. Sometimes a bull goes for long periods without being ridden at all because of his ability to throw his rider.

An occasional cow may disrupt the entire herd. When cattle in the herd are being driven into a pen, one or more will bolt from the herd and head back to the pasture or range. When this happens the entire herd may go with them. The same cow or cows cause this difficulty time after time. Removing such cows from the herd will help solve this problem.

Even in the milking barn certain cows have learned the bad

habit of kicking. Once this habit is developed, a cow seldom gets over it. The legs must be tied, or a special kick-preventing apparatus must be placed on the legs each time before milking. A kicking cow becomes more of a problem when a milking machine is used for milking.

According to old-time cowboys in the trail-drive days, one or a few steers were the ringleaders in starting a stampede. A bad stampede sometimes resulted in the death of horses, cowboys, or other cattle. Much labor was also required to gather the stampeded animals again into a single herd. Once cattle stampeded, they would stampede again at a later time with less provocation.

Cattle producers are either born with the ability to handle cattle correctly or they learn it from someone else. Good advice given by older to younger cattle producers is to think like a cow. In so doing, trouble may be avoided before it has the opportunity to occur.

STUDY QUESTIONS

1. What is meant by behavior of beef cattle? Why is it important?
2. What is responsible for beef cattle behaving the way they do?
3. Define habituation, conditioning, and insight learning.
4. Why are beef cattle less likely to be accustomed to humans than dairy cattle?
5. What is meant by gregarious? Are cattle gregarious?
6. Why is dominant-subordinate behavior of importance in beef cattle?
7. What characteristics are related to the dominance of some individuals in a herd?
8. What is meant by a *buller steer*? How can the problem be solved? Why is it important?
9. What is meant by agonistic behavior? Why is it important to know something about this kind of behavior in cattle?
10. What may be responsible for cattle becoming belligerent?
11. How do bulls detect cows in heat? What determines sexual behavior?
12. Describe sexual behavior in the cow.
13. What type of behavior is noticed in cows when they near parturition?
14. Do cows and heifers ever fail to claim their calves? Why is this important? How can a cow be induced to claim her calf?
15. How does a cow recognize her calf? How do a cow and calf communicate?
16. Describe the nursing behavior of a newborn calf. Why must a calf nurse soon after birth?

17. How do cattle ingest food? What factors may interfere with the ability of a cow to ingest food? What may be the consequences in cows that cannot ingest food normally?

18. Will cattle eat snow? How can this type of behavior be of practical importance?

19. What is meant by eliminative behavior? What uses have been or could be made of cow manure?

20. Do cattle have good powers of reasoning? Can cattle learn?

INDEX